Electromyography Signal Acquisition and Processing for Movement Analysis

Electromyography Signal Acquisition and Processing for Movement Analysis

Editors

Francesco Di Nardo
Valentina Agostini
Silvia Conforto

MDPI • Basel • Beijing • Wuhan • Barcelona • Belgrade • Manchester • Tokyo • Cluj • Tianjin

Editors

Francesco Di Nardo
Università Politecnica delle Marche
Italy

Valentina Agostini
Politecnico di Torino
Italy

Silvia Conforto
University Roma Tre
Italy

Editorial Office
MDPI
St. Alban-Anlage 66
4052 Basel, Switzerland

This is a reprint of articles from the Special Issue published online in the open access journal *Sensors* (ISSN 1424-8220) (available at: https://www.mdpi.com/journal/sensors/special_issues/emg_signal).

For citation purposes, cite each article independently as indicated on the article page online and as indicated below:

LastName, A.A.; LastName, B.B.; LastName, C.C. Article Title. *Journal Name* **Year**, *Volume Number*, Page Range.

ISBN 978-3-0365-7204-8 (Hbk)
ISBN 978-3-0365-7205-5 (PDF)

Contents

Preface to "Electromyography Signal Acquisition and Processing for Movement Analysis"

The assessment of muscle recruitment is acknowledged as one of the main issues of movement analysis. Muscle activity is typically monitored using surface electromyography (sEMG), a non-invasive technique widely adopted both in research and clinical settings. Recent advancements in commercial EMG signal acquisition technologies and sensors, the development of high-density surface EMG systems, the introduction of sensor fusion, and the availability of data storage and file sharing systems have changed the perspectives surrounding the measuring, capturing, and analysing of EMG signals, specifically in movement analysis. The areas of application are also increasing and differentiating. Besides typical fields such as basic research, clinics, and sports, EMG analysis is increasingly proposed in novel scenarios related to robotics, exoskeleton technology, prosthetics, assistive devices, electrical stimulation, and ergonomics.

The present book is designed to comprehensively cover the open research issues related to the improvement of classic approaches and the development of innovative technology and methodology for EMG-signal acquisition and processing in the domain of movement analysis. Furthermore, the scientific articles included in the current book also aim to focus on different fields of application of EMG analysis, including in clinics, physiology, rehabilitation, sports, and ergonomics. Computational intelligence methods, such as machine and deep learning, have recently emerged as promising tools for the development and application of intelligent systems in interpreting EMG signals. Contributions in this field are also included.

Francesco Di Nardo , Valentina Agostini , and Silvia Conforto
Editors

Article

A Simulation Study to Assess the Factors of Influence on Mean and Median Frequency of sEMG Signals during Muscle Fatigue

Giovanni Corvini * and Silvia Conforto

Department of Industrial, Electronics and Mechanical Engineering (DIIEM), Roma Tre University, 00146 Rome, Italy
* Correspondence: giovanni.corvini@uniroma3.it

Abstract: Mean and Median frequency are typically used for detecting and monitoring muscle fatigue. These parameters are extracted from power spectral density whose estimate can be obtained by several techniques, each one characterized by advantages and disadvantages. Previous works studied how the implementation settings can influence the performance of these techniques; nevertheless, the estimation results have never been fully evaluated when the power density spectrum is in a low-frequency zone, as happens to the surface electromyography (sEMG) spectrum during muscle fatigue. The latter is therefore the objective of this study that has compared the Welch and the autoregressive parametric approaches on synthetic sEMG signals simulating severe muscle fatigue. Moreover, the sensitivity of both the approaches to the observation duration and to the level of noise has been analyzed. Results showed that the mean frequency greatly depends on the noise level, and that for Signal to Noise Ratio (SNR) less than 10dB the errors make the estimate unacceptable. On the other hand, the error in calculating the median frequency is always in the range 2–10 Hz, so this parameter should be preferred in the tracking of muscle fatigue. Results show that the autoregressive model always outperforms the Welch technique, and that the 3rd order continuously produced accurate and precise estimates; consequently, the latter should be used when analyzing severe fatiguing contraction.

Keywords: power spectral density; spectral estimation techniques; Welch method; Burg method; autoregressive model

Citation: Corvini, G.; Conforto, S. A Simulation Study to Assess the Factors of Influence on Mean and Median Frequency of sEMG Signals during Muscle Fatigue. *Sensors* **2022**, *22*, 6360. https://doi.org/10.3390/s22176360

Academic Editor: Georg Fischer

Received: 2 May 2022
Accepted: 19 August 2022
Published: 24 August 2022

Publisher's Note: MDPI stays neutral with regard to jurisdictional claims in published maps and institutional affiliations.

1. Introduction

In the last years, the use of surface electromyography (sEMG) exponentially increased in a variety of contexts and applications such as clinical assessment [1], sport performance evaluation [2], gesture recognition [3], classification [4], and prosthesis control [5,6]. In fact, this non-invasive technique provides useful information on the state of muscles [7]. For example, by variables in the time domain related to the signal amplitude, such as envelope or Root Mean Square (RMS), information on timing of muscle activation and on muscular force can be obtained [8,9], while by frequency parameters, information on muscle physiology and on muscular fatigue [10] can be derived. Among the various pieces of information, the one related to muscle fatigue is certainly of extreme interest. Since muscular fatigue has been associated with electrical signs, such as an increase of the amplitude of the sEMG signal and a compression of its spectrum toward the low-frequency area [11,12], attention has been devoted to the detection of parameters able to outline this behavior. Thus, to investigate the variation in the frequency content of the power spectrum, Mean Frequency (MNF) and Median Frequency (MDF) have been proposed [13] because they have been demonstrated to be related to alterations of firing rate and recruitment patterns of Motor Units (MUs) [14] that occur due to metabolic changes during fatigue. These important spectral features can be extracted from the Power Spectral Density (PSD) of the sEMG signals. Nevertheless, due to the finiteness of the real signals, the power spectrum cannot be computed, but only estimated; hence, several estimation techniques have been developed, each with its own advantages and disadvantages [15]. In general, it is possible to assess the quality of the estimates by studying the bias and the variance of the estimators [16]; however, the estimates also seem to be affected by the specific implementation settings of the estimators, such as the length and the shape of the signal segmentation window [17,18], the number of segments used for estimation [18], the frequency distribution of the spectrum [19] and the model order in parametric approaches [18,20]. Farina and Merletti [20], as well as Clancy et al. [10], compared the performance of different estimation methods on the basis of the epoch length used to process the sEMG signals. However, to our knowledge, comparisons have not been extended to muscle fatigue conditions in which the spectrum of the sEMG signal may take on shapes quite different from those typical of non-fatigue protocols [21].

In fact, the last methodological works that focused on the comparison of different methods in assessing the spectral parameters is the one of Farina and Merletti [20], in which useful recommendations for the spectral estimation with the autoregressive model were provided (i.e., the use of the 10th order of the Burg method). As a result, all subsequent studies exploit such recommendations without considering that changes in the frequency content of signals might affect the spectral estimates. For example, the study by Zhang et al. [22] investigated the PSD estimation of non-stationary signals with a time-varying autoregressive model, but still using a fixed order for the parametric approach. Moreover, a recent study [19] showed that the spectral estimates extracted from sEMG are affected by the frequency content of the signals.

The aim of this work, thus, is to test whether and to what extent the results of previous studies [19,20] can be considered valid in the case of fatiguing contractions associated with frequency distributions different from those found in the case of no fatigue. Two techniques for the Power Spectral Density estimation are considered: the Welch method and a parametric approach based on the AutoRegressive (AR) model.

The methods are applied on several synthetic sEMG time-series, each with its own Time duration (T), but all with the same compressed spectral shape. Different amounts of white Gaussian noise, indicated by the Signal to Noise Ratio (SNR) variable, are added to the signals to simulate as much as possible real acquisition conditions [23].

The performance of the estimators is assessed on the ability to determine the spectral parameters (i.e., MNF and MDF) and is quantified through the Mean Absolute Error (MAE) and its variance. This error is used as the criterion to determine the most robust estimation approach with respect to (i) the spectrum content and shape, (ii) the time duration of the signals, and (iii) the level of noise. For the parametric approach, the order of the model is also studied as a factor of influence of the results.

The paper is organized as follows: the estimation techniques for the Power Spectral Density estimation are described; the experimental design is presented; the statistical analysis is explained. Then, the performance of the two compared techniques is presented in terms of the error committed in the extraction of spectral parameters. Finally, the discussion section comments the findings, and in the conclusion section, some guidelines are provided.

2. Materials and Methods

In this section, the model used for generating the signals is illustrated. The two estimation techniques, which have been used for the performance comparison, are fully described, as well as the spectral parameters that are calculated from the estimated power spectra. Then, the error that was used for the assessment of the performance is explained, and finally the ANOVA tests for the statistical analyses are presented.

2.1. Simulation Procedure

The model proposed by Stulen and De Luca [24] was used to generate a set of synthetic sEMG signals. This model takes as input a zero mean white process with unit variance that is filtered by a band-pass filter of which the square modulus of the transfer function is:

$$P_{xx}(f) = \frac{k^2 f_h^4 f^2}{\left(f^2 + f_l^2\right)\left(f^2 + f_h^2\right)^2} \tag{1}$$

In this way, P_{xx} represents the ideal PSD, k^2 is a scaling factor, f_l and f_h are the low and high cut-off frequencies, respectively, and f is the frequency that ranges from zero to half of the sampling frequency ($f_s/2$), because only the positive part of the spectrum is considered. The number of spectral lines (L) in the range 0–$f_s/2$ depends on the duration of the analyzed signal. The model parameters were set as follows: $k = 1$ and $f_s = 1024$ Hz. The two cut-off frequencies, $f_l = 20$ Hz and $f_h = 40$ Hz have been selected such that the ideal MNF and MDF had a value of 39.84 Hz and 30.95 Hz, respectively. This specific pair of cut-off frequencies was chosen as low as possible to generate consistent myoelectric signals presenting a substantial compression of the power spectrum shape towards the low-frequency area, thus simulating a strong level of muscle fatigue, as highlighted in [25,26]. Eight different types of sEMG signals were generated considering eight different durations as in [20]: (a) T = 250 ms, (b) T = 500 ms, (c) T = 750 ms, (d) T = 1000 ms, (e) T = 1250 ms, (f) T = 1500 ms, (g) T = 1750ms, (h) T = 2000 ms. For each type, 1000 realizations were generated. A further random 1000 realizations of white Gaussian noise were added to the sEMG signals. Four typical SNR conditions were simulated, from 5 to 20dB, as in [23,27]. The resulting myoelectric signals have the following form:

$$x_n = \sum_{j=0}^{N} g_n h_{n-j} + q_n \qquad\qquad n = 0, 1, \ldots, N \tag{2}$$

where N is the number of samples, g_n is a realization of white Gaussian noise used as input of the shaping filter h_n, and q_n is a further realization of white Gaussian noise. The two processes of noise, g_n and q_n were assumed to be independent. The filter h_n was obtained by taking the real part of the inverse Fourier Transform of the amplitude spectrum, that is the square root of $P_{xx}(f)$, and its phase was reconstructed as the imaginary part in the Hilbert transformation of the logarithm of the magnitude, as explained in [24].

2.2. Mean and Median Frequency

Mean and median frequency were computed from the power spectral densities. MNF is an average frequency, which is computed as:

$$\text{MNF} = \frac{\sum_{l=1}^{L} P_l f_l}{\sum_{l=1}^{L} P_l} \tag{3}$$

where f_l is the l-th frequency, P_l is the l-th line of the power spectrum, and L represents the total number of spectral lines in the positive part of the spectrum.

MDF, instead, is the frequency that splits the sEMG power spectrum into two regions exactly equivalent in power [13], and it is defined as:

$$\sum_{l=1}^{l_{\text{MDF}}} P_l = \sum_{l_{\text{MDF}}}^{L} P_l = \frac{1}{2} \sum_{l=1}^{L} P_l \tag{4}$$

where P_l, f_l, and L are the same as above. When the spectrum is symmetric with respect to its center line (e.g., Gaussian), MNF and MDF coincide, but typically, when dealing with myoelectrical signals, the distribution of the power in the frequency domain is left skewed and therefore the MDF is lower than the MNF.

2.3. Techniques to Estimate and Compute the Power Spectral Density

In the following, the Welch method, which is a non-parametric technique that estimates the power spectral density directly from the data, and the autoregressive model for PSD estimation are presented [28].

2.3.1. Non-Parametric Estimation

The periodogram is one of the most known non-parametric estimation techniques but, unfortunately, this estimator is not consistent because the variance of its estimate does not tend to zero as the number of samples increases. Consequently, improved versions, which aimed to solve this issue, have been proposed, such as the Bartlett [29] and the Welch method [30]. The first one solves the inconsistency problem dividing the total length of the signal into S segments, computing the periodogram in each segment and then averaging the results to obtain the final PSD estimate; the second method works in a similar way, but it further improves the resulting PSD estimate because it allows overlapping windows. In this way, the improvement comes from the greater number of windows (thus decreasing the variance of the estimate), as well as the reduction of the loss of information at the extremities of the window due to the effect of the Fourier transform. The S segments are obtained by multiplying the signal to a window function (whose length is smaller than the total length of the signal), which is translated over the entire signal with a fixed overlap of samples. Hence, the resulting power spectral density can be estimated as:

$$\hat{P}_{xx}(f) = \frac{1}{S} \sum_{s=1}^{S} I_m^{(s)}(f) \quad \text{where } f = 0 : \frac{f_s}{m} : \frac{f_s}{2} \tag{5}$$

where S is the total number (13) of segments, and $I_m^{(s)}$ indicates the s-th periodogram that is estimated on m samples according to the following equation:

$$I_m^{(s)}(f) = \frac{1}{U} \left| \sum_{m=0}^{M} w_m x_m e^{-j2\pi m f} \right|^2 \quad \text{where } U = \frac{1}{M} \sum_{m=0}^{M} w_m^2 \tag{6}$$

with M being the total number of samples of the window, w_m the window function, x_m the signal, and U a gain factor.

In this work, according to the results showed in Figure A1, the length of the window function was set to 25% of the total length of the signal, while the overlap was set to 25% of the length of the segment, so the total number of segments S was equal to 13. The zero padding technique was applied to all the windows such that each periodogram was estimated on a total number of samples equal to the length of the entire signal. Results from a previous study have shown that the Tukey window function, also known as tapered cosine, outperformed other window functions in MNF and MDF assessment [19]; for this reason, this window was selected for the implementation of the Welch algorithm.

2.3.2. Parametric Estimation

For the parametric estimation, Autoregressive-Moving Average (ARMA) models are the most known. This parametric approach allows to estimate parameters of a mathematical model that can generate (and forecast) future samples by a linear combination of present and past inputs, and its past output. Autoregressive is a special case of the ARMA model and it is the most widely used for spectral estimation [20]. We used the Burg method [31], which estimates the model parameters directly from the measured data minimizing the prediction error that is generated by the difference of the actual output of the model and the real value of the signals analyzed. Given an order of the model p, the Burg technique estimates only the reflection coefficients a_{pp} to predict future samples of a signal according to the following equation:

$$\hat{x}(n) = - \sum_{z=1}^{p} a_{pz} x(n-z) \tag{7}$$

where p is the order of the model, a_{pz} are the prediction coefficients that can be computed using the iterative Levinson–Durbin algorithm, and a_{pp} are the reflection coefficients (obtained when the index z is equal to the order p) that can be obtained by the minimization of the forward and backward errors of the estimation [15]. As a result, the Burg method aims to simultaneously minimize the sum of both the forward and the backward errors via the Least Mean Square Error (LSME) criterion. The power spectral density is thus computed as:

$$P_{xx}(f) = \frac{\sigma_z^2}{\left|1 + \sum_{z=1}^{p} a_{pz}e^{-j2\pi z f}\right|^2} \tag{8}$$

where σ_z^2 is the total error and a_{pz} are defined as before. In this work, six different orders, heuristically selected between the 3rd and the 30th order to compare their performance in the implementation of the Burg method, were analyzed. Different orders have been compared because the results of a previous work [19] demonstrated that the optimal order differs for the MNF and MDF computation, especially when a compression of the spectrum started to be visible in the frequency domain.

2.4. Statistical Analysis

For the statistical analyses, the MAE is computed as following:

$$\text{MAE} = \frac{1}{C} \sum_{c=1}^{C} y_d - y_c \qquad \qquad with\ c = 0,\ 1,\ \ldots,\ 1000 \tag{9}$$

where y_d is the ideal value of MNF (or MDF), y_c is the MNF (or MDF) value computed from the estimation technique, and c is the total number of generated signals. Descriptive statistics (mean and standard deviation) were computed for both parameters. Interaction effects among factors were investigated by performing three-way ANOVA considering the following factors:

- *method*, 6 levels (Welch, Burg 3rd, 4th, 7th, 10th, 15th and 30th order)
- *duration*, 8 levels (250 ms, 500 ms, 750 ms, 1000 ms, 1250 ms, 1500 ms, 1750 ms, 2000 ms)
- *SNR*, 4 levels (5 dB, 10 dB, 15 dB, 20 dB)

When the interaction effect among the three factors was significant, we set the values of *SNR* and we computed a two-way ANOVA for each level of the *SNR* factor; in turn, if the interaction effect between the two other factors (*duration* and *method*) was significant, we set the values of the *duration* factor, and then performed one-way ANOVA on *method* for each level of the *duration* factor. On the other hand, when the three- and two-way ANOVA were not significant, the main effect with one-way ANOVA on the *method* factor was directly studied. In each case, when the main effect of *method* was significant, the Tukey's HSD post-hoc test was applied. Statistical analyses were conducted in MATLAB and the significance levels were set at: * $p < 0.05$, ** $p < 0.01$, *** $p < 0.001$.

3. Results

In Figure 1, the ideal PSD as well as those estimated with the Welch and Burg methods are shown. In this study, the difference between the ideal and the estimated shape was not assessed because we are interested in the values of the spectral parameters for fatigue detection. In Figure 1a,b, the PSDs estimated from signals with time duration equal to 250 ms and 2000 ms, respectively, when the level of noise is very high (SNR = 5 dB) are shown; in Figure 1c,d, instead, the PSD come from signals with SNR = 20dB and duration 250 and 2000 ms, respectively.

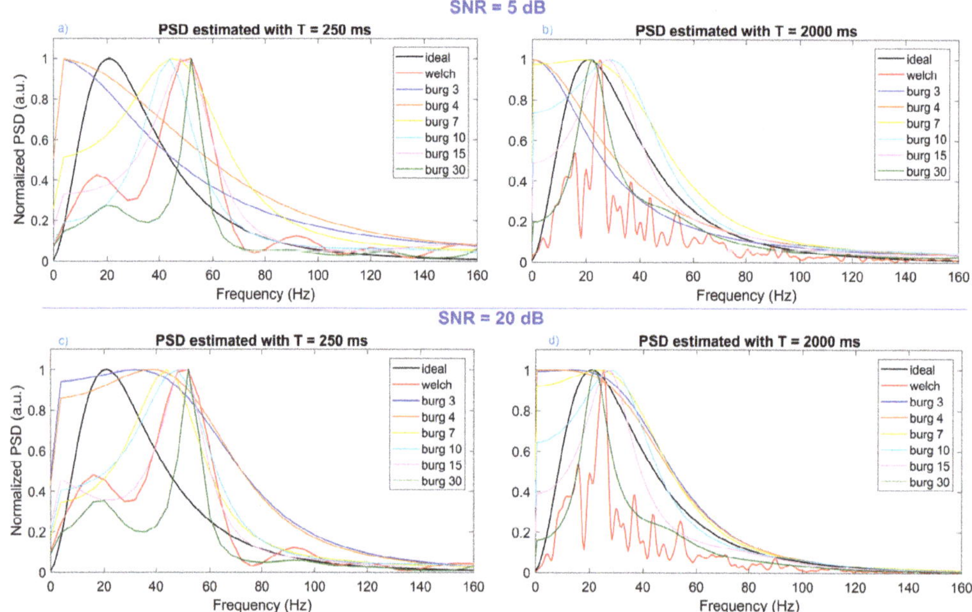

Figure 1. Representation of the ideal Power Spectral Density (PSD) together with power spectra estimated with Welch method and six different orders of the Burg method. (**a,b**) show the spectra estimated from signals whose Time duration (T) is equal to 250 and 2000 ms, respectively. These two figures were generated when Signal-to-Noise-Ratio (SNR) was low, that means there was a high level of noise. In the same way, in (**c,d**) power spectral densities, which were extracted from signals with time duration equal to 250 and 2000 ms, respectively, can be seen, but the SNR of signals was high, that is there was a low level of noise.

In Figure 1a, it can be noticed that, when dealing with brief signals (T = 250 ms), neither the spectrum obtained with Welch nor those computed with Burg succeed in the approximation of the ideal spectrum shape (in black). In fact, low orders of Burg produced a spectrum shape truncated around 0–5 Hz, and thus they were not able to approximate the ideal shape. In the same way, orders too high (30th) and Welch failed to generate a well-shaped spectrum shape because their spectra contained one small peak in correspondence of the high peak of the ideal spectrum, while the largest peak could be found around 50–60 Hz. It seems that the 10th order of Burg had the most similar shape, even if its peak (around 45 Hz) did not coincide with the ideal one around 20–25 Hz.

When time duration (T ≥ 1000 ms) of the signal increased, as shown by one example in Figure 1b, some high orders of Burg method (15th and 30th) approximated the spectrum shape well, having the central peak in the same frequency range of the ideal one. The spectrum estimated by the Welch method shifted the peak towards the low-frequency area, but it started to exhibit more oscillation. On the other hand, Figure 1c,d showed the power spectra estimated from brief (T = 250 ms) and long (T = 2000 ms) signals, respectively, with low level of noise (SNR = 20 dB). As can be seen, these two figures are similar to those corresponding to low level of SNR, indicating that the SNR did not substantially influence the estimation of the power spectrum shape. The main difference could be seen in the approximation of the spectrum shape obtained by the 3rd and 4th order of the AR model with signals of brief duration (T = 250 ms): when the SNR was equal to 20dB, the shape started to approximate the ideal one with a smoothed peak (Figure 1b) instead of having a sharp peak (Figure 1a).

Then, a three-way ANOVA was computed on both MNF and MDF, and the result of the test are summarized in Table 1. While no significant three-way interaction among *method*, *duration*, and *SNR* was visible in the study of the MNF, a statistically significant three-way interaction effect ($p < 0.05$) among these three factors could be seen when dealing with the MDF.

Table 1. Three-way ANOVA tests were performed to investigate the effects of the three factors on both Mean (MNF) and Median frequency (MDF). The three factors used were *method* (Welch and orders of Burg), *duration* (T varying from 250 to 2000 ms), and *SNR* (from 5 to 20 dB). The two-way and three-way interaction effects are shown starting from the 4th row of the table up to the 7th one, respectively. The significance levels were set at: * $p < 0.05$, ** $p < 0.01$, *** $p < 0.001$. Mean square, F-statistic, and *p*-value are provided. All the values less than 2×10^{-16} were indicated as 0, with the corresponding significance indicated by *** $p < 0.001$.

Source	Mean Frequency			Median Frequency		
	Mean Sq.	F	Prob > F	Mean Sq.	F	Prob > F
method	2.22×10^4	2.95×10^3	0 ***	1.52×10^4	2.11×10^3	0 ***
duration	5.67×10^3	754.21	0 ***	1.79×10^4	2.48×10^3	0 ***
SNR	2.74×10^7	3.65×10^6	0 ***	3.39×10^5	4.68×10^4	0 ***
*method*duration*	10.05	1.33	0.07	89.06	12.30	0 ***
*method*SNR*	1.93×10^3	257.06	0 ***	9.02×10^3	1.24×10^3	0 ***
*duration*SNR*	293.79	39.02	0 ***	277.71	38.36	0 ***
*method*duration*SNR*	1.74	0.23	1	21.57	2.98	0 ***
Error	7.52			7.23		

3.1. Mean Frequency

For the MNF, no significant three-way interaction effect among *method*, *duration*, and *SNR* was found, but there were significant two-way interaction effects between the following pairs of factors: *method* and *SNR*, and *duration* and *SNR*. The SNR influenced the estimate of the MNF producing substantial errors that were significantly different from one level to another independently from the estimation method, passing from an error of about 50 Hz when SNR was equal to 5 dB to an error of about 3 Hz when SNR was equal to 20 dB. The *duration*, instead, influenced the precision of the estimate: as the duration increased, the variance of the error decreased. However, since we are interested in finding the more robust method for the estimation, one-way ANOVA was performed on the *method* factor for each level of the *SNR* and for each level of the *duration* factor. Each of the test results was significant ($p < 0.0001$) and thus post-hoc tests were performed on the *method* factor. The results in Figure 2 show that the 3rd order of Burg outperformed Welch method and all the other orders of Burg ($p < 0.01$), except for one case: when SNR was equal to 20 dB, the mean of the MAE between the 3rd and all other orders were not statistically different ($p = 0.99$).

Results in Figure 3 are similar to those in Figure 2, but they represent the case when signals had T = 2000 ms. By comparing these results with those in Figure 2, it can be seen how the increase in the time duration of signals reduced the variance of the error, improving the precision of each method. Also in this case, the 3rd order of Burg outperformed Welch and all the other orders of Burg ($p < 0.01$) except when SNR was equal to 20dB: in this case, the difference between the 3rd and the 4th order was not significant ($p = 0.99$).

In general, Burg outperformed Welch especially when SNR was very low (SNR = 5 and 10 dB). As soon as the SNR increased, the difference between the two methods decreased, and even if significant, the difference between the best order of Burg and Welch was less than 1 Hz when SNR = 20 dB, for both, brief and long signals (T = 250 and T = 2000 ms, respectively). These results can be easily visualized in Figures 2 and 3, where the MNF errors are reported for six orders of Burg and for Welch method. Each subplot in the figures corresponds to the analysis performed by setting one value of the *SNR*.

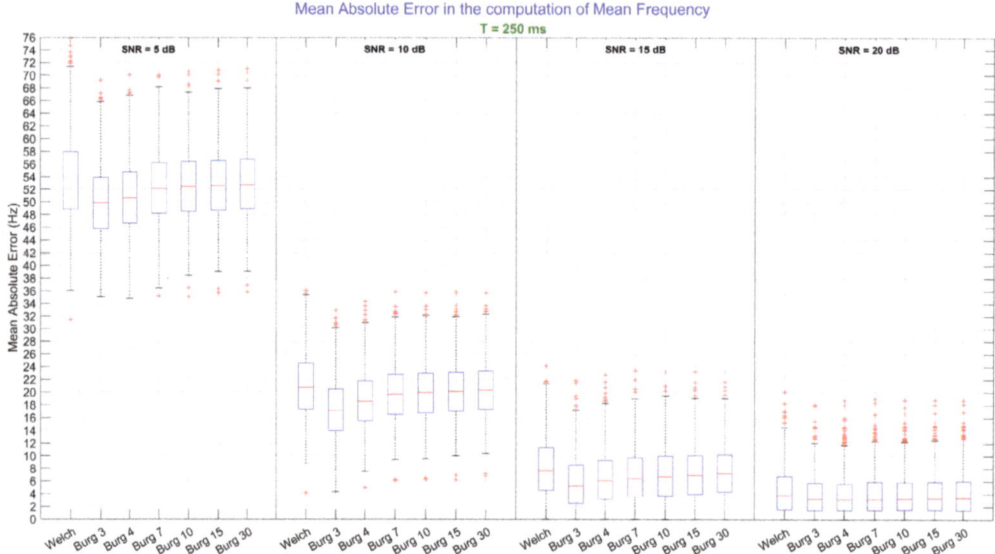

Figure 2. Mean frequency values computed from the power spectral density, which were estimated by Welch and Burg methods. The analysis performed on brief signals (T = 250 ms) is shown. In each subplot, a specific level of Signal-to-Noise Ratio (SNR) is represented. Since one-way ANOVA on the *method* factor was significant, post-hoc tests were performed: the significance level was set at $p < 0.05$.

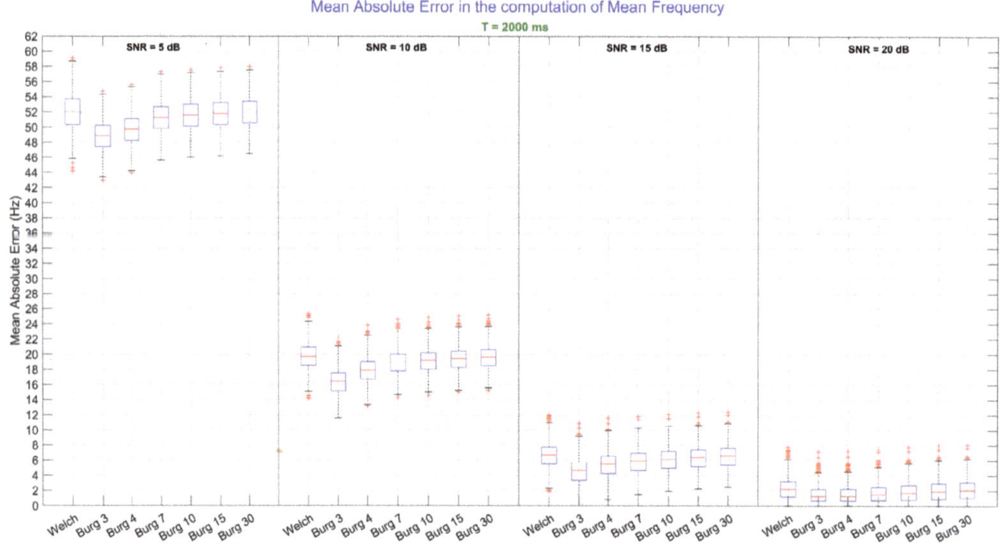

Figure 3. Mean frequency values computed from the power spectral density, which were estimated by Welch and Burg methods. The analysis performed on long signals (T = 2000 ms) is shown. In each subplot, a specific level of Signal-to-Noise Ratio (SNR) is represented. Since one-way ANOVA on the *method* factor was significant, post-hoc tests were performed: the significance level was set at $p < 0.05$.

3.2. Median Frequency

For the MDF, three-way ANOVA revealed that there was a significant interaction effect ($p < 0.001$) among *method*, *duration*, and *SNR*, as shown in Table 1. Therefore, we set the value of the *SNR* factor, and a two-way ANOVA was performed, considering the interactions between the *method* and the *duration* for each level of *SNR*. The results, summarized in Table 2, suggest that there was always a significant interaction effect ($p < 0.001$) between the *method* and the *duration* factor.

Table 2. Two-way ANOVA for the analysis of interaction effects between the *method* and the *duration* factors. Each ANOVA test was performed considering one level of *SNR* at time. The significance level is set at: * $p < 0.05$, ** $p < 0.01$ and *** $p < 0.001$. Mean square, F-statistic, and *p*-value are provided. All the values less than 2×10^{-16} are indicated as 0, with the corresponding significance indicated by *** $p < 0.001$.

	SNR = 5 dB		SNR = 10 dB		SNR = 15 dB		SNR = 20 dB	
Source	F	Prob > F	F	Prob > F	F	Prob > F	F	Prob > F
method	3.17×10^3	0 ***	357.65	0 ***	71.34	0 ***	60.41	0 ***
duration	186.11	0 ***	629.21	0 ***	1.17×10^3	0 ***	1.35×10^3	0 ***
*method*duration*	5.48	0 ***	8.09	0 ***	3.75	0 ***	2.35	0 ***

Therefore, one-way ANOVA tests were performed on the *method* factor for each level of the *duration* factor, in turn computed for each level of *SNR*. All the one-way ANOVAs were statistically significant ($p < 0.001$), and thus post-hoc tests were executed to find out which level of the *method* factor produced the minimum error and had the best performance. In Figure 4, results obtained from signals with SNR = 5 dB are shown.

We can see that the lowest order of the Burg model outperformed the Welch method and all the other orders ($p < 0.05$) for every time duration of the signals; the same difference with Welch and all other orders was still present for the 4th order. For brief signals (T = 250 ms), the Welch method produced similar error to the 7th and 10th order ($p > 0.05$), while by increasing the time duration of the signals, it slightly reduced the error, producing comparable results ($p > 0.05$) to those obtained with higher orders (15th and 30th). Quantitatively, the mean difference between the 3rd and the 4th order was about 2–3 Hz, while between the 3rd order and higher ones and Welch was about 4–5 Hz. In general, the error produced by the best method (3rd order) was about 5 Hz when signals were very brief (T = 250 ms) and it decreased as well as the duration of signals increasing, reaching the initial error of about 2.5 Hz. This decrease (2.5 Hz) was found for each considered technique, indicating that longer signals allow to have a better frequency resolution.

In Figure 5, instead, it is possible to see the results obtained when SNR was equal to 20 dB. In general, the minimum error was produced by the 15th order of the AR model. When the signals had brief duration (T = 250 ms), the difference between the 15th order and the other levels of *method* factor were not significant, except when compared to the 30th order ($p < 0.05$). When the time duration of signals started to increase (T > 500 ms), the 15th order produced the minimum error, whose difference was statistically significant ($p < 0.05$) with respect to Welch and all other orders except for the 3rd and 4th order. Quantitatively, the mean difference between the 15th order and the other levels of the *method* factor was about 0.7–1 Hz for brief signals and it decreased to 0.2–0.5 for longer signals (T > 1000). In general, the error produced by the best method (15th order) was about 3 Hz when signals were very brief (T = 250 ms), but when the duration of signals increased, the total error reduced to 1.5 Hz. A decrease of about 1.5–3 Hz was found for each considered technique, confirming that better frequency resolution was obtained when working with longer signals.

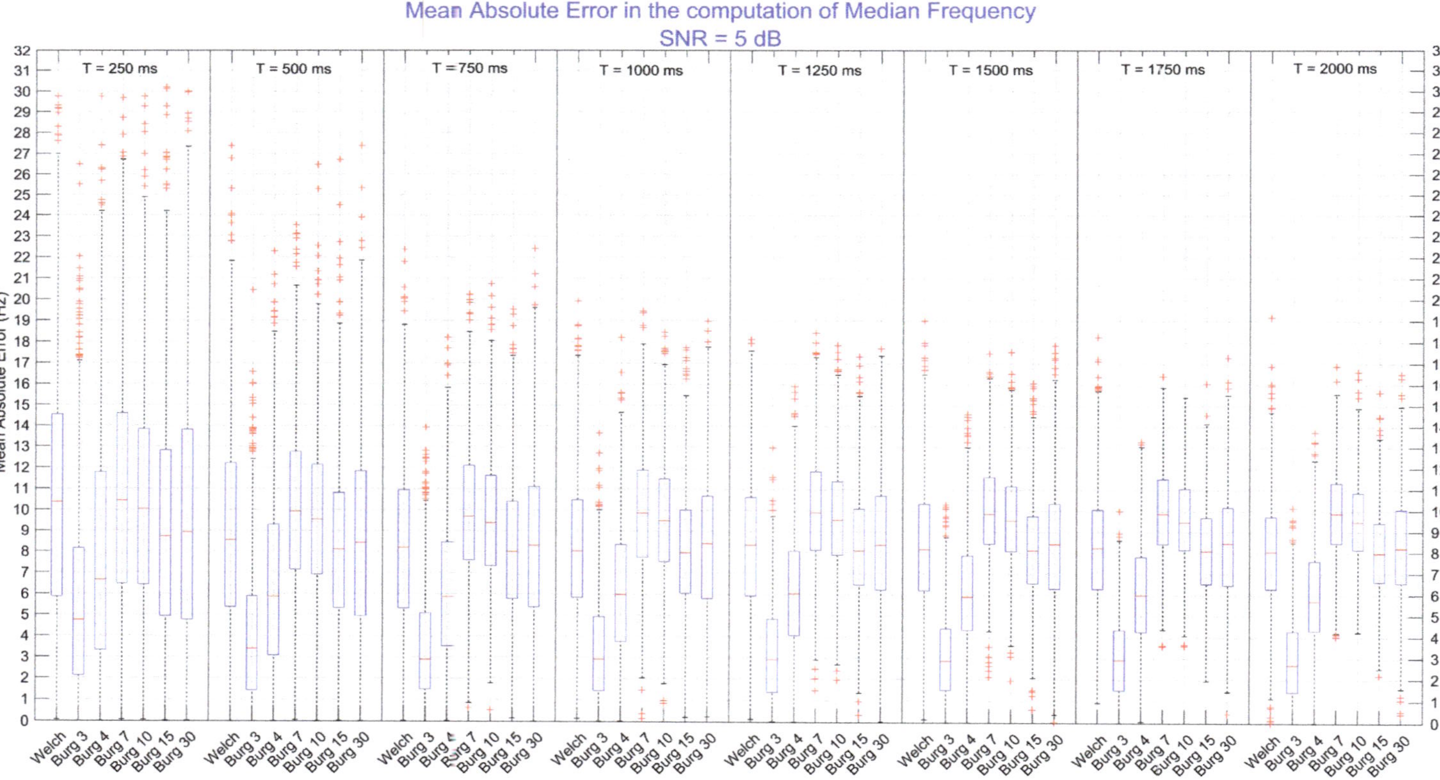

Figure 4. Mean absolute error computed on the median frequency parameters. All values were extracted from power spectral densities that have been estimated by Welch and Burg methods. The figure is divided into 8 subgroups (separated by vertical lines) representing the levels of the *duration* factor (e.g., T = 250 ms, . . . , and T = 2000 ms). Each represented subgroup was computed when the level of noise was high (SNR equal to 5 dB). The significance level was set at $p < 0.05$. The mean differences between the 3rd order and all other levels of *method* were statistically significant ($p < 0.05$).

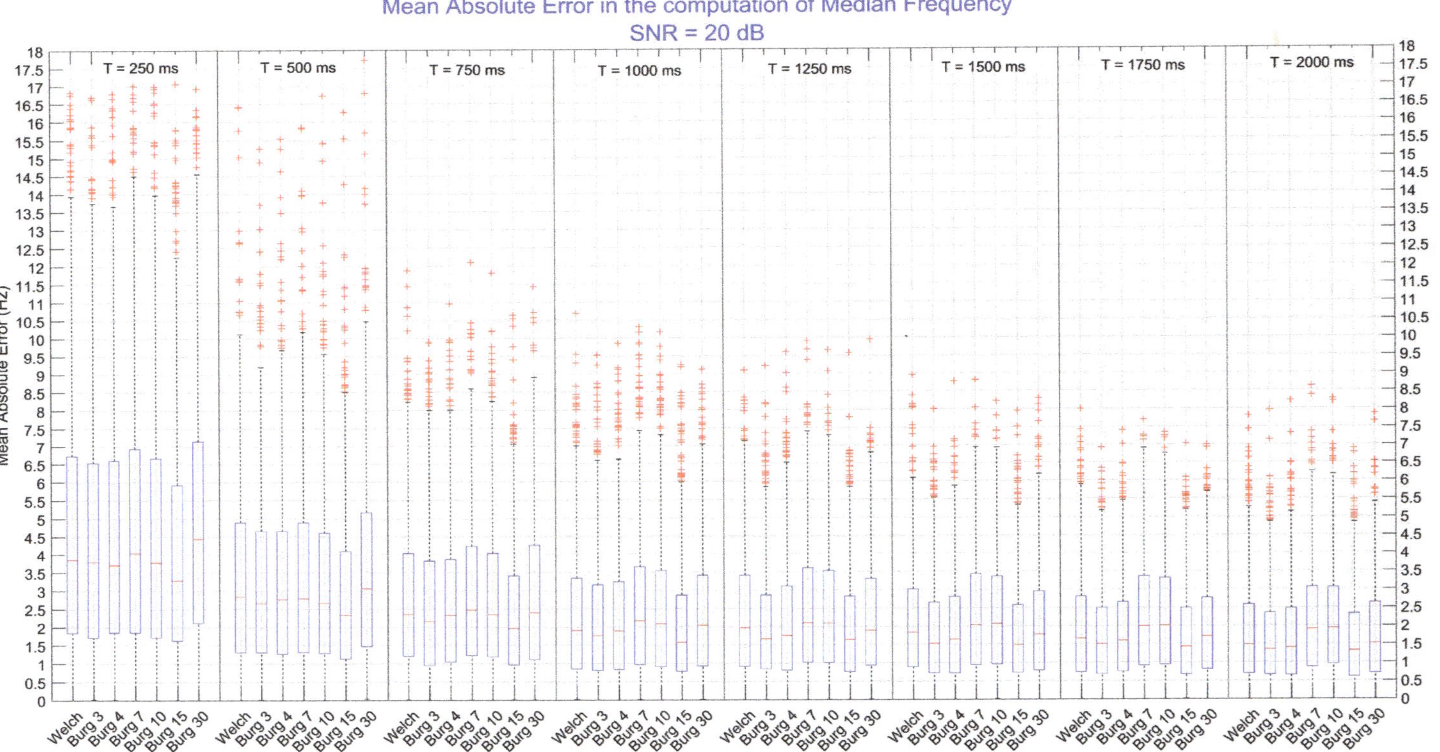

Figure 5. Mean absolute error computed on the median frequency parameters. All values were extracted from power spectral densities that were estimated by Welch and Burg methods. The figure is divided in 8 subgroups (separated by vertical lines) representing the levels of the *duration* factor. Each represented subgroup was computed when the signal-to-noise ratio was high (20 dB). The significance level was set at $p < 0.05$. The 15th order produces the minimum error but the mean differences with the error produced by the other orders and Welch were not significant ($p < 0.05$).

4. Discussion

This study aimed to investigate the effects produced by the compression of the power spectral density, due to muscle fatigue, on the computation of the spectral parameters.

Although the results of this work will give some suggestion in the choice of the method to be used for the extraction of the spectral parameters, a few considerations need to be highlighted to correctly interpret the results. First, this study focused on a single power spectral shape representing an extreme case, that is severe muscle fatigue, which usually might be found in real data when analyzing muscle contractions until failure. Second, these findings, which have been extracted on synthetic sEMG, cannot be validated on real signals because the real value of the spectral parameters is unknown. As a result, the suggestions provided for the choice of the method could only ensure that the error in the spectral estimates will be limited depending on the analyzed condition.

Analyzing the spectra computed with the two estimation techniques, we noticed that the shapes of the spectra were not substantially influenced by the level of SNR (see comparison Figure 1a,c, or Figure 1b,d), except for the low orders (3rd and 4th) of the Burg method: this happens because a few parameters of the model are influenced by the high level of noise and are not able to approximate the shape of the spectrum to the ideal one being truncated at very low frequency. On the other hand, the time duration of the signal influenced the resulting shape: in fact, by increasing the length of the signal, we increased the frequency resolution. Both methods benefit from this increase in frequency resolution, but the Welch method still presented a lot of oscillations over the spectrum, which then affected the computation of the spectral parameter reducing the goodness of the estimates. Although there were no visible effects produced by the SNR on the shape, this factor highly influences the estimate of Mean and Median Frequency, as it can be seen in Table 1.

By analyzing the error produced in computing the Mean Frequency, we noticed that the time duration of the signal had no significant influence on the error, while the SNR had a great significant effect. When the level of noise was high (SNR = 5 and 10 dB), the error generated by the computation of the Mean Frequency was around 50 Hz and 19 Hz, respectively. Therefore, these huge errors are not acceptable, and we suggest avoiding the Mean Frequency use when dealing with noisy signals. Instead, if the level of SNR was high (SNR = 15 and 20 dB), the error around 5–7 Hz and 2–3 Hz, respectively, is still acceptable: results showed that the 3rd order of the Burg model is always the most performant in comparison with Welch and high orders of Burg. These specific results can only be considered valid when dealing with fatiguing contraction that are producing a harsh compression of the power spectrum. This finding is in contrast with the suggestion of always using a 10th order of the autoregressive model given by Farina and Merletti [20], but this is due to the fact that they considered a spectrum shape with the peak around 70–80 Hz that a 3rd order model is not able to approximate well. In contrast, if we need to analyze a compressed spectrum, the truncated shape obtained by the 3rd order (see Figure 1) produced a lower value of the Mean Frequency that was closer to the simulated ideal value. Therefore, in agreement with [19], we recommend decreasing the order of the autoregressive model to compute the Mean Frequency for tracking the development of muscle fatigue. However, the user should be very careful in using the Mean Frequency as an indicator of fatigue because it is highly influenced by the noise of signals, and the high level of error could lead to misleading results.

The analysis performed on the computation of the Median Frequency, instead, revealed that the obtained value was influenced by the interaction effects of the estimation technique, the time duration, and the amount of noise of the signals. As can be seen from Figure 3, even when the level of noise was very high (SNR = 5 dB), the error in the computation of Median Frequency was around 5–10 Hz, depending on the used estimation technique. As the time duration of the signal increased, the dispersion of the error around its mean, instead, was greatly mitigated. These findings confirm that the Median Frequency is more robust because it is less sensitive to noise than the mean frequency [24]. In fact, when the level of noise was low (SNR = 20dB), the error decreased to 4 Hz, and there were no significant differences between the errors produced by the different techniques. Moreover, as the time duration increased, with a consequent increase in the frequency resolution, there was a further reduction in the error down to 2 Hz. These results indicate that high accuracy in the computation of the Median Frequency can be obtained with both Welch and Burg techniques. On the other hand, the precision of the measure mainly depends on the time duration of the signals. For all these reasons, this study proposes to use a low order of the autoregressive model (3rd–4th) to estimate the Median Frequency when high level of muscle fatigue is to be assessed. Median Frequency should be preferred to the Mean Frequency if accurate measures are required even in presence of noise. In general, a normal sEMG shape of the spectrum could be estimated by high orders, (i.e., the 10th or the 6th, as stated in [20] and [19], respectively), but the order of the Burg methods need to be decrease to 3rd or 4th order as soon as muscle fatigue is approached.

5. Conclusions

This study aimed to investigate the effects produced by the compression of the power spectral density toward the low-frequency area; this variation in the frequency content is caused by the progression of muscle fatigue and influences the calculation of the Mean and Median Frequency. Two estimation techniques, Welch and Burg, were compared for the estimation of the power spectral density and the extraction of the spectral parameters. The purpose of this study, moreover, was to describe how the time duration and the level of noise of the signals affect the estimate of the power spectrum when it is harshly compressed in the low-frequency area.

The main finding of this work is that the Median Frequency should be preferred as indicator of muscle fatigue because it is less sensitive to noise than the Mean Frequency [24], always producing errors in the range of 2–10 Hz, according to the specific case. In fact, the use of Mean Frequency should be avoided when dealing with noisy signals (SNR <= 10 dB) because it produced enormous errors that are unacceptable.

In general, by increasing the time duration, and thus increasing the frequency resolution, improvements are produced in precision of the estimation, while increasing the SNR produces improvements in the accuracy of the estimates. Results suggested that the 3rd order of the autoregressive model produced accurate estimates analyzing fatiguing contractions, and therefore it is not necessary to use a high order (the 10th) as stated in [20], that will also increase the complexity and the time computation of the algorithm. These results, though, are valid when we are dealing with a power spectrum very compressed towards the low-frequency area due to the progression of high level of muscle fatigue; however, as stated in [19], the order of the autoregressive model for estimating the spectral parameter is not fixed, but it should be properly changed according to the frequency content of the spectrum that is examined, ranging from the 3rd order in presence of severe muscle fatigue to the 6th/8th order in normal conditions.

Author Contributions: Conceptualization, G.C. and S.C.; Data curation, G.C.; Formal analysis, G.C.; Funding acquisition, S.C.; Investigation, G.C. and S.C.; Methodology, G.C. and S.C.; Project administration, S.C.; Resources, G.C. and S.C.; Software, G.C.; Supervision, S.C.; Validation, G.C.; Visualization, G.C.; Writing—original draft, G.C.; Writing—review & editing, G.C. and S.C. All authors have read and agreed to the published version of the manuscript.

Funding: The research presented in this article was carried out as part of the program BRIC 2019-ID48 funded by INAIL. The funders had no role in study design, data collection and analysis, decision to publish, or preparation of the manuscript.

Conflicts of Interest: The authors declare no conflict of interest.

Abbreviations

The following abbreviations are used in this manuscript:

AR	AutoRegressive model
ARMA	AutoRegressive Moving Average model
LMSE	Least Mean Square Error
MDF	Median Frequency
MAE	Mean Absolute Error
MNF	Mean Frequency
MUs	Motor Units
OV	Overlap
PSD	Power Spectral Density
RMS	Root Mean Square
sEMG	surface Electromyography
SNR	Signal to Noise Ratio
T	Time duration
WL	Window Length

Appendix A

In the following, some additional statistical analyses are showed in order to justify the choices of the settings for the Welch window (length equal to 25% of the original signal and overlap equal to 25% of the length of the window).

Welch Window Length and Overlap

We compared the results in the computation of Mean (MNF) and Median Frequency (MDF) obtained by using several implementations of the Welch method. Since there is the possibility of varying the length of the window used for the computation of the spectrum and the amount of overlap between one window and the next adjacent one, we investigated the effect of these two factors on the accuracy of the spectral parameters estimate. The statistical analyses were performed with a three-ways ANOVA to study the interaction effects among the factor *method*, which identifies the different combination of epoch length and overlap, the factor *duration*, which represents the total length of the analyzed signal, and the *SNR* factor that indicate the level of noise in the signal. For simplicity, we use the abbreviation WL for window length (where the percentage refers to the length of the original signal) and OV for the overlap (where the percentage refers to the length of the window).

Factors for ANOVA analyses:

Method:

1. WL = 50%, OV = 50%
2. WL = 33%, OV = 50%
3. WL = 25%, OV = 50%
4. WL = 50%, OV = 25%
5. WL = 33%, OV = 25%
6. WL = 25%, OV = 25%
7. WL = 50%, OV = 10%
8. WL = 33%, OV = 10%
9. WL = 25%, OV = 10%

Duration:

1. 250 ms
2. 500 ms
3. 750 ms
4. 1000 ms
5. 1250 ms
6. 1500 ms
7. 1750 ms

8. 2000 ms

   SNR:

1. 5 dB
2. 10 dB
3. 15 dB
4. 20 dB

First, we performed a three-way ANOVA, shown in Table A1 considering the mean frequency as response variable.

Table A1. Three-way ANOVA tests were performed to investigate the effects of the three factors on both, mean and median frequency. The three factors used are *method* (Welch and orders of Burg), *duration* (T varying from 250 to 2000 ms) and *SNR* (from 5 to 20 dB). Their interaction effects are showed from the 4th to the 7th row of the table. The significance levels were set at: * $p < 0.05$, ** $p < 0.01$, *** $p < 0.001$. All the values less than 2×10^{-16} were indicated as 0, with the corresponding significance indicated by *** $p < 0.001$.

Source	Sum Sq.	d.f.	Mean Sq.	F	Prob > F
method	688.1	8	86	6.43	0 ***
duration	71,280	7	10,182.9	761	0 ***
SNR	108,516,793.5	3	36,172,264.5	2,703,252.41	0 ***
*method*duration*	817.4	56	14.6	1.09	0.2983
*method*SNR*	123.9	24	5.2	0.39	0.997
*duration*SNR*	6694.1	21	318.8	23.82	0 ***
*method*duration*SNR*	291.2	168	1.7	0.13	1
Error	3,849,879	287,712	13.4		
Total	112,446,567.4	287,999			
Constrained (Type III) sums of squares					

As we can see in the table, there was no interaction effect among the three factors, neither between *method* and *duration* or between *method* and *SNR*, which are the two interaction effects we wanted to investigate. Therefore, we directly analyzed the main effect of the *method* factor to study how different combinations of window length and overlap influence the estimate of MNF. A Tukey post-hoc test was performed to find out which combination of WL and OV performed better, and the results are summarized in the following figure.

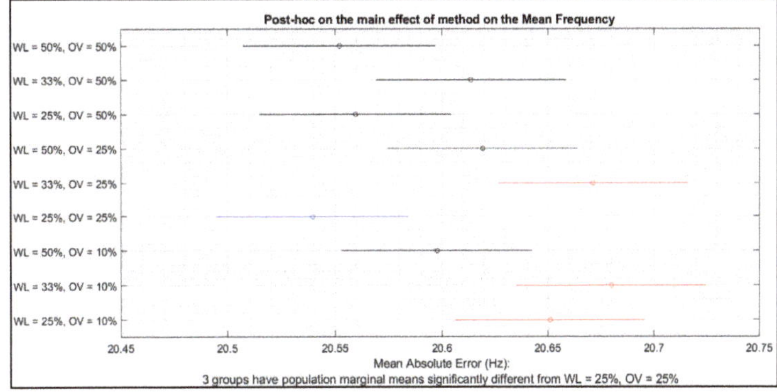

Figure A1. Post-hoc test on the main effect of the *method* factor on the Mean Frequency. Significance was set at 5%. Blue color shows the Window Length (WL)-Overlap (OV) combination that we want to compare with the others. In black are represented all the combinations whose mean is not statistically different ($p > 0.05$) from that one in blue. In red are represented all the combinations whose mean is statistically different ($p < 0.05$) from the one in blue.

We noticed that the combination window length equal to 25% of the total signal length and overlap equal to 25% of the window length itself produced the minimum error that was significantly different from other three combinations. Since similar results were obtained in the analysis of the median frequency, we selected this particular combination (WL = 25%, OV = 25%) of settings for the Welch method and compared this one with the orders of the autoregressive model.

References

1. Campanini, I.; Disselhorst-Klug, C.; Rymer, W.Z.; Merletti, R. Surface EMG in Clinical Assessment and Neurorehabilitation: Barriers Limiting Its Use. *Front. Neurol.* **2020**, *11*, 934. [CrossRef]
2. Felici, F.; Del Vecchio, A. Surface Electromyography: What Limits Its Use in Exercise and Sport Physiology? *Front. Neurol.* **2020**, *11*, 578504. [CrossRef]
3. Nasri, N.; Orts-Escolano, S.; Cazorla, M. An sEMG-Controlled 3D Game for Rehabilitation Therapies: Real-Time Time Hand Gesture Recognition Using Deep Learning Techniques. *Sensors* **2020**, *20*, 6451. [CrossRef]
4. Toledo-Pérez, D.; Martínez-Prado, M.; Gómez-Loenzo, R.; Paredes-García, W.; Rodríguez-Reséndiz, J. A Study of Movement Classification of the Lower Limb Based on up to 4-EMG Channels. *Electronics* **2019**, *8*, 259. [CrossRef]
5. Stein, J.; Narendran, K.; McBean, J.; Krebs, K.; Hughes, R. Electromyography-Controlled Exoskeletal Upper-Limb–Powered Orthosis for Exercise Training After Stroke. *Am. J. Phys. Med. Rehabil.* **2007**, *86*, 255–261. [CrossRef]
6. Parajuli, N.; Sreenivasan, N.; Bifulco, P.; Cesarelli, M.; Savino, S.; Niola, V.; Esposito, D.; Hamilton, T.J.; Naik, G.R.; Gunawardana, U.; et al. Real-Time EMG Based Pattern Recognition Control for Hand Prostheses: A Review on Existing Methods, Challenges and Future Implementation. *Sensors* **2019**, *19*, 4596. [CrossRef]
7. Farina, D.; Stegeman, D.F.; Merletti, R. Biophysics of the generation of EMG signals. In *Surface Electromyography: Physiology, Engineering, and Applications*; Merletti, R., Farina, D., Eds.; John Wiley & Sons, Inc.: Hoboken, NJ, USA, 2016; Chapter 2; pp. 30–53. [CrossRef]
8. Ranaldi, S.; Corvini, G.; De Marchis, C.; Conforto, S. The Influence of the SEMG Amplitude Estimation Technique on the EMG–Force Relationship. *Sensors* **2022**, *22*, 3972. [CrossRef]
9. Merletti, R.; Muceli, S. Tutorial. Surface EMG Detection in Space and Time: Best Practices. *J. Electromyogr. Kinesiol.* **2019**, *49*, 102363. [CrossRef]
10. Clancy, E.A.; Farina, D.; Merletti, R. Cross-Comparison of Time- and Frequency-Domain Methods for Monitoring the Myoelectric Signal during a Cyclic, Force-Varying, Fatiguing Hand-Grip Task. *J. Electromyogr. Kinesiol.* **2005**, *15*, 256–265. [CrossRef] [PubMed]
11. Cifrek, M.; Medved, V.; Tonković, S.; Ostojić, S. Surface EMG Based Muscle Fatigue Evaluation in Biomechanics. *Clin. Biomech.* **2009**, *24*, 327–340. [CrossRef] [PubMed]
12. González-Izal, M.; Malanda, A.; Gorostiaga, E.; Izquierdo, M. Electromyographic Models to Assess Muscle Fatigue. *J. Electromyogr. Kinesiol.* **2012**, *22*, 501–512. [CrossRef] [PubMed]
13. Phinyomark, A.; Thongpanja, S.; Hu, H.; Phukpattaranont, P.; Limsakul, C. The Usefulness of Mean and Median Frequencies in Electromyography Analysis. In *Computational Intelligence in Electromyography Analysis—A Perspective on Current Applications and Future Challenges*; Naik, G.R., Ed.; IntechOpen: London, UK, 2012. [CrossRef]
14. Corvini, G.; Holobar, A.; Moreno, J.C. On Repeatability of MU Fatiguing in Low-Level Sustained Isometric Contractions of Tibialis Anterior Muscle. In *Converging Clinical and Engineering Research on Neurorehabilitation IV*; Torricelli, D., Akay, M., Pons, J.L., Eds.; Biosystems & Biorobotics; Springer International Publishing: Cham, Switzerland, 2022; Volume 28, pp. 909–913. [CrossRef]
15. Kay, S.M.; Marple, S.L. Spectrum Analysis—A Modern Perspective. *Proc. IEEE* **1981**, *69*, 1380–1419. [CrossRef]
16. Hof, A.L. Errors in Frequency Parameters of EMG Power Spectra. *IEEE Trans. Biomed. Eng.* **1991**, *38*, 1077–1088. [CrossRef]
17. Merletti, R.; Balestra, G.; Knaflitz, M. Effect of FFT Based Algorithms on Estimation of Myoelectric Signal Spectral Parameters. In *Images of the Twenty-First Century, Proceedings of the Annual International Engineering in Medicine and Biology Society, Seattle, WA, USA, 9–12 November 1989*; IEEE: Piscataway, NJ, USA, 1989; Volume 3, pp. 1022–1023. [CrossRef]
18. Mañanas, M.A.; Jané, R.; Fiz, J.A.; Morera, J.; Caminal, P. Influence of Estimators of Spectral Density on the Analysis of Electromyographic and Vibromyographic Signals. *Med. Biol. Eng. Comput.* **2002**, *40*, 90–98. [CrossRef] [PubMed]
19. Corvini, G.; D'Anna, C.; Conforto, S. Estimation of Mean and Median Frequency from Synthetic SEMG Signals: Effects of Different Spectral Shapes and Noise on Estimation Methods. *Biomed. Signal Process. Control* **2022**, *73*, 103420. [CrossRef]
20. Farina, D.; Merletti, R. Comparison of Algorithms for Estimation of EMG Variables during Voluntary Isometric Contractions. *J. Electromyogr. Kinesiol.* **2000**, *10*, 337–349. [CrossRef]
21. Lowery, M.M.; Vaughan, C.L.; Nolan, P.J.; O'Malley, M.J. Spectral Compression of the Electromyographic Signal Due to Decreasing Muscle Fiber Conduction Velocity. *IEEE Trans. Rehab. Eng.* **2000**, *8*, 353–361. [CrossRef] [PubMed]
22. Zhang, Z.G.; Liu, H.T.; Chan, S.C.; Luk, K.D.K.; Hu, Y. Time-Dependent Power Spectral Density Estimation of Surface Electromyography during Isometric Muscle Contraction: Methods and Comparisons. *J. Electromyogr. Kinesiol.* **2010**, *20*, 89–101. [CrossRef]
23. Bonato, P.; D'Alessio, T.; Knaflitz, M. A Statistical Method for the Measurement of Muscle Activation Intervals from Surface Myoelectric Signal during Gait. *IEEE Trans. Biomed. Eng.* **1998**, *45*, 287–299. [CrossRef]

24. Stulen, F.B.; De Luca, C.J. Frequency Parameters of the Myoelectric Signal as a Measure of Muscle Conduction Velocity. *IEEE Trans. Biomed. Eng.* **1981**, *BME-28*, 515–523. [CrossRef]
25. Cao, L.; Wang, Y.; Hao, D.; Rong, Y.; Yang, L.; Zhang, S.; Zheng, D. Effects of Force Load, Muscle Fatigue, and Magnetic Stimulation on Surface Electromyography during Side Arm Lateral Raise Task: A Preliminary Study with Healthy Subjects. *BioMed Res. Int.* **2017**, *2017*, e8943850. [CrossRef] [PubMed]
26. Puce, L.; Pallecchi, I.; Marinelli, L.; Mori, L.; Bove, M.; Diotti, D.; Ruggeri, P.; Faelli, E.; Cotellessa, F.; Trompetto, C. Surface Electromyography Spectral Parameters for the Study of Muscle Fatigue in Swimming. *Front. Sports Act. Living* **2021**, *3*, 644765. [CrossRef] [PubMed]
27. Rinaldi, M.; D'Anna, C.; Schmid, M.; Conforto, S. Assessing the Influence of SNR and Pre-Processing Filter Bandwidth on the Extraction of Different Muscle Co-Activation Indexes from Surface EMG Data. *J. Electromyogr. Kinesiol.* **2018**, *43*, 184–192. [CrossRef]
28. Oppenheim, A.V.; Schafer, R.W. *Digital Signal Processing*; Prentice-Hall: Hoboken, NJ, USA, 1975; pp. 548–554. ISBN 0-13-214635-5.
29. Bartlett, M.S. Periodogram analysis and continuous spectra. *Biometrika* **1950**, *37*, 1–16. [CrossRef]
30. Welch, P. The use of fast Fourier transform for the estimation of power spectra: A method based on time averaging over short, modified periodograms. *IEEE Trans. Audio Electroacoust.* **1967**, *15*, 70–73. [CrossRef]
31. Burg, J.P. Maximum Entropy Spectral Analysis. In Proceedings of the 37th Meeting Society of Exploration Geophysics, Oklahoma City, OK, USA, 31 October 1967.

Article

Muscle Co-Contraction Detection in the Time–Frequency Domain

Francesco Di Nardo *, Martina Morano, Annachiara Strazza and Sandro Fioretti

Department of Information Engineering, Università Politecnica delle Marche, Via Brecce Bianche,
60131 Ancona, Italy; s1107637@studenti.univpm.it (M.M.); a.strazza@pm.univpm.it (A.S.);
s.fioretti@staff.univpm.it (S.F.)
* Correspondence: f.dinardo@staff.univpm.it

Abstract: Background: Muscle co-contraction plays a significant role in motion control. Available detection methods typically only provide information in the time domain. The current investigation proposed a novel approach for muscle co-contraction detection in the time–frequency domain, based on continuous wavelet transform (CWT). Methods: In the current study, the CWT-based cross-energy localization of two surface electromyographic (sEMG) signals in the time–frequency domain, i.e., the CWT coscalogram, was adopted for the first time to characterize muscular co-contraction activity. A CWT-based denoising procedure was applied for removing noise from the sEMG signals. Algorithm performances were checked on synthetic and real sEMG signals, stratified for signal-to-noise ratio (SNR), and then validated against an approach based on the acknowledged double-threshold statistical algorithm (DT). Results: The CWT approach provided an accurate prediction of co-contraction timing in simulated and real datasets, minimally affected by SNR variability. The novel contribution consisted of providing the frequency values of each muscle co-contraction detected in the time domain, allowing us to reveal a wide variability in the frequency content between subjects and within stride. Conclusions: The CWT approach represents a relevant improvement over state-of-the-art approaches that provide only a numerical co-contraction index or, at best, dynamic information in the time domain. The robustness of the methodology and the physiological reliability of the experimental results support the suitability of this approach for clinical applications.

Keywords: surface EMG signal; co-contraction detection; muscular synergies; the time–frequency domain; wavelet transform

Citation: Di Nardo, F.; Morano, M.; Strazza, A.; Fioretti, S. Muscle Co-Contraction Detection in the Time–Frequency Domain. *Sensors* **2022**, *22*, 4886. https://doi.org/10.3390/s22134886

Academic Editor: Enrico G. Caiani

Received: 17 May 2022
Accepted: 24 June 2022
Published: 28 June 2022

Publisher's Note: MDPI stays neutral with regard to jurisdictional claims in published maps and institutional affiliations.

1. Introduction

Muscle co-contraction is defined as the concurrent activation of agonist and antagonist muscles crossing a targeted joint [1]. It plays a significant role in motion control during physiological activities related to motor learning. Specifically, in able-bodied subjects, co-contraction is a mechanism that occurs to achieve a homogeneous pressure on a joint's surface, preserving the articular stability and controlling its mechanical impedance [2]. Thus, it occurs frequently in everyday activities such as learning a motor task or handling a tool or an object [3]. In the elderly, injured, and pathological individuals, co-activations play a key role in developing compensation strategies by enhancing joint stability [4,5]. Furthermore, increased co-contraction levels have been detected in orthopedic and neuromuscular patients in order to generate additional joint stiffness to improve joint stability [6,7]. Thus, a quantitative assessment of muscle co-contraction could be considered a meaningful tool that could help us to gain a deeper understanding of how pathology can affect the muscle recruitment strategies during different tasks, including walking.

Several different techniques have been introduced to assess muscle co-contraction, however, a gold standard is not available yet. Advanced mathematical models based on muscle moment assessment have been adopted [8]. Furthermore, simple indexes based on the analysis of electromyography (EMG) signals are potentially very suitable for clinical application [9–13]. Falconer and Winter [13] introduced a co-contraction index (CI) based on the computation of areas under the curve of rectified EMG signals from antagonist muscles. Similar formulations were also reported later [9,10]. Although it is a long-standing index, CI is still commonly held nowadays to provide a rough clinical evaluation of co-contraction [14,15]. However, CI suffers from one main limitation: it provides a single numerical value, which only represents the intensity of the simultaneous muscle activation (the co-contraction level) of each pair of muscles [8]. No information is provided on how long the co-contraction lasts and when it occurs. This information is not clinically negligible, indeed, it has been demonstrated that the modified duration of the co-contraction could be a marker of impairment or could affect the metabolic cost [5,16]. Rudolph et al. tried to overcome this limitation by developing a dynamic co-contraction index that describes the temporal and magnitude components of the EMG signals from antagonist muscles [11]. However, a recent study of young adults showed that Rudolph's index may present poor reliability during gait [17].

Further studies, including those conducted by the present group of researchers, have quantified muscle co-contraction as the time range where the EMG activity of the two muscles is superimposed [5,18,19]. In this procedure, the onset and offset of a single muscle activity were typically assessed by applying reliable algorithms, such as [20,21]. The overlapping of the activation intervals of the two antagonist muscles provided the time range of the co-contraction. The beginning and the end of this time range have been acknowledged as the onset and the offset of the co-contraction, respectively. All these algorithms reported a good accuracy in assessing the onset and offset of a single muscle's activity. However, for assessing the co-contraction interval, the algorithm must be applied twice, once for each muscle. Thus, the identification of the co-contraction onset and offset could suffer from the propagation of uncertainty.

To the best of our knowledge, no study is reported in the literature that tries to characterize the frequency content of muscle co-contraction. Typically, frequency analysis is used to quantify the muscle fatigue process in stationary isometric contraction by means of traditional frequency-based techniques, such as Fourier transform [22]. Advanced time–frequency techniques, such as wavelet transform, have been adopted to study the spectral properties of EMG signals over time, trying to characterize the frequency content of muscle activation [23,24]. However, no attempt was made to assess the frequency content of muscle co-contraction.

The purpose of the current work was to propose a novel approach for assessing muscle co-contraction in the time–frequency domain, using the cross-energy localization of the surface electromyographic (sEMG) signals provided by the continuous wavelet transform (CWT) analysis. Specifically, this approach aims to assess a single co-contraction signal and is able to quantify muscular co-contraction in terms of the time interval (onset/offset), the frequency band (maximum–minimum), and the amplitude. CWT is an advanced signal processing technique that maps a time waveform into the time–frequency domain, providing a lossless representation of non-stationary signals in the time and frequency domains. This multiresolution analysis allowed us to define the time–frequency energy density of a signal and to provide a localized statistical assessment of the time–frequency cross-energy density between two signals, i.e., the CWT coscalogram function. The CWT coscalogram was successfully adopted to test the cross-correlation between two different bio-signals [25]. As far as we know, except for a preliminary effort made by this same research group [26], this is the first essay to interpret the CWT coscalogram function between the sEMG signals from two antagonist muscles as the muscular co-contraction activity in the time–frequency domain.

2. Materials and Methods

2.1. Co-Contraction Detection

A methodology based on continuous wavelet transform (CWT) was adopted to assess muscular co-contraction. CWT is a flexible approach to signal decomposition. CWT is a time–frequency approach, which provides a lossless representation of non-stationary signals in time and frequency domains concomitantly, as represented below:

$$CWT_{sEMG}(t,a,b) = \int sEMG(t) \cdot \psi^*_{a,b}(t)\, dt \quad a \neq 0 \tag{1}$$

where $sEMG(t)$ is the input signal and $\psi_{a,b}(t)$ is the so-called mother wavelet at scale a and time b, this is represented by Equation (2), as follows:

$$\psi_{a,b}(t) = \frac{1}{\sqrt{a}}\, \psi\left(\frac{t-b}{a}\right) \tag{2}$$

In the present study, CWT analysis was achieved by using the Daubechies of order 4 (factorization in 6 levels) as the mother wavelet. This choice was based on the similarity of Daubechies mother wavelet to the shape of motor unit action potentials [27].

A process for removing noise from the sEMG was performed by applying CWT denoising [28]. The soft Donoho threshold was employed to this aim. The sEMG signal was reconstructed by revised CWT coefficients. Efficacy of the denoising procedure was tested by evaluating the signal-to-noise ratio (SNR) before and after the application of the denoising procedure. SNR was determined by Equation (3), as follows:

$$SNR(\sigma_s, \sigma_n) = 10 \times \log \frac{(\sigma_s^2)}{(\sigma_n^2)} \tag{3}$$

where σ_s and σ_n are the signal and noise standard deviation (SD), respectively. The scalogram function $P_{sEMG}(a,b)$ has been used to represent the energy localization in the time–frequency domain, according to Equation (4), as follows:

$$P_{sEMG}(a,b) = |W_{sEMG}(a,b)|^2 \tag{4}$$

where $W_{sEMG}(a,b)$ is the matrix of CWT coefficients at time b and scale a for the $sEMG(t)$ signal. In the present study, the CWT scalogram function in the time–frequency domain of denoised $sEMG$ signal was adopted for assessing muscular activation.

Local cross-correlation between two signals could be identified by computing CWT cross-energy density between signals by means of CWT coscalogram function [25]. For two sEMG signals ($sEMG_1(t)$ and $sEMG_2(t)$), CWT coscalogram function, $PW_{sEMG}(a,b)$, was computed as follows:

$$PW_{sEMG}(a,b) = W_{sEMG_1}(a,b) \cdot W^*_{sEMG_2}(a,b) \tag{5}$$

where $W_{sEMG_1}(a,b)$ and $W_{sEMG_2}(a,b)$ represent the matrices of CWT coefficients of the two denoised signals, at scale a and time b, the operator * represents the conjugate complex. In the present study, CWT coscalogram function in the time–frequency domain of denoised sEMG signals was adopted for assessing the co-contraction signal between the selected muscles. Co-contraction timing was computed in a single stride as the beginning (onset) and the end (offset) of the time interval when the coscalogram function surpassed 1% of the cross-energy-density peak in the selected stride [29]. Once the co-contraction interval was detected in the time domain, the correspondent co-contraction content in the frequency domain was computed as the frequency range associated with the coscalogram function to that specific time interval. The maximum and minimum of the frequency content were, thus, quantified for each one of the co-contractions assessed in time domain. A block diagram describing CWT procedure for the assessment of the muscle co-contraction signal is presented in Figure 1.

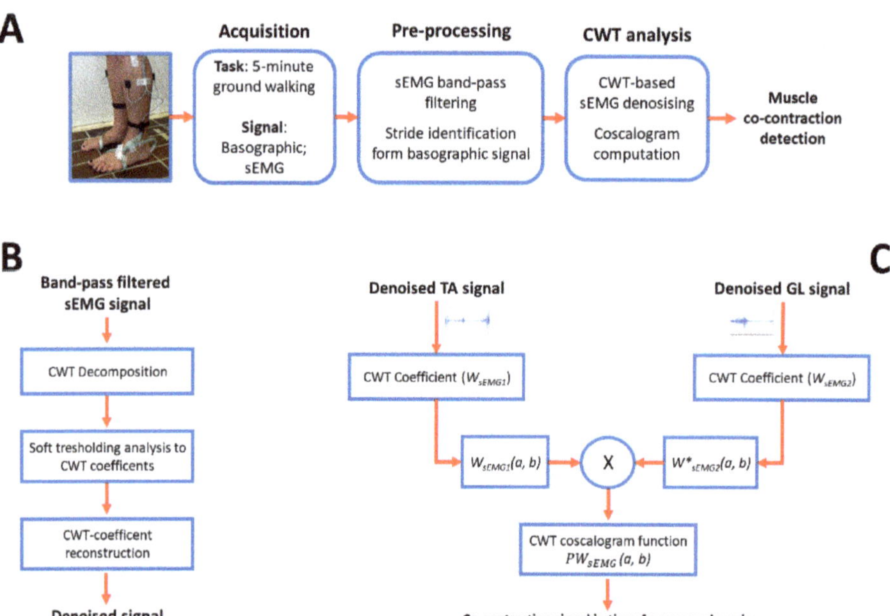

Figure 1. (**A**) Scheme of the entire experimental processing, from data acquisition to muscle co-contraction detection; (**B**) Block diagram representing CWT denoising procedure; (**C**) Block diagram representing the CWT-based procedure for the assessment of muscle co-contraction signals. TA and GL mean tibialis anterior and gastrocnemius lateralis, respectively.

2.2. Simulation Study

A simulation study was conducted for evaluating the performance of the CWT-based procedure in the assessment of the co-contraction signal. It is acknowledged that simulated sEMG signals can be modeled as a process by adding background uncorrelated noise to a bandlimited stochastic process, with zero-mean Gaussian distribution of amplitude and fixed power level, according to [30]. In this study, this distribution was obtained by bandpass filtering a Gaussian series of uncorrelated samples. The bandpass filter cutoff frequencies were 80 and 120 Hz, respectively. This Gaussian distribution was truncated to simulate the sEMG activity due to the muscle activation [30]. Background uncorrelated noise was achieved by a further independent zero-mean Gaussian distribution. Each simulated sEMG signal was generated with a sampling frequency of fs = 1000 Hz, and a time window of =1 s. Different simulated sEMG signals were created by varying the standard deviation, σ, and the time support, $2 \times \alpha \times \sigma$, of the Gaussian distribution to simulate the physiological variability associated with the recruitment of different muscles. The variation in σ was achieved according to the desired value of the SNR. The following four different SNR values were considered: 5 dB, 10 dB, 15 dB, and 20 dB. Each synthetized signal passed the Anderson's whiteness test ($p < 0.05$). An example of simulated sEMG signals representing the activity of two muscles is reported in Figure 2.

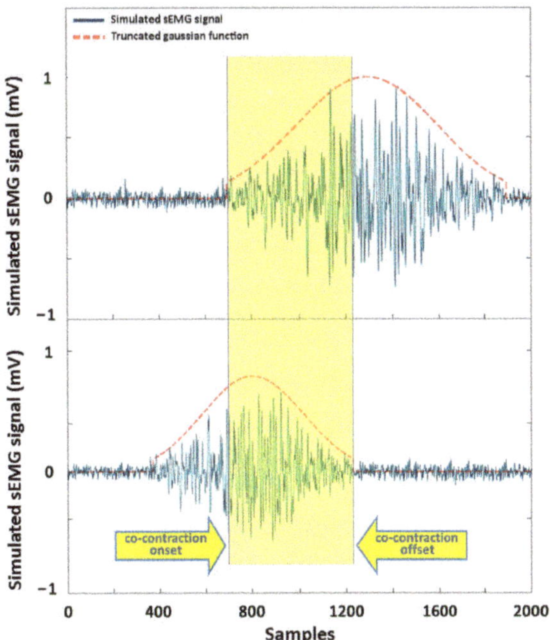

Figure 2. Example of two simulated sEMG signals (in blue), suitable for co-contraction detection. Truncated Gaussian function used to simulate the sEMG activity of each muscle is depicted in dashed red lines. Yellow area represents the co-contraction interval between the two signals. The two yellow arrows indicate the onset and the offset of simulated co-contraction, adopted as ground truth.

Performance of the CWT approach for co-contraction detection was evaluated in the time domain by a direct comparison with the ground truth. Specifically, the start and the end of the truncated Gaussian function used to model the simulated signal (Figure 2) were adopted as the ground-truth events for the onset and offset of the activation of a single muscle, respectively. Next, the ground truth for the onset and offset of muscle co-contraction was computed as the beginning and the end of the time interval, where the concomitant presence of the two simulated sEMG signals was detected (yellow area in Figure 2), following the indication reported in [5,18,19]. The beginning sample and the end sample of this co-contraction interval were the ground-truth onset and offset of the co-contraction, respectively. Figure 2 depicts this procedure. Co-contraction interval assessed by the proposed CWT approach was directly compared with the ground-truth co-contraction interval, in terms of onset and offset events, to evaluate the performance of the approach. Results are reported as absolute error (AE) and time delay (TD). AE was computed as the absolute value of the time distance between the predicted event and the corresponding reference event. TD was computed as the relative value (with sign) of the same time distance. Signs "−" and "+" were adopted to indicate that the predicted event occurred earlier and later than the corresponding value in the reference signal, respectively.

2.3. Experimental Study

Experimental data included the foot–floor contact and sEMG signals collected during the walking of 30 healthy young adults, which was retrospectively taken from the dataset built up at Movement Analysis Lab, Università Politecnica delle Marche, Ancona, Italy. The data are freely available by consulting the public repository of medical research data PhysioNet [31–33]. Subjects with a body mass index (BMI) lower than 18 kg/m^2 (underweight) and higher than 25 kg/m^2 (overweight) were ruled out from the investigation. Patients who communicated manifest disorders, diseases, pain, or who had undergone surgical intervention were kept out of the considered population.

The acquisition system Step32 (Medical Technology, Turin, Italy) was used to acquire all the walking signals. A sampling rate of 2000 Hz and a resolution of 12 bit were adopted for the acquisition procedure. Differential sensors were employed to measure sEMG signals over the following two ankle muscles of each leg during 5 min of ground walking: the gastrocnemius lateralis (GL) and the tibialis anterior (TA). The procedure of positioning the sEMG probes was conducted according to the SENIAM recommendation for sensor locations on muscles in the lower leg and foot [34]. Characteristics of the probes were as follows: single differential probes with fixed geometry, constituted by Ag/Ag-Cl disks; size—7 mm × 27 mm × 19 mm; electrode diameter—4 mm; interelectrode distance—12 mm; gain—1000; high-pass filter—10 Hz; and input impedance, >1.5 GΩ, CMRR > 126 dB. The sEMG signals were bandpass filtered (20–450 Hz) and then processed as reported in Section 2.1. CWT scalogram and coscalogram were computed separately in every single stride.

Footswitches, applied on the sole of each foot (heel, first, and fifth metatarsal heads), were utilized to synchronously capture basographic data during the same task. An eight-level coded basographic signal was acquired from three foot switches (size: 11 mm × 11 mm × 0.5 mm; activation force: 3 N), which were beneath the heel, first, and fifth metatarsal heads of each foot. Foot-switch signals were converted to the following four levels: heel contact (H), flat foot contact (F), push off (P), and swing (S). They were processed to identify the beginning and the end of each stride, i.e., the gait cycle [35]. Subjects walked barefoot on level ground for approximately 5 min at their natural speed and pace, following an eight-shaped path, which included rectilinear segments and curves. This study was fulfilled observing the ethical principles of the Helsinki Declaration and was approved by the local ethical committee.

Unlike the simulated signals, no ground truth was available for the experimental dataset. Thus, the direct computation of detection accuracy, AE, and TD was not possible. Thus, the performances of the proposed CWT approach have been evaluated in the time domain by a direct comparison with the outcomes achieved by using a reference approach. Reference values were assessed as follows: The onset and offset timing of TA and GL activity was identified on denoised sEMG signals by applying the double-threshold statistical detector (DT), an approach particularly suitable for application in the walking task [36]. Onset and offset of co-contraction intervals were computed as the beginning and the end of the superimposition between the activation intervals, assessed by DT, for TA and GL in the same stride. These values were adopted as reference values of the co-contraction onset and offset.

2.4. Statistics

The significant difference of the parameter distribution was determined through the following statistical tests. The normality of each distribution was tested by adopting the Shapiro–Wilk test. The possible significance of the statistical difference was tested for the following: (1) for normal distributions by means of the two-tailed, non-paired Student's *t*-test, between two distributions and using analysis of variance (ANOVA) for multi-group comparison, and (2) for non-normal distributions through the Mann–Whitney test, between two distributions and through Kruskal–Wallis' test for multi-group comparison. 5% was the threshold adopted for detecting the test significance.

3. Results

3.1. Simulation Study

Figure 3 shows the two-dimensional color representation of the CWT scalogram function for the two simulated sEMG signals (panel A and B, respectively). Panel C of the same figure depicts the cross-energy density in the time–frequency domain between the denoised simulated sEMG signals, represented by the CWT coscalogram function, i.e., the estimated co-contraction signal.

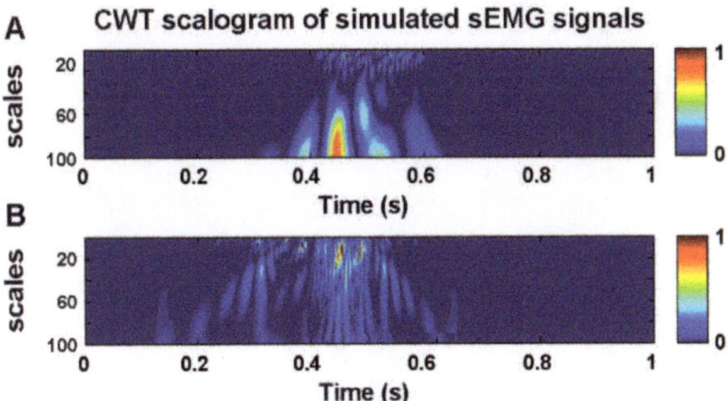

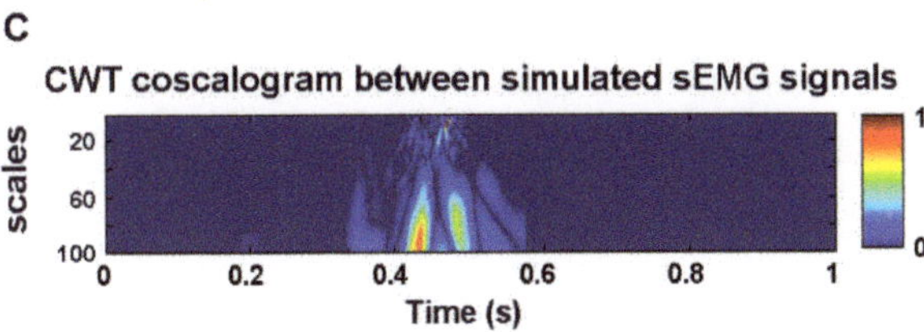

Figure 3. Two-dimensional color representation of CWT scalogram function for two simulated sEMG signals (**A**,**B**). (**C**) shows the CWT coscalogram between the simulated sEMG signals in (**A**,**B**).

The accuracy in predicting the co-contraction onsets and offsets in the whole dataset of the simulated sEMG signals was 100%. This means that the proposed CWT approach was able to identify all the simulated co-contraction intervals (recall = 100%) and that no false positives were detected (precision = 100%). The prediction errors for the co-contraction onset and offset were quantified in terms of AE and TD. The mean, SD, median, 25th percentile, and 75th percentile for each SNR value are depicted in Figure 4 and further reported for the TD in Table 1. The AE values decreased with an increase in the SNR from 5 to 20 dB. The mean and median TD values for both the onset and offset fell in the [−5 ms; 5 ms] range for all the SNR values. The AE values computed for SNR = 5 dB were significantly higher ($p < 0.05$) than all the AE values computed in the other three SNR conditions for the detection of both the onset and offset instants of the co-contraction.

Table 1. Time delay (TD) ± SD computed over simulated sEMG co-contractions per SNR value, expressed in dB.

	SNR (dB)	Mean (ms)	SD (ms)	Median (ms)	25th Percentile (ms)	75th Percentile (ms)
Onset	5	0.52	14.6	−4.0	−9.1	9.2
	10	0.96	6.9	−1.2	−4.0	8.1
	15	0.10	6.4	2.1	−4.3	5.2
	20	−1.16	4.3	−1.4	−3.2	2.4
Offset	5	−1.80	7.9	−3.1	−7.0	5.4
	10	0.76	4.3	−2.2	−4.2	3.3
	15	0.52	4.0	−1.2	−4.1	3.0
	20	0.20	2.5	−1.0	−2.1	1.4

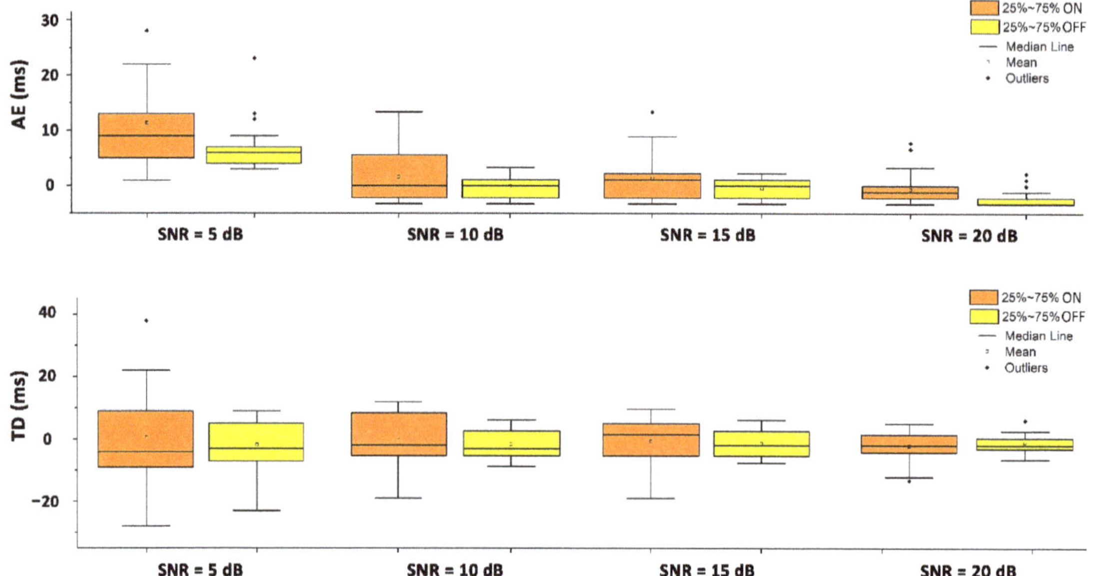

Figure 4. Whisker plots of absolute error (AE, upper panel) and time delay (TD, lower panel) computed over simulated sEMG co-contractions per SNR value expressed in dB. Orange bars represent the onset values. Yellow bars represent the offset values. The AE values computed for SNR = 5 dB were significantly higher ($p < 0.05$) than all the AE values computed in the other three SNR conditions for onset and offset instants of the co-contraction.

3.2. Experimental Study

Detailed values of the SNR for raw sEMG signals (SNR raw) and for sEMG signals after CWT denoising (SNR denoised) for the whole population are reported in Table 2. The mean, SD, median, 25th percentile, and 75th percentile are reported at the bottom of the table.

An increase in the SNR value after the denoising procedure was reported in each single signal. An SNR improvement ($p < 0.05$) was detected for both the TA and the GL in terms of the mean, SD, median, 25th percentile, and 75th percentile value. The two-dimensional color representation of the CWT scalogram function for the TA (panel A) and the GL (panel B) denoised sEMG signals is reported in Figure 5, for a representative stride from subject five (low SNR). Panel C of the same figure shows the cross-energy density in the time–frequency domain between the denoised TA and GL signals, represented by the CWT coscalogram function, i.e., the estimated co-contraction signal in a representative stride in the walking task. The 3D representation of the CWT coscalogram function for the same stride is depicted in Figure 6.

The average co-contraction intervals achieved in each subject by the application of the proposed CWT approach are illustrated as horizontal bars in the percentage of gait cycle, in Figure 7. The co-contractions detected in the stance and swing phase are highlighted in blue and red, respectively.

The performance of the CWT approach in providing the co-contraction onset and offset was assessed by a direct comparison with the DT algorithm in the experimental sEMG signals from thirty subjects, including a total of 16,315 strides, resulting in at least 100 co-contraction bursts per subject. The results for each subject are reported in Figure 8. No significant differences were identified between the average values over the whole population, achieved by the two approaches for the co-contraction onset and offset. Furthermore, no significant difference was detected in each single subject ($p > 0.05$, Figure 8).

Table 2. SNR of the raw sEMG signals (SNR raw) and after CWT denoising (SNR denoised) for tibialis anterior and gastrocnemius lateralis.

	Tibialis Anterior		Gastrocnemius Lateralis	
Subject	SNR Raw	SNR Denoised	SNR Raw	SNR Denoised
1	8.9	17.7	13.1	39.1
2	4.7	8.5	5.8	17.6
3	13.3	17.5	16.4	17.5
4	9.8	12.7	5.0	7.7
5	7.3	9.1	8.7	10.5
6	12.6	19.2	16.1	18.5
7	24.7	30.8	10.7	15.4
8	23.5	32.3	20.5	33.0
9	4.5	9.3	3.3	7.3
10	2.4	4.3	6.0	8.2
11	27.3	35.4	16.3	21.7
12	22.1	29.7	11.3	16.8
13	17.2	21.7	13.1	19.4
14	16.4	19.3	18.5	20.7
15	5.5	9.8	13.9	18.5
16	18.1	24.3	14.9	15.4
17	25.3	32.9	17.4	20.5
18	27.1	31.2	25.8	30.7
19	11.5	18.1	12.7	15.1
20	16.2	18.9	17.0	20.1
21	14.2	16.7	18.1	19.4
22	12.2	16.4	13.9	15.8
23	20.2	25.6	12.9	17.4
24	11.8	20.9	18.9	21.7
25	13.1	18.8	21.0	24.4
26	17.8	19.9	20.5	23.7
27	14.6	17.6	15.6	18.3
28	5.4	9.8	3.9	8.5
29	17.3	20.5	12.5	15.1
30	10.6	15.7	16.8	23.1
Mean	14.5	19.5	14.0	18.7
SD	7.0	8.0	5.5	7.1
Median	13.8	18.9	14.4	18.4
25th percentile	9.8	15.7	11.3	15.4
75th percentile	18.1	24.3	17.4	21.7

All the values are expressed in dB. SNR—signal-to-noise ratio. SNR distributions after denoising (SNR denoised) were significantly different ($p < 0.05$) from the corresponding value of the SNR raw, for both TA and GL.

The current CWT algorithm was also able to provide the frequency content of each one of the co-contractions detected in the time domain, as the frequency of the coscalogram signal in the specific time range where the co-contraction was detected. The values of the maximum frequency of each co-contraction signal are shown in Figure 9. Specifically, each bar in Figure 9 represents the maximum frequency of each one of the co-contractions represented in Figure 7. This was computed as the average value over all the maximum frequency values assessed in each stride, where the co-contraction was detected in that specific subject. For example, the first blue bar in Figure 9 is the average maximum frequency computed in the first co-contraction of subject one (in Figure 7); the second blue bar is the average maximum frequency computed in the first co-contraction of subject one (in Figure 7), and so on. The white space between the colored bars represents the absence of a specific co-contraction in a subject (in stance or in swing). The values of the minimum frequency of each co-contraction signal, computed as for the maximum frequency, are shown in Figure 10. The separation between the co-contractions in stance and swing is still respected here, as in Figure 7.

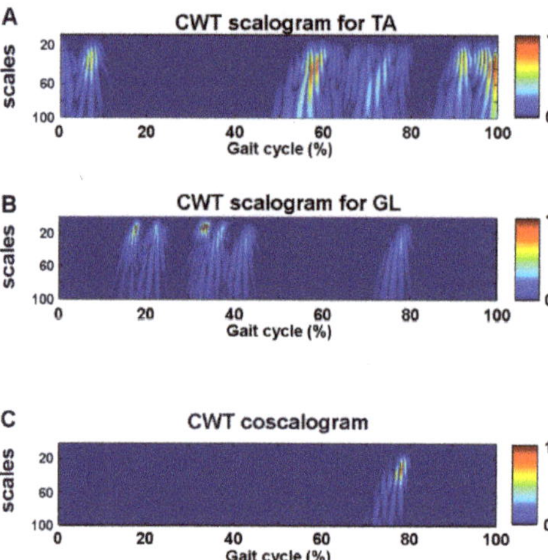

Figure 5. Two-dimensional color representation of CWT scalogram function for TA (**A**) and GL (**B**) denoised sEMG signals and CWT coscalogram between TA and GL (**C**) for a representative stride from subject five.

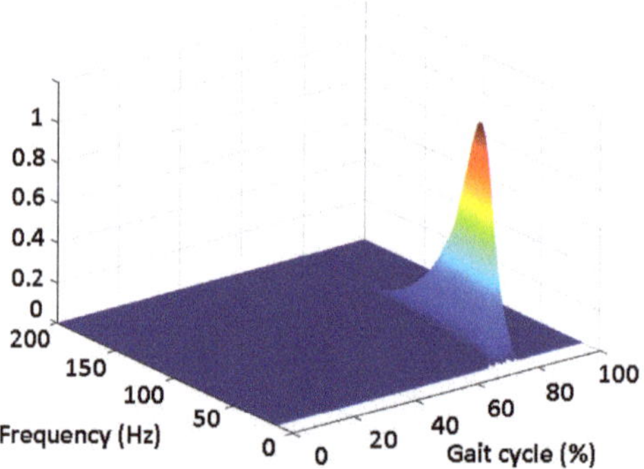

Figure 6. Three-dimensional color representation of CWT coscalogram between TA and GL for the same representative stride of subject five, depicted in Figure 5.

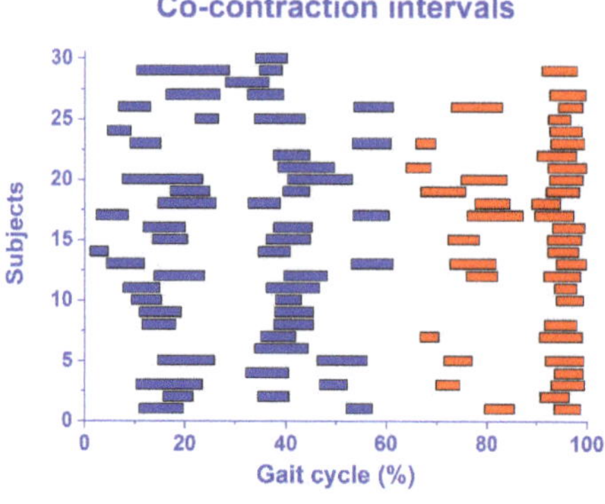

Figure 7. All the average co-contraction intervals detected in stance (blue bars) and in swing (red bars) in each one of the thirty subjects of the population. Values are expressed in percentage of gait cycle.

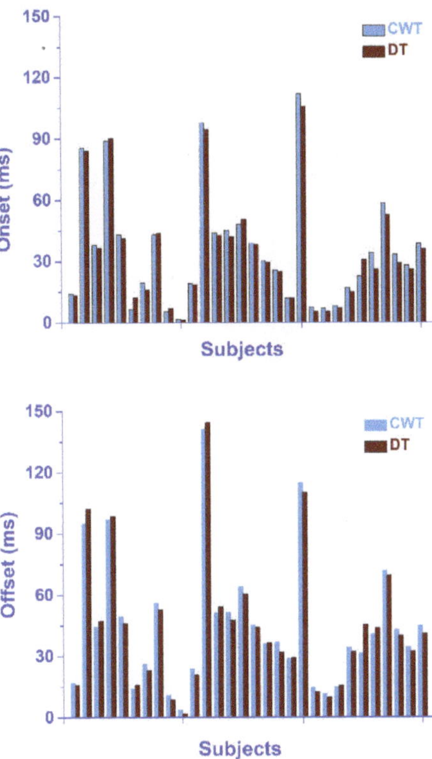

Figure 8. Average co-contraction onset and offset detected in early stance by CWT (cyan bars) vs. DT (brown bars) on real sEMG signals collected during walking. Values are expressed in ms, as the time–distance from the previous heel strike.

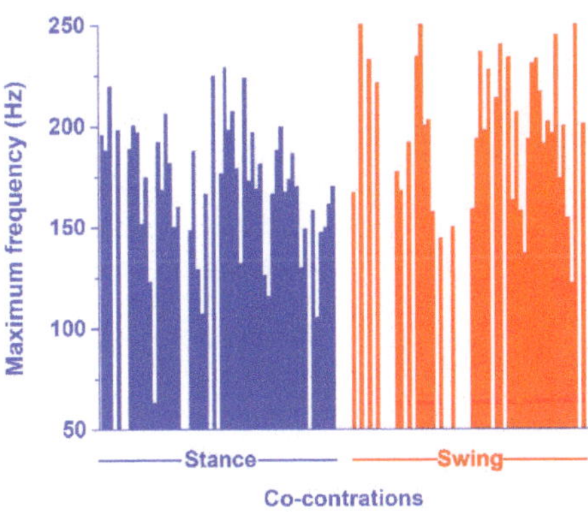

Figure 9. Average values of the maximum frequency computed in every co-contraction detected in time domain, as reported in Figure 7. Blue and red bars represent the maximum frequency of the co-contraction detected in stance and swing, respectively. Each single bar represents the average value of the maximum frequency detected for each co-contraction in a single subject.

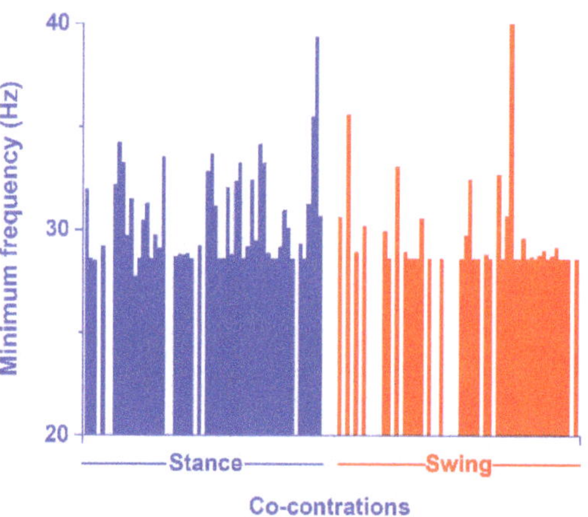

Figure 10. Average values of the minimum frequency computed in every co-contraction detected in time domain, as reported in Figure 7. Blue and red bars represent the minimum frequency of the co-contraction detected in stance and in swing, respectively. Each single bar represents the average value of the minimum frequency detected for each co-contraction in a single subject.

4. Discussion

The current investigation was designed to propose a novel approach for muscle co-contraction detection, based on the CWT cross-energy localization of two sEMG signals in the time–frequency domain, i.e., the CWT coscalogram. As far as we know, this is the first thorough study that has tried to adopt the CWT coscalogram between two sEMG signals from antagonist muscles in order to characterize the muscular co-contraction activity in the time–frequency domain. The CWT coscalogram allowed us to compute a single co-contraction signal, by which it was possible to concomitantly assess the timing of the co-contraction occurrence, its frequency content, and its amplitude.

In the field of sEMG data analysis, sEMG signals have been processed in the time–frequency domain, based on wavelet transform technology. This has also been conducted with the aim of assessing muscle co-contractions and synergies. Lee et al. [37] achieved an sEMG-based estimation of lower limb muscle co-contractions in patients with incomplete spinal cord injury. A wavelet analysis was used to process the signal in order to compute the sEMG signal intensity. Furthermore, the principal component analysis was adopted to characterize the co-contraction. The outcomes were reported in terms of correlation, however, the time-duration and frequency content were not computed and the detection accuracy was not reported. Du et al. [38] attempted to discriminate the possible difference of lumbar muscles' co-contraction in patients affected by lumbar disc herniation. A wavelet analysis was performed to filter the sEMG signals, while classic index-based time–domain methods were used to assess the co-contraction between filtered sEMG signals. The results were provided in terms of a simple index and a co-contraction ratio, however, the time-duration and frequency content were not assessed and the detection accuracy was not reported. Frere et al. tried to assess the synergy of upper-limb muscles in gymnasts [39]. A wavelet analysis was only used to process the signal in order to extract the sEMG envelopes suitable to compute the muscle synergies. The actual computation of the synergies was achieved by adopting the traditional non-negative matrix decomposition method. Similarly, Xie et al. [40] identified muscle synergy in wrist motion by means of a time–frequency approach. A wavelet packet transform was used to decompose and characterize the sEMG signals in each specific frequency band. Synergy modules were extracted by applying the non-negative matrix factorization method in each frequency band. In all these approaches, the wavelet technique was only adopted with the aim of processing and filtering the sEMG signals. Other techniques (e.g., non-negative matrix factorization, NMF, principal component analysis, PCA, and classic index-based approaches) have been used to provide quantitative information on co-contraction. Moreover, NMF methods are only appropriate in assessing the synergies between a certain number of muscles, as they do not work for the identification of the synergy between two muscles, i.e., the co-contraction. Thus, to the best of our knowledge, none of the wavelet-based attempts to characterize muscle co-contraction have been able to provide a quantitative concomitant assessment of co-contraction time duration and frequency content. Conversely, this present study provides the timing, magnitude, and frequency content of muscle co-contraction concomitantly. In addition, the current approach not only allowed us to achieve the frequency content of the whole co-contraction signal, but also of the frequency band of each single occurrence of co-contraction in the gait cycle. These considerations summarize the main contributions of the study.

Although no attempt has been made to identify the frequency content, some approaches in the time domain are available to assess co-contraction timing [5,9–13,18,19]. However, the only studies able to provide co-contraction timing (i.e., the onset and offset of muscle co-contraction) are those based on the identification of muscle co-contraction as the time range where the sEMG activity of the two muscles is superimposed [5,18,19]. The overlapping of the activation intervals of the two antagonist muscles provides the time range of co-contraction. The beginning and the end of this time range is acknowledged as the onset and the offset of the co-contraction, respectively. Thus, the proposed CWT approach has been validated against those studied by a direct comparison of the real and simulated co-contraction onset and offset identified in the time domain.

The performances of the CWT approach in the detection of muscular co-contraction were directly estimated in both the synthetic and real sEMG signals. For the simulation study, since no reference data or technique were available in the time–frequency domain, validation was performed in the time domain, where co-contraction is typically quantified as the temporal interval where the sEMG activity of the two muscles is superimposed [5,18,19]. Essentially, the direct comparison of the co-contraction interval provided by the CWT approach was performed with the time interval, where the two clean, simulated sEMG signals were superimposed (ground truth). The accuracy in predicting the co-contraction onsets and offsets in the whole dataset of the simulated sEMG signal was 100%. Furthermore, the mean error with respect to the ground truth (as defined in Section 2.3) was computed in terms of AE and TD ($\pm$SD) (Figure 4). These results of the assessment of the co-contraction between two muscles are comparable with those achieved for the detection of the activation of a single muscle in the aforementioned recent studies [20,21,29,36,41]. However, it is worth highlighting that each one of the aforementioned studies reported the performance of the algorithm in the assessment of the activity of a single muscle. For assessing co-contraction onset and offset, the algorithm should be applied twice, once for each muscle. Thus, a problem could arise with the propagation of the error and a larger bias could occur. This issue does not concern the present CWT approach because it works on a single co-contraction signal. Moreover, the accuracy of the CWT approach could be considered widely satisfactory within the tested SNR range of the simulated sEMG signals. A statistical analysis showed that, for SNR values >5 dB, the performance of the CWT approach in the detection of both onset and offset instants did not significantly change (Figure 4). This result matched with [20]. The significant increase in the AE detected for SNR = 5 dB ($p < 0.05$, upper panel in Figure 4) indicated a worsening of the detector performance for low SNR values. This worsening, however, did not affect the reliability of the detection, since the AE values are comparable with values reported by others [20,41] for similar or higher SNRs and for the activity of a single muscle. The TD values lower than 5 ms in the whole SNR range support the trustworthiness of the prediction and show that the estimates are not polarized (lower panel in Figure 4).

The CWT approach was also tested on a dataset of the experimental sEMG signals from ankle antagonist muscles (walking data from 30 able-bodied subjects). As indicated in [42], a wavelet approach guarantees a reliable denoising of the sEMG signals. In the present population, the denoising procedure allowed a significant improvement ($p < 0.05$) in the average SNR of 34.5% for TA and 33.6% for GL (Table 2), leading to an average increase in the SNR from 14.5 ± 7.0 dB to 19.5 ± 8.0 dB for TA, and from 14.0 ± 5.5 dB to 18.7 ± 7.1 dB for GL. This provided more reliable ankle muscle sEMG signals (reduced noise) for the subsequent co-contraction quantification. Figure 5 shows that the application of the CWT approach to real signals was able to represent the co-contraction between the TA and GL activity in the functions of time (x-axis), frequency (y-axis), and magnitude (colored scale). In particular, this representation allowed a simple and direct identification of co-contraction in the time domain (% of gait cycle). A 3D graph (Figure 6) appeared to be more suitable for an overall description of the co-contractions, including the magnitude and frequency bands. A graphical representation of all the co-contractions detected in the 30-subject population used in this study is reported in Figure 7. This graph allows us to identify the main zones where co-contraction occurs and the occurrence rate of this co-contraction. The outcomes are consistent with the results reported by the referenced studies on ankle muscle co-contractions, detected in numerous strides (more than ten thousand), with a more traditional approach (i.e., the superimposition of activation intervals between two muscles) [19]. Indeed, the occurrence of co-contraction in early stance, during weight acceptance, and in the final swing were detected both in [19] and in the current analysis, supporting the suitability of the current detector for experimental applications. The high occurrence rate during the late swing (Figure 7) is also in agreement with the results reported in [19]. A further validation of the results provided by the CWT approach in the time domain was performed by means of a direct comparison with a reference co-contraction interval and assessed as the temporal interval where the sEMG activity of the two muscles were superimposed, as described in Section 2.3. No gold standard is available for the detection of muscular onset and offset during walking. In the present study, Bonato's double-threshold (DT) statistical detector was used as a reference [36], due to its acknowledged reliability in assessing muscular onset and offset [43]. Since no ground truth was available for this dataset, the computation of the detection accuracy was not possible. However, the reliability of the present approach is supported by the fact that the two approaches (CWT and DT) identified the same number of co-contraction intervals in the whole experimental sEMG dataset. The outcomes of the co-contraction onset and offset that were assessed in real sEMG signals with SNR (after denoising), ranged from 4.3 dB to 39.1 dB (as shown in Table 2) and are reported in Figure 8. The performance of each of the two detectors in providing the co-contraction onset and offset were comparable, since the small difference highlighted in Figure 8 was not statistically significant ($p > 0.05$). This suggests the trustworthiness of the present results in the time domain.

However, the actual novel contribution of the current CWT approach consists of providing the frequency values of each muscle co-contraction detected in the time domain, as the frequency of the coscalogram signal in the specific time range where the co-contraction is detected. To the best of our knowledge, this is the first attempt that have been made to achieve this. This represents a relevant improvement over the state-of-the-art approaches that provide only a numerical co-contraction index [8–10,13,14] or, at best, dynamic information in the time domain [4,11,17–19], since it has been shown that frequency content could be used to reveal changes in the electrophysiological characteristics associated to specific disorders of the neuromotor system [44]. Figure 9 shows the distribution of the maximum values of the frequency content provided by the CWT algorithm in the present population, separately for the stance and swing phases. The figure highlights a wide variability in the frequency maximum between the subjects and within the strides. In particular, co-contraction seems to assume higher average and peak values of frequency in swing, compared to stance. Minimum frequency variability was also detected, mainly during the stance phase (Figure 10). A plausible interpretation of this phenomenon could be related to the functional/control tasks of the different co-contraction occurrences during the gait cycle. Even though the quantitative analysis of co-contraction frequency content is beyond the purpose of the current research project, these outcomes raise a novel question that deserves to be investigated. New and specific tools, such as the approach introduced by the current study, could be beneficial to this goal.

In conclusion, the current study proposes the CWT coscalogram function between the sEMG signals from antagonist muscles as a suitable tool to assess the muscular co-contraction activity in the time–frequency domain. This CWT approach successfully provided an overall characterization of the co-contraction phenomenon with a 3D range of variability (time, magnitude, and frequency), allowing us to monitor possible changes of this range to correlate the relative role of each one of the 3D components (time, magnitude, and frequency) in the phenomenon. This approach could be particularly valuable in clinical and rehabilitation environments, since changes in the co-contraction picture of a subject was acknowledged as a marker of neurological impairment [5], meaning that patients could be adequately encouraged to obtain positive long-term clinical outcomes [45]. The robustness of the methodology, the satisfactory accuracy, the precision in co-contraction detection, and the physiological reliability of the experimental results further support the suitability of the present tool for clinical applications. In the current paper, the performance of the CWT approach was validated on simulated signals and on real sEMG signals, which were acquired during walking. Future studies should focus on a direct validation of different motor tasks, such as jumping, squatting, and running.

Author Contributions: Conceptualization, F.D.N. and A.S.; methodology, F.D.N. and A.S.; software, M.M. and A.S.; validation, A.S., M.M. and F.D.N.; investigation, A.S. and F.D.N.; resources, F.D.N. and S.F.; data curation, F.D.N., A.S. and M.M.; writing—original draft preparation, F.D.N.; writing—review and editing, F.D.N., A.S. and S.F.; supervision, S.F. All authors have read and agreed to the published version of the manuscript.

Funding: This research received no external funding.

Institutional Review Board Statement: The study was conducted in accordance with the Declaration of Helsinki. An ethical review and approval were waived for this study because only simulated data and data available in the literature and already employed in previous studies were used.

Informed Consent Statement: Informed consent was obtained from all subjects involved in the study. Written informed consent has been obtained from the patients to publish this paper.

Data Availability Statement: Data are freely available by consulting the public repository of medical research data, PhysioNet, at the following link: https://physionet.org/content/semg/1.0.0/ (accessed on 30 May 2022).

Conflicts of Interest: The authors declare no conflict of interest.

References

1. Piche, E.; Chorin, F.; Zory, R.; Duarte Freitas, P.; Guerin, O.; Gerus, P. Metabolic cost and co-contraction during walking at different speeds in young and old adults. *Gait Posture* **2022**, *91*, 111–116. [CrossRef] [PubMed]
2. Babadi, S.; Vahdat, S.; Milner, T.E. Neural Substrates of Muscle Co-contraction during Dynamic Motor Adaptation. *J. Neurosci.* **2021**, *41*, 5667–5676. [CrossRef] [PubMed]
3. Milner, T.E. Adaptation to destabilizing dynamics by means of muscle cocontraction. *Exp. Brain Res.* **2002**, *143*, 406–416. [CrossRef] [PubMed]
4. Nagai, K.; Yamada, M.; Mori, S.; Tanaka, B.; Uemura, K.; Aoyama, T.; Ichihashi, N.; Tsuboyama, T. Effect of the muscle coactivation during quiet standing on dynamic postural control in older adults. *Arch. Gerontol. Geriatr.* **2013**, *56*, 129–133. [CrossRef]
5. Rosa, M.C.; Marques, A.; Demain, S.; Metcalf, C.D.; Rodrigues, J. Methodologies to assess muscle co-contraction during gait in people with neurological impairment. A systematic literature review. *J. Electromyogr. Kinesiol.* **2014**, *24*, 179–191. [CrossRef]
6. Lamontagne, A.; Richards, C.L.; Malouin, F. Coactivation during gait as an adaptive behavior after stroke. *J. Electromyogr. Kinesiol.* **2000**, *10*, 407–415. [CrossRef]
7. McGinnis, K.; Snyder-Mackler, L.; Flowers, P.; Zeni, J. Dynamic joint stiffness and co-contraction in subjects after total knee arthroplasty. *Clin. Biomech.* **2013**, *28*, 205–210. [CrossRef]
8. Souissi, H.; Zory, R.; Bredin, J.; Gerus, P. Comparison of methodologies to assess muscle co-contraction during gait. *J. Biomech.* **2017**, *24*, 141–145. [CrossRef]
9. Frost, G.; Dowling, J.; Dyson, K.; Bar-Or, O. Cocontraction in three age groups of children during treadmill locomotion. *J. Electromyogr. Kinesiol.* **1997**, *7*, 179–186. [CrossRef]

10. Hesse, S.; Brandl-Hesse, B.; Seidel, U.; Doll, B.; Gregoric, M. Lower limb muscle activity in ambulatory children with cerebral palsy before and after the treatment with Botulinum toxin A. *Restor. Neurol. Neurosci.* **2000**, *17*, 1–8.

11. Rudolph, K.S.; Axe, M.J.; Snyder-Mackler, L. Dynamic stability after ACL injury: Who can hop? *Knee Surg Sports Traumatol. Arthrosc.* **2000**, *8*, 262–269. [CrossRef] [PubMed]

12. Marin-Pardo, O.; Laine, C.M.; Rennie, M.; Ito, K.L.; Finley, J.; Liew, S.-L. A Virtual Reality Muscle–Computer Interface for Neurorehabilitation in Chronic Stroke: A Pilot Study. *Sensors* **2020**, *20*, 3754. [CrossRef] [PubMed]

13. Falconer, K.; Winter, D.A. Quantitative assessment of co-contraction at the ankle joint in walking. *Electromyogr. Clin. Neurophysiol.* **1985**, *25*, 135–149. [PubMed]

14. Iwamoto, Y.; Takahashi, M.; Shinkoda, K. Differences of muscle co-contraction of the ankle joint between young and elderly adults during dynamic postural control at different speeds. *J. Physiol. Anthropol.* **2017**, *36*, 32. [CrossRef]

15. Nam, C.; Zhang, B.; Chow, T.; Ye, F.; Huang, Y.; Guo, Z.; Li, W.; Rong, W.; Hu, X.; Poon, W. Home-based self-help telerehabilitation of the upper limb assisted by an electromyography-driven wrist/hand exoneuromusculoskeleton after stroke. *J. Neuroeng. Rehabil.* **2021**, *18*, 137. [CrossRef]

16. Peterson, D.S.; Martin, P.E. Effects of age and walking speed on coactivation and cost of walking in healthy adults. *Gait Posture* **2018**, *31*, 355–359. [CrossRef]

17. Mohr, M.; Lorenzen, K.; Palacios-Derflingher, L.; Emery, C.; Nigg, B.M. Reliability of the knee muscle co-contraction index during gait in young adults with and without knee injury history. *J. Electromyogr. Kinesiol.* **2018**, *38*, 17–27. [CrossRef]

18. Strazza, A.; Mengarelli, A.; Fioretti, A.; Burattini, L.; Agostini, V.; Knaflitz, M.; Di Nardo, F. Surface-EMG analysis for the quantification of thigh muscle dynamic co-contractions during normal gait. *Gait Posture* **2017**, *51*, 228–233. [CrossRef]

19. Di Nardo, F.; Mengarelli, A.; Maranesi, E.; Burattini, L.; Fioretti, S. Assessment of the ankle muscle co-contraction during normal gait: A surface electromyography study. *J. Electromyogr. Kinesiol.* **2015**, *25*, 347–354. [CrossRef]

20. Vannozzi, G.; Conforto, S.; D'Alessio, T. Automatic detection of surface EMG activation timing using a wavelet transform based method. *J. Electromyogr. Kinesiol.* **2010**, *20*, 767–772. [CrossRef]

21. Severini, G.; Conforto, S.; Schmid, M.; D'Alessio, T. Novel formulation of a double threshold algorithm for the estimation of muscle activation intervals designed for variable SNR environments. *J. Electromyogr. Kinesiol.* **2012**, *22*, 878–885. [CrossRef] [PubMed]

22. González-Izal, M.; Malanda, A.; Gorostiaga, E.; Izquierdo, M. Electromyographic models to assess muscle fatigue. *J. Electromyogr. Kinesiol.* **2012**, *22*, 501–512. [CrossRef] [PubMed]

23. Sacco, I.C.; Hamamoto, A.N.; Onodera, A.N.; Gomes, A.A.; Weiderpass, H.A.; Pachi, C.G.; Yamamoto, J.F.; von Tscharner, V. Motor strategy patterns study of diabetic neuropathic individuals while walking—A wavelet approach. *J. Biomech.* **2014**, *47*, 2475–2482. [CrossRef]

24. Mohr, M.; von Tscharner, V.; Emery, C.A.; Nigg, B.M. Classification of gait muscle activation patterns according to knee injury history using a support vector machine approach. *Hum. Mov. Sci.* **2019**, *66*, 335–346. [CrossRef] [PubMed]

25. Sukiennik, P.; Białasiewicz, J.T. Cross-correlation of bio-signals using continuous wavelet transform and genetic algorithm. *J. Neurosci. Methods* **2015**, *247*, 13–22. [CrossRef] [PubMed]

26. Strazza, A.; Verdini, F.; Mengarelli, A.; Cardarelli, S.; Burattini, L.; Fioretti, S.; Di Nardo, F. A time-frequency approach for the assessment of dynamic muscle co-contractions. *IFMBE Proc.* **2018**, *68*, 223–226. [CrossRef]

27. Rafiee, J.; Rafiee, M.A.; Prause, N.; Schoen, M.P. Wavelet basis functions in biomedical signal processing. *Expert Syst. Appl.* **2011**, *38*, 6190–6201. [CrossRef]

28. Reaz, M.B.I.; Hussain, M.S.; Mohd-Yasin, F. Techniques of EMG signal analysis: Detection, processing, classification and applications (Correction). *Biol. Proced. Online* **2006**, *8*, 163. [CrossRef]

29. Di Nardo, F.; Basili, T.; Meletani, S.; Scaradozzi, D. Wavelet-Based Assessment of the Muscle-Activation Frequency Range by EMG Analysis. *IEEE Access* **2022**, *10*, 9793–9805. [CrossRef]

30. Ghislieri, M.; Cerone, G.L.; Knaflitz, M.; Agostini, V. Long short-term memory (LSTM) recurrent neural network for muscle activity detection. *J. Neuroeng. Rehabil.* **2021**, *18*, 153. [CrossRef]

31. Di Nardo, F.; Morbidoni, C.; Fioretti, S. Surface electromyographic signals collected during long-lasting ground walking of young able-bodied subjects (version 1.0.0). *PhysioNet* **2022**. [CrossRef]

32. Goldberger, A.; Amaral, L.; Glass, L.; Hausdorff, J.; Ivanov, P.C.; Mark, R.; Mietus, J.E.; Moody, G.B.; Peng, C.K.; Stanley, H.E. PhysioBank, PhysioToolkit, and PhysioNet: Components of a new research resource for complex physiologic signals. *Circulation* **2000**, *101*, e215–e220. [CrossRef]

33. Di Nardo, F.; Morbidoni, C.; Cucchiarelli, A.; Fioretti, S. Influence of EMG-Signal Processing and Experimental Set-up on Prediction of Gait Events by Neural Network. *Biomed. Signal Process. Control* **2021**, *63*, 102232. [CrossRef]

34. Freriks, B.; Disselhorst-Klug, C.; Rau, G. Development of recommendations for sEMG sensors and sensor placement procedures. *J. Electromyogr. Kinesiol.* **2000**, *10*, 361–374. [CrossRef]

35. Agostini, V.; Balestra, G.; Knaflitz, M. Segmentation and Classification of Gait Cycles. *IEEE Trans. Neural. Syst. Rehabil. Eng.* **2013**, *22*, 946–952. [CrossRef]

36. Bonato, P.; D'Alessio, T.; Knaflitz, M. A statistical method for the measurement of muscle activation intervals from surface myoelectric signal during gait. *IEEE Trans. Biomed. Eng.* **1998**, *45*, 287–299. [CrossRef]

37. Lee, S.S.M.; Lam, T.; Pauhl, K.; Wakeling, J.M. Quantifying muscle coactivation in individuals with incomplete spinal cord injury using wavelets. *Clin. Biomech.* **2020**, *73*, 101–107. [CrossRef]
38. Du, W.; Li, H.; Omisore, O.M.; Wang, L.; Chen, W.; Sun, X. Co-contraction characteristics of lumbar muscles in patients with lumbar disc herniation during different types of movement. *Biomed. Eng. Online* **2018**, *17*, 8. [CrossRef]
39. Frère, J.; Göpfert, B.; Slawinski, J.; Tourny-Chollet, C. Shoulder muscles recruitment during a power backward giant swing on high bar: A wavelet-EMG-analysis. *Hum. Mov. Sci.* **2012**, *31*, 472–485. [CrossRef]
40. Xie, P.; Chang, Q.; Zhang, Y.; Dong, X.; Yu, J.; Chen, X. Estimation of Time-Frequency Muscle Synergy in Wrist Movements. *Entropy* **2022**, *24*, 707. [CrossRef]
41. Merlo, A.; Farina, D.; Merletti, R. A fast and reliable technique for muscle activity detection from surface EMG signals. *IEEE Trans. Biomed. Eng.* **2003**, *50*, 316–323. [CrossRef]
42. Sharma, T.; Veer, K. Comparative study of wavelet denoising in myoelectric control applications. *J. Med. Eng. Technol.* **2016**, *40*, 80–86. [CrossRef]
43. Staude, G.; Flachenecker, C.; Daumer, M.; Wolf, W. Onset detection in surface electromyographic signals: A systematic comparison of methods. *EURASIP J. Appl. Signal. Process.* **2001**, *2*, 67–81. [CrossRef]
44. Angelova, S.; Ribagin, S.; Raikova, R.; Veneva, I. Power frequency spectrum analysis of surface EMG signals of upper limb muscles during elbow flexion-A comparison between healthy subjects and stroke survivors. *J. Electromyogr. Kinesiol.* **2018**, *38*, 7–16. [CrossRef]
45. Preece, S.J.; Jones, R.K.; Brown, C.A.; Cacciatore, T.W.; Jones, A.K. Reductions in co-contraction following neuromuscular re-education in people with knee osteoarthritis. *BMC Musculoskelet. Disord.* **2016**, *17*, 372. [CrossRef]

Article

The Influence of the sEMG Amplitude Estimation Technique on the EMG–Force Relationship

Simone Ranaldi [1], Giovanni Corvini [1], Cristiano De Marchis [2] and Silvia Conforto [1,*]

[1] Department of Industrial, Electronics and Mechanical Engineering, Roma Tre University, 00154 Roma, Italy; simone.ranaldi@uniroma3.it (S.R.); giovanni.corvini@uniroma3.it (G.C.)

[2] Department of Engineering, University of Messina, 98158 Messina, Italy; cristiano.demarchis@unime.it

* Correspondence: silvia.conforto@uniroma3.it

Abstract: The estimation of the sEMG–force relationship is an open problem in the scientific literature; current methods show different limitations and can achieve good performance only on limited scenarios, failing to identify a general solution to the optimization of this kind of analysis. In this work, this relationship has been estimated on two different datasets related to isometric force-tracking experiments by calculating the sEMG amplitude using different fixed-time constant moving-window filters, as well as an adaptive time-varying algorithm. Results show how the adaptive methods might be the most appropriate choice for the estimation of the correlation between the sEMG signal and the force time course. Moreover, the comparison between adaptive and standard filters highlights how the time constants exploited in the estimation strategy is not the only influence factor on this kind of analysis; a time-varying approach is able to constantly capture more information with respect to fixed stationary approaches with comparable window lengths.

Keywords: sEMG processing; force estimation; isometric contractions

Citation: Ranaldi, S.; Corvini, G.; De Marchis, C.; Conforto, S. The Influence of the sEMG Amplitude Estimation Technique on the EMG–Force Relationship. *Sensors* **2022**, *22*, 3972. https://doi.org/10.3390/s22113972

Academic Editor: Victor Sysoev

Received: 13 April 2022
Accepted: 20 May 2022
Published: 24 May 2022

Publisher's Note: MDPI stays neutral with regard to jurisdictional claims in published maps and institutional affiliations.

1. Introduction

Surface ElecroMyoGraphy (sEMG) has been widely used as a means for measuring muscle activity during force generation, typically relating the amplitude of the sEMG signal to the amount of force exerted during a particular movement [1–5]. In most adopted models of the relationship between muscle activity and force, the capability of the contraction in generating a torque at a joint is dependent on the length of the muscle itself and its rate of change [6,7]; for this reason, the sEMG–force relationship is typically studied only in isometric conditions [8]. Even in this limited scenario, considering the different influencing factors on sEMG amplitude, as well as the numerosity of the muscles acting on each joint, the relationship between sEMG amplitude and generated force is approximately linear only below a certain level of force [9]. These models apply to static or quasi-static contractions, during which the time relationship is not considered, but when there is a need to model time-varying behaviors, and electromechanical delay (i.e., the delay between EMG and force onsets) must be taken into account. Considering the multitude of sources of noise in the investigation of the sEMG–force relationship, it is crucial to have a robust and thorough processing schema for information extraction from the biological signals, to be able to isolate fine characteristics that can be assigned to the correct influence factor, considering the particular experimental scenario [10]. Although it can be argued that this is a very general problem in sEMG-related experiments, the scientific literature still fails to have a simple and powerful model relating single-muscle sEMG recordings and force output during different conditions, suggesting that a more detailed characterization of the models and methods at the core of these analyses is needed.

Estimating the amplitude of the sEMG signal is a critical step in most of the well-established clinical and research analyses [1,11]; this computation is however often carried out by means of highly subjective methods, and the effect of different estimation strategies on the target parameters is unknown and uncharacterized. In the particular case of force estimation, since sEMG amplitude has been related to motor-unit synchronization [12], the envelope extraction phase is the main analytical tool to be applied on the signal. Moreover, synaptic noise is also present at the source of the signal (i.e., the neural drive) [13], so that optimal filtering is essential in managing different components of the noise.

sEMG envelope estimation is performed by low-pass filtering the rectified version of the raw signal, at cutoff frequencies that vary approximately in the range 2–20 Hz for general applications [14–17]; for isometric and quasi-static contractions, the optimal value has been typically considered to be in the very low-frequency portion of this range, considering that the sEMG amplitude can be hypothesized to have a frequency content below 5–10 Hz in these scenarios [18], with optimal cutoff frequencies around 2–3 Hz [5]. Filtering is achieved either by IIR filtering or by moving-window algorithm, the latter technique being more general in terms of its applicability also in online processing [19]; although the particular filtering strategy can reasonably be supposed to have small effects on outcomes, the different cutoff frequencies (or window lengths) strongly affect the smoothness and the responsiveness of the amplitude estimator, potentially affecting the correlations among estimated sEMG amplitude and any target signal, such as force variations.

In this work, different sEMG amplitude-estimation methods are tested on two different datasets to investigate whether choices in the processing schema and particularly in the filters used for sEMG envelope extraction yield results that are characterized by different correlation levels with the force signal, hence being more or less suitable to be used in sEMG–force relationship analyses.

2. Materials and Methods

2.1. Datasets

For this study, data coming from two different isometric force-tracking experiments have been used. In detail, the two experiments are related to isometric contractions of both upper limb (*triceps brachii lateral head*) and lower limb (*tibialis anterior*) muscles. Both tasks have been selected to have an experimental condition in which most of the force is generated by the recorded muscle alone, with minimal contribution from muscles belonging to different anatomical groups. Both experimental procedures have been approved by the local ethics committee of Roma Tre University.

2.1.1. Experimental Protocol for the *Triceps* Experiment

For these datasets, 16 healthy subjects (all males, 28 ± 2 years old, height 179 ± 10 cm, weight 82.4 ± 10.2 kg), righthanded were enrolled. Subjects were asked to track a force signal composed by the cyclical repetition of 10 s contractions at 20% and 40% of their maximum voluntary contraction (MVC), separated by a 5 s resting period at a very low force level (10% MVC). Visual feedback was realized by showing a colored line that the subjects had to keep within two limits corresponding to the desired value ±5% MVC (Figure 1).

During the whole task, sEMG signals are recorded with a wired StepPC system (DEMItalia) and synchronized with the force values recorded with a custom-made load cell (sensitivity: 10 mV/N, full-scale range: 2000 N). The visual stimulation signal is provided and recorded through a custom LabView panel that manages the whole data acquisition. Sampling frequency has been set to 1 kHz for both sEMG and force signals.

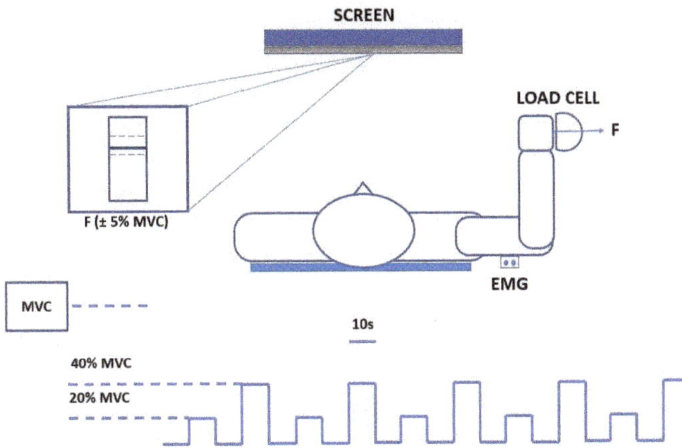

Figure 1. A visual representation of the experimental setup and the *triceps brachii* force-tracking protocol.

2.1.2. Experimental Protocol for the *Tibialis* Experiment

Ten healthy subjects were enrolled for the *tibialis* experimental protocol (8 females and 2 males, 27 ± 3 years old, height 170 ± 8 cm, weight 64.3 ± 13.9 kg). Participants were asked to avoid any kind of fatiguing activity the day before the measurements.

A visual representation of the experimental protocol is given in Figure 2. Subjects were comfortably seated with the knee at a 90 degrees flexion angle. The dominant foot was placed under a fixed structure containing the force sensor. Heel was kept fixed to the ground during the whole experiment. The experiment consisted of a series of 5 contractions lasting 2 min at a 50% MVC force level. Appropriate resting periods between contractions were inserted between trials to avoid the presence of fatigue at the beginning of the successive trial. Visual feedback was provided to the subject by showing on a screen the time-varying force trace superimposed to the reference values.

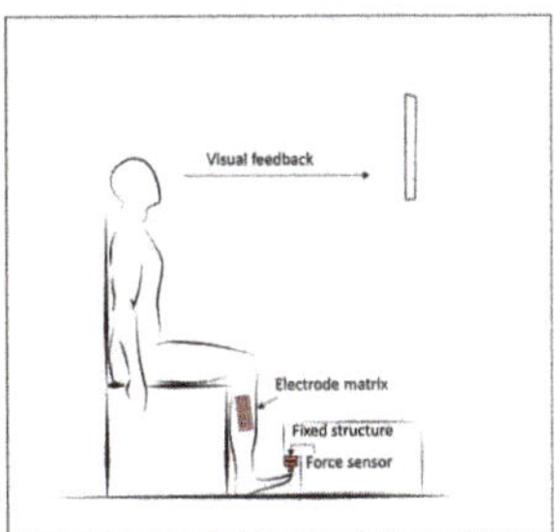

Figure 2. A visual representation of the experimental setup for the *tibialis anterior* force-tracking protocol.

Surface electromyography signals of the *tibialis anterior* muscle have been recorded through a high-density EMG (HD sEMG) sensor (SESSANTAQUATTRO, OTBioelettronica, Turin, Italy); from the HD sEMG data, a bipolar signal was selected to mimic in the best way SENIAM recommendations for sEMG recordings. Both sEMG and force signals were acquired at a sampling frequency of 2 kHz. Force signals were recorded with a compression load cell with a full-scale range of 220 N (FC2231-50L, TE Connectivity, Schaffhausen, Switzerland).

2.2. sEMG Processing

The processing strategy for the sEMG signal was standardized for both the experimental protocols. In more detail, as a first step the sEMG signal was pre-processed using standard denoising techniques (3rd-order Butterworth band-pass filter between 25 and 450 Hz and 3rd-order Butterworth notch filter at 50 Hz with a 1 Hz bandwidth), prior to envelope extraction.

The sEMG envelope has then been extracted via rectification and a moving RMS filter with different window lengths, namely 500, 200, 100, 80, 66 and 44 ms (MW_{500}, MW_{200}, MW_{100}, MW_{80}, MW_{66} and MW_{44}). In addition, the adaptive envelope computed with the algorithm in [20] has been inserted in the analysis (MW_{ADA}), to test the effect of a time-dependent time constant in the estimation of the sEMG–force relationship.

The adaptive algorithm works via iteratively adapting a sample-by-sample window length for an RMS filter. The optimization target is the minimization in the RMS error in the estimation of the amplitude of the sEMG signal, and the convergence criterion is based on the evaluation of the estimation entropy.

The SNR for each event has been calculated via the inverse of the coefficient of variation of the sEMG amplitude during the contraction phase. Given its optimality in terms of information extraction, the results from the adaptive algorithm have been used for this estimation

$$SNR_{dB} = 10 \log_{10} \frac{\mu_{ADA}}{\sigma_{ADA}}$$

The signal and noise power has been calculated by considering the particular problem to be solved; in this sense, μ_{ADA} is the mean value of the envelope during the contraction, and σ_{ADA} is its variability.

This estimation has been carried out by analyzing a 1 s window starting 4 s after the contraction onset, for each event of each subject. Average values across all events and all subjects were taken as a general estimation of the SNR in the different tested conditions.

2.3. sEMG–Force Relationship Measures

The quality of the estimated relationship has been tested by analyzing the contraction and release phases separately.

The phases have been defined differently for the two datasets:

- *Triceps*. For this dataset, the events have been defined starting from the signal related to the visual stimuli that has been given to the subject during the experiment.
- *Tibialis anterior*. For this dataset, events have been defined directly from the force signal, defining a threshold based on the noise level.

In both datasets, the segments related to onsets and offsets have been defined ranging from 0.5 s before the event, with a 2.5 s total duration.

Considering that most of the models for sEMG–force relationship are linear in nature [4], for both phases the correlation coefficient between the force signal and the sEMG envelope has been calculated as a quality parameter. The correlation coefficient has been calculated at its peak value from the cross-correlation function, in order not to take into account the delays induced by the filtering and the physiological electromechanical delay, which might insert trends into the results that are not indicative of the quality of the sEMG–force relationship.

In addition to the correlation measures, the root mean square error (RMSE) between the normalized time course of the force and the sEMG has been computed [21] as the RMS value of the difference between the two signals. Before calculating this parameter, the two segments were aligned to the delay corresponding to the maximum correlation, to compensate for physiological or instrumental time delays between the two quantities, such as the electromechanical delay [5].

The statistical significance of the differences between algorithms in the performance indicators has been tested by means of a one-way ANOVA test, with the algorithm as a factor. ANOVA assumptions were tested by checking the normality of the residuals of the model with a Shapiro–Wilk test. Three different tests were carried out separately—two for both force levels of the *triceps* dataset and one for the *tibialis* signals. Post hoc analysis was carried out with a Tukey test.

3. Results

3.1. Signal to Noise Ratio

The SNR values for the three conditions are shown in Figure 3. SNR levels for data coming from the *tibialis* dataset are significantly lower than all the values recorded for the *triceps* experiment ($p < 0.05$, Wilcoxon Rank-Sum test with Bonferroni correction for multiple comparisons). No statistical difference has been recorded for the two force levels of the *triceps* experiment.

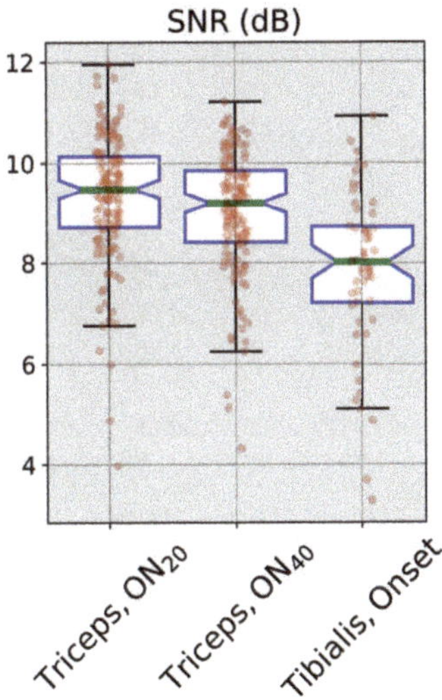

Figure 3. SNR values for the three tested conditions. Green line represents the median, blue lines are the inter-quartile range. Red dots refer to the single values.

3.2. Time-Varying Filter Time Constant

The average behavior, over all repetitions and all subjects, of the time-varying filter window length is shown in Figure 4 for both datasets. For the *triceps* dataset, a complete cycle is shown (both 20% and 40% force levels). For the *tibialis* data, before calculating the average behavior, all the trials have been interpolated over a fixed number of points (30,000 samples), considering the variable duration of the experimental trial; this step has been performed only for visualization purposes, not for the calculation of the actual correlation parameters.

The experimental protocols described in the Methods section resulted in an average of eight events for each *triceps* subject and five for the *tibialis* participants.

In both datasets, the time-varying filter time constant shows pronounced minima in correspondence to the onset of the contractions. Moreover, in both scenarios, the filter window stabilizes at a value of 90 ms during static or quasi-static contraction phases. Dataset-related differences that can be identified are:

- For the *tibialis* dataset, the minimum window length is around 60 ms, different to the 70 ms value that is recorded on the *triceps* signal;
- The variability of the *tibialis* window length is higher, with a noise level that is close to the actual range of the *triceps* dataset;
- For the *triceps* dataset, in which also a contraction offset phase is present, the same local minima for the window length can be found;

- In the *triceps* dataset, the local minima are more pronounced for higher-level contractions, both for onset and offset phases.

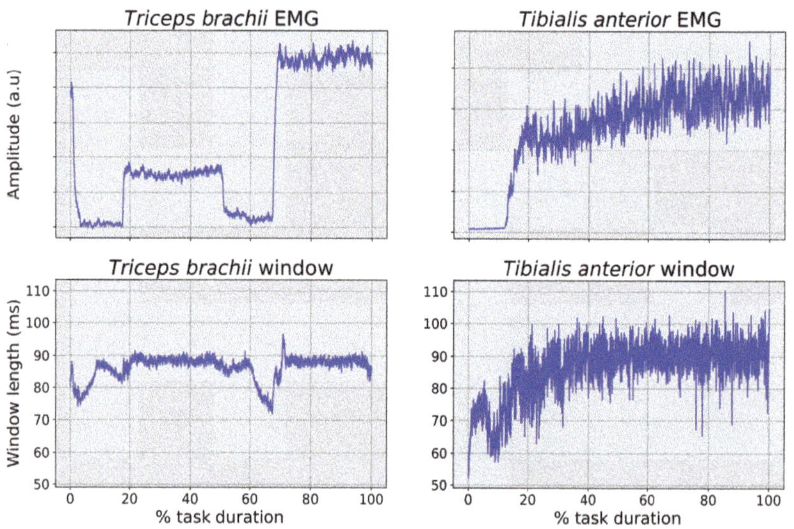

Figure 4. Average sEMG envelope and time-varying time constant behavior for the two datasets.

3.3. Correlation Metrics

For all the tested conditions, there is a significant worsening of the performance of the fixed-window algorithms for shorter time windows. In general, results on the *tibialis* dataset show lower correlation values and higher variability, while still maintaining the same trend. The adaptive algorithm performs with comparable correlation values ($p > 0.05$) to the best-performing algorithms (MW_{500} and MW_{200}) on the *tibialis* dataset and on the high force level of the *triceps* data; for lower force value, MW_{ADA} performs significantly worse than MW_{500} ($p = 0.03$).

Similar results can be found on the RMSE values coming from the same analysis, as reported in Figure 5. For this parameter, the same significant differences can be identified in the *triceps* data, while no differences are present on the *tibialis* results.

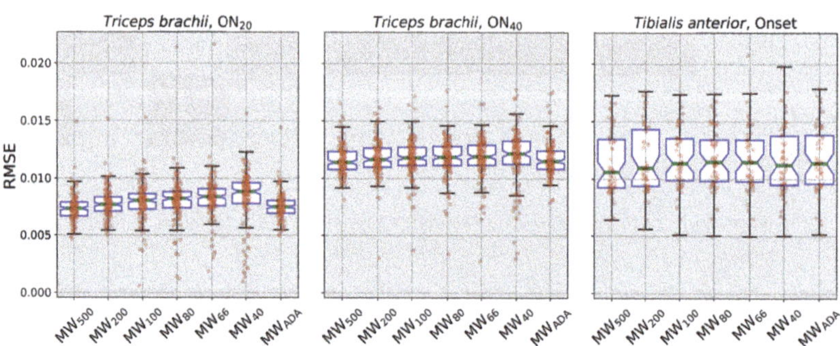

Figure 5. RMSE values. Colors have the same meaning as in Figure 3

4. Discussion

In this study, two simple force-tracking experiments have been exploited to test the effect of the processing choices (i.e., the amplitude-estimation strategy) on the evaluation of the sEMG–force relationship. Data from submaximal isometric contractions have been processed and tested in terms of the correlation values between the estimated amplitude envelope of the sEMG signal and the force curve during the onset phases and of the RMSE between the two signals. The set of classical fixed-time approaches presented here represents a subset of the methods that have been applied in the literature; the results coming from this analysis can be reasonably extended to other methods that need to be selected with some optimality criteria before being applied to sEMG–force estimation analyses.

In this work, we designed two experiments to have a condition in which all the force is generated by a single muscle; this condition is far from being often encountered in typical experimental scenarios. However, the multi-muscle force-generation condition requires some mathematical models to be built on top of the envelope-estimation procedure, so that these results presented here can be generalized to these wider conditions.

The tested methods are divided into standard, fixed-time constant moving-window procedures, and the adaptive procedure introduced in [20]. It has been shown that this latter algorithm can capture information not only in its main output (i.e., the sEMG envelope), but also in the time-dependent behavior of the estimated optimal time constant for the moving-window filter. In both the scenarios shown in Figure 4, it is evident how, for these slowly varying signals, this feature of the algorithm is also preserved. From a purely signal-feature point of view, the data from the two datasets are different in terms of rate of force change (i.e., the time to reach the target force from the baseline condition), sampling frequency and noise level; however, even with this inhomogeneous characteristic of the dataset, it is possible to identify a strongly repeatable minimum in the window length when the force level (and consequently the signal amplitude) abruptly changes.

In addition, the difference between the steady-state value for the time-varying window and the local minima at the onsets are different as a function of the force output that is generated by the contraction. When analyzing data only coming from the *triceps* dataset, this is true for both the onset and offset phases; data in the *tibialis* dataset are characterized by a higher force output (expressed in terms of fractions of MVC), thus justifying the shorter time windows that are recorded at the onset.

As an additional difference between the two experiments, it can be noted from Figure 3 that data from the *tibialis* dataset have a higher noise power with respect to *triceps*; the difference in SNR is clearly visible in the *plateau* phase of the contraction in Figure 6. Even in this high-noise scenario, the same trend for the correlation parameters can be identified, supporting the optimality hypothesis for the adaptive estimation of the sEMG envelope. RMS values do not show any difference across the two algorithms as a consequence of the high noise that is present close to the plateau region for the contraction. However, this high variability in the results is also present for the slower filter, which tends to filter out any high-frequency noise; considering this, it is reasonable to suppose that the absence of the same trend for RMSE values over the *tibialis* dataset is not to be ascribed to worse performance of the algorithms, but only to a more challenging condition in the dataset itself. It should be noted that this difference in SNR is not only to be ascribed to the recording and experimental settings, but it is indeed coming from the signal-dependent nature of the noise itself [22,23]; when the effects of the signal-dependent noise model are not negligible, the adoption of an optimal algorithm for the estimation of the sEMG amplitude has an even higher importance for ensuring the correctness of the results.

In all cases, the time dynamics identified by the adaptive algorithm is enclosed in the range 60–100 ms, which is completely included in the range that has been tested using the fixed-window methods (44–500 ms). However, in terms of correlation with the force, values coming from the fixed-window approach with time dynamics closer to MW_{ADA} (i.e., MW_{66}, MW_{80} and MW_{100}) are slightly or significantly lower than the ones relative to the adaptive approach. Although for the dataset coming from the *triceps brachii* contractions the values are still very high (>0.95 for all the methods except for MW_{80}), on a noisier and generally less standardized dataset, such as the one coming from the *tibialis* experiment, the differences are more evident, with the only methods that are close to the 0.95 correlation value being the two slowest fixed filters (MW_{500} and MW_{200}) and the adaptive algorithm.

All these trends in the results strongly suggest that, when analyzing the sEMG–force relationship, even if the task has slow dynamics, the time constant of the filters is not the only processing choice that influences the results.

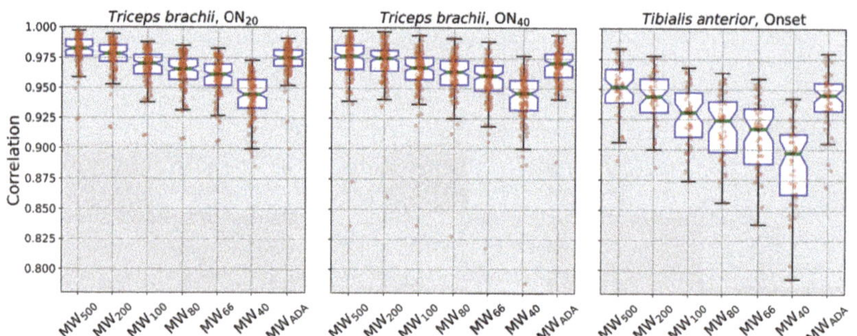

Figure 6. Correlation values. Colors have the same meaning as in Figure 3.

As a general consideration, the slowly varying nature of the force-tracking experiments that have been analyzed here puts the slower filters in a clear advantage with respect to the shorter time windows; moreover, the nature of the quality parameter (i.e., the correlation) is intrinsically higher for time signals that come from very long time windows that are consequently very smooth. Even with this advantage, and even by estimating an envelope that is less smooth, MW_{ADA} is comparable to those optimal solutions; as a consequence of this result, it is reasonable to suppose that the behavior of the adaptive algorithm is more consistent across different scenarios, in which the amplitude of the sEMG signal is varying with faster dynamics.

In this paper, the analysis has been focused on simple yet crucial quality parameters for the estimation of the sEMG–force relationship, namely the maximal correlation value and the RMSE, which has already been proven to be effective in characterizing the quality of force estimation from sEMG [24]. These parameters are focused on the characterization of the shape identification capabilities of the different methods, without quantifying the information about the timing error in the onset detection. Although it is possible to also quantify this feature of the sEMG relationship (possibly also exploiting the signal related to the time-varying window length from MW_{ADA}), such an analysis requires that several influence factors such as the electromechanical delay are taken into account, requiring complex and controlled experimental scenarios. Moreover, onset identification is a well-known problem in the scientific literature on sEMG signal processing, which already has optimal and widely accepted solutions. In this work, the focus has been put on identifying general results in a highly uncontrolled scenario using very simple experimental procedures and quality parameters, to identify general trends that can pave the way to an optimization of the sEMG processing choices for force-estimation applications.

Considering the fact that, for the *tibialis* dataset, the signals were extracted from an HD sEMG recording, the results presented here can be considered to be valid also in the case in which the sEMG–force relationship is exploited by this recording technology [3,21,25]. Estimating the amplitude of the different HD sEMG channel signals via an adaptive procedure can reasonably improve the outcomes of any processing algorithm that estimate force-related measures starting from the sEMG amplitude, yielding analogous performance differences with respect to the compact correlation measure that has been tested in this work using single-channel recordings.

An accurate estimation of the sEMG–force relationship can yield relevant parameters not only during the onset and offset phases. For example, it has been demonstrated that the force output is characterized by an increasing variability in the presence of neuromuscular fatigue [26–28], so that, with very high correlations between the force output and the sEMG signal, it can theoretically be possible to also identify this feature from the sEMG alone, giving rise to the definition of novel fatigue indicators. When neuromuscular fatigue is present, the sEMG signal has been shown to have increased amplitude and a spectrum that is more focused in the low-frequency region; although the presented results show that the performance of MW_{ADA} are consistent across different force level, its mathematical assumptions and the presence of a pre-whitening filter in the algorithm ensure that the shift in the main frequency components have little or no effect in the final amplitude estimation.

In addition to improve fatigue detection, a repeatable, stable and reliable estimation of the sEMG–force relationship is also able to highlight more advanced and detailed information on force-generation mechanisms and force control in general, such as frequency and coherence behavior [28–30] or responses to visual stimuli in force-tracking procedures [31]. Both these scenarios are typically investigated in terms of very low-frequency oscillations (less than 1 Hz); the adoption of accurate and optimal amplitude-estimation algorithms might ensure that this information is captured in a stable manner even if no ad hoc very steep and narrow low-pass filters are adopted. Here, we focused on the analysis of consistent changes to the force level (i.e., the onsets); however, it is reasonable to suppose that the same advantage of the adaptive algorithm can also be recorded in the isotonic phase of the contraction, during which different mechanisms (e.g., fatigue itself) might result in small changes to the force level.

Although the results presented here are relative to a well-established research question (i.e., sEMG-based force estimation and tracking) and have strong physiological models underlying the experimental design [7], the considerations that have been made in this work are easily generalized to any application in which the key objective is to track dynamic changes to sEMG amplitude to estimate biomechanically relevant time-varying quantities. In all those cases, the adoption of an adaptive optimal procedure is reasonably the safest choice in being able to capture the wider portion of the relevant variability in the sEMG amplitude.

Author Contributions: Conceptualization, S.R., C.D.M. and S.C.; methodology, S.R. and C.D.M.; software, S.R., G.C. and C.D.M.; validation, S.R. and C.D.M.; formal analysis, S.R. and C.D.M.; investigation, S.R., G.C. and C.D.M.; resources, S.C.; data curation, G.C. and C.D.M.; writing—original draft preparation, S.R. and C.D.M.; writing—review and editing, S.R., G.C., C.D.M. and S.C.; visualization, S.R. and G.C.; supervision, C.D.M. and S.C.; project administration, S.C.; funding acquisition, S.C. All authors have read and agreed to the published version of the manuscript.

Funding: This research received no external funding.

Institutional Review Board Statement: The study was conducted according to the guidelines of the Declaration of Helsinki.

Informed Consent Statement: Informed consent was obtained from all subjects involved in the study.

Data Availability Statement: Data used in this study are available from the corresponding author under reasonable request.

Conflicts of Interest: The authors declare no conflict of interest.

Abbreviations

The following abbreviations are used in this manuscript:

sEMG	Surface ElectroMyoGraphy
HD sEMG	High-Density Surface ElectroMyoGraphy
MW	Moving Window
MVC	Maximum Voluntary Contraction
RMS	Root Mean Square
RMSE	Root Mean Square Error

References

1. Hogrel, J.Y. Clinical applications of surface electromyography in neuromuscular disorders. *Neurophysiol. Clin. Neurophysiol.* **2005**, *35*, 59–71. [CrossRef] [PubMed]
2. Mobasser, F.; Eklund, J.M.; Hashtrudi-Zaad, K. Estimation of elbow-induced wrist force with EMG signals using fast orthogonal search. *IEEE Trans. Biomed. Eng.* **2007**, *54*, 683–693. [CrossRef] [PubMed]
3. Johns, G.; Morin, E.; Hashtrudi-Zaad, K. Force modelling of upper limb biomechanics using ensemble fast orthogonal search on high-density electromyography. *IEEE Trans. Neural Syst. Rehabil. Eng.* **2016**, *24*, 1041–1050. [CrossRef] [PubMed]
4. Bida, O.; Rancourt, D.; Clancy, E. Electromyogram (EMG) amplitude estimation and joint torque model performance. In Proceedings of the IEEE 31st Annual Northeast Bioengineering Conference, Hoboken, NJ, USA, 2–3 April 2005; pp. 229–230.
5. Staudenmann, D.; Roeleveld, K.; Stegeman, D.F.; Van Dieën, J.H. Methodological aspects of SEMG recordings for force estimation–a tutorial and review. *J. Electromyogr. Kinesiol.* **2010**, *20*, 375–387. [CrossRef]
6. Huxley, A.F.; Simmons, R.M. Proposed mechanism of force generation in striated muscle. *Nature* **1971**, *233*, 533–538. [CrossRef]
7. Hill, A.V. The heat of shortening and the dynamic constants of muscle. *Proc. R. Soc. Lond. Ser. B Biol. Sci.* **1938**, *126*, 136–195.
8. Clancy, E.A.; Hogan, N. Relating agonist-antagonist electromyograms to joint torque during isometric, quasi-isotonic, nonfatiguing contractions. *IEEE Trans. Biomed. Eng.* **1997**, *44*, 1024–1028. [CrossRef]
9. Disselhorst-Klug, C.; Schmitz-Rode, T.; Rau, G. Surface electromyography and muscle force: Limits in sEMG–force relationship and new approaches for applications. *Clin. Biomech.* **2009**, *24*, 225–235. [CrossRef]
10. Clancy, E.A.; Bida, O.; Rancourt, D. Influence of advanced electromyogram (EMG) amplitude processors on EMG-to-torque estimation during constant-posture, force-varying contractions. *J. Biomech.* **2006**, *39*, 2690–2698. [CrossRef]
11. De Marchis, C.; Schmid, M.; Bibbo, D.; Bernabucci, I.; Conforto, S. Inter-individual variability of forces and modular muscle coordination in cycling: A study on untrained subjects. *Hum. Mov. Sci.* **2013**, *32*, 1480–1494. [CrossRef]
12. Yao, W.; Fuglevand, R.J.; Enoka, R.M. Motor-unit synchronization increases EMG amplitude and decreases force steadiness of simulated contractions. *J. Neurophysiol.* **2000**, *83*, 441–452. [CrossRef]
13. Farina, D.; Negro, F. Common synaptic input to motor neurons, motor unit synchronization, and force control. *Exerc. Sport Sci. Rev.* **2015**, *43*, 23–33. [CrossRef]
14. D'Alessio, T.; Conforto, S. Extraction of the envelope from surface EMG signals. *IEEE Eng. Med. Biol. Mag.* **2001**, *20*, 55–61. [CrossRef]
15. Burden, A. How should we normalize electromyograms obtained from healthy participants? What we have learned from over 25 years of research. *J. Electromyogr. Kinesiol.* **2010**, *20*, 1023–1035. [CrossRef]
16. Shuman, B.R.; Schwartz, M.H.; Steele, K.M. Electromyography data processing impacts muscle synergies during gait for unimpaired children and children with cerebral palsy. *Front. Comput. Neurosci.* **2017**, *11*, 50. [CrossRef]
17. Hug, F.; Turpin, N.A.; Dorel, S.; Guével, A. Smoothing of electromyographic signals can influence the number of extracted muscle synergies. *Clin. Neurophysiol. Off. J. Int. Fed. Clin. Neurophysiol.* **2012**, *123*, 1895–1896. [CrossRef]
18. Clancy, E.A.; Farry, K.A. Adaptive whitening of the electromyogram to improve amplitude estimation. *IEEE Trans. Biomed. Eng.* **2000**, *47*, 709–719. [CrossRef]
19. Clancy, E.A.; Hogan, N. Single site electromyograph amplitude estimation. *IEEE Trans. Biomed. Eng.* **1994**, *41*, 159–167. [CrossRef]
20. Ranaldi, S.; De Marchis, C.; Conforto, S. An automatic, adaptive, information-based algorithm for the extraction of the sEMG envelope. *J. Electromyogr. Kinesiol.* **2018**, *42*, 1–9. [CrossRef]
21. Hajian, G.; Etemad, A.; Morin, E. Automated channel selection in high-density sEMG for improved force estimation. *Sensors* **2020**, *20*, 4858. [CrossRef]
22. Harris, C.M.; Wolpert, D.M. Signal-dependent noise determines motor planning. *Nature* **1998**, *394*, 780–784. [CrossRef]
23. Jones, K.E.; Hamilton, A.F.d.C.; Wolpert, D.M. Sources of signal-dependent noise during isometric force production. *J. Neurophysiol.* **2002**, *88*, 1533–1544. [CrossRef]
24. Ma, R.; Zhang, L.; Li, G.; Jiang, D.; Xu, S.; Chen, D. Grasping force prediction based on sEMG signals. *Alex. Eng. J.* **2020**, *59*, 1135–1147. [CrossRef]
25. Al Harrach, M.; Carriou, V.; Boudaoud, S.; Laforet, J.; Marin, F. Analysis of the sEMG/force relationship using HD-sEMG technique and data fusion: A simulation study. *Comput. Biol. Med.* **2017**, *83*, 34–47. [CrossRef]
26. Contessa, P.; Adam, A.; De Luca, C.J. Motor unit control and force fluctuation during fatigue. *J. Appl. Physiol.* **2009**, *107*, 235–243. [CrossRef]
27. Masuda, K.; Masuda, T.; Sadoyama, T.; Inaki, M.; Katsuta, S. Changes in surface EMG parameters during static and dynamic fatiguing contractions. *J. Electromyogr. Kinesiol.* **1999**, *9*, 39–46. [CrossRef]
28. Castronovo, A.M.; Negro, F.; Conforto, S.; Farina, D. The proportion of common synaptic input to motor neurons increases with an increase in net excitatory input. *J. Appl. Physiol.* **2015**, *119*, 1337–1346. [CrossRef]
29. Moon, H.; Kim, C.; Kwon, M.; Chen, Y.T.; Onushko, T.; Lodha, N.; Christou, E.A. Force control is related to low-frequency oscillations in force and surface EMG. *PLoS ONE* **2014**, *9*, e109202. [CrossRef]
30. Lodha, N.; Christou, E.A. Low-frequency oscillations and control of the motor output. *Front. Physiol.* **2017**, *8*, 78. [CrossRef]
31. Park, S.H.; Kwon, M.; Christou, E.A. Motor output oscillations with magnification of visual feedback in older adults. *Neurosci. Lett.* **2017**, *647*, 8–13. [CrossRef]

 *sensors*

Article

Machine Learning for Detection of Muscular Activity from Surface EMG Signals

Francesco Di Nardo [1,*]**, Antonio Nocera** [1]**, Alessandro Cucchiarelli** [1]**, Sandro Fioretti** [1] **and Christian Morbidoni** [2]

[1] Department of Information Engineering, Università Politecnica delle Marche, Via Brecce Bianche, 60131 Ancona, Italy; s1102518@studenti.univpm.it (A.N.); a.cucchiarelli@staff.univpm.it (A.C.); s.fioretti@staff.univpm.it (S.F.)

[2] Department of Management and Business Administration, University of Chieti-Pescara, 65127 Pescara, Italy; christian.morbidoni@unich.it

* Correspondence: f.dinardo@staff.univpm.it

Abstract: Background: Muscular-activity timing is useful information that is extractable from surface EMG signals (sEMG). However, a reference method is not available yet. The aim of this study is to investigate the reliability of a novel machine-learning-based approach (DEMANN) in detecting the onset/offset timing of muscle activation from sEMG signals. Methods: A dataset of 2880 simulated sEMG signals, stratified for signal-to-noise ratio (SNR) and time support, was generated to train a hidden single-layer fully-connected neural network. DEMANN's performance was evaluated on simulated sEMG signals and two different datasets of real sEMG signals. DEMANN was validated against different reference algorithms, including the acknowledged double-threshold statistical algorithm (DT). Results: DEMANN provided a reliable prediction of muscle onset/offset in simulated and real sEMG signals, being minimally affected by SNR variability. When directly compared with state-of-the-art algorithms, DEMANN introduced relevant improvements in prediction performances. Conclusions: These outcomes support DEMANN's reliability in assessing onset/offset events in different motor tasks and the condition of signal quality (different SNR), improving reference-algorithm performances. Unlike other works, DEMANN's adopts a machine learning approach where a neural network is trained by only simulated sEMG signals, avoiding the possible complications and costs associated with a typical experimental procedure, making this approach suitable to clinical practice.

Keywords: onset detection; muscle activation; machine learning; neural networks; surface EMG

Citation: Di Nardo, F.; Nocera, A.; Cucchiarelli, A.; Fioretti, S.; Morbidoni, C. Machine Learning for Detection of Muscular Activity from Surface EMG Signals. *Sensors* **2022**, *22*, 3393. https://doi.org/10.3390/s22093393

Academic Editor: Georg Fischer

Received: 18 March 2022
Accepted: 27 April 2022
Published: 28 April 2022

Publisher's Note: MDPI stays neutral with regard to jurisdictional claims in published maps and institutional affiliations.

1. Introduction

Assessing muscle-recruitment timing is relevant in different fields, including clinical gait analysis and electromyography-driven assistive devices [1,2]. Traditionally, onset/offset events are detected by visual inspection of surface electromyographic (sEMG) signals by trained experts [3]. However, visual inspection may be time-consuming, not completely reproducible/repeatable, and not suitable for large datasets [4]. A further classical approach is represented by threshold-based automatic methods [5]. Among these, the double-threshold statistical algorithm (DT) is a robust approach, and nowadays it is still widely adopted for clinical and research purposes [6,7]. Further approaches are typically developed based on time-frequency analysis [8–11] and signal filtering by a Teager-Kaiser energy operator (TKEO) [12]. As reported [4,13], performances of the above-mentioned approaches could be significantly affected by the relative amount of background noise compared to the magnitude of the actual sEMG signal, i.e., low values of the signal-to-noise ratio (SNR). A further issue to consider is that the majority of these approaches do not take into account those conditions where SNR is not constant throughout the signal acquisition, such as during prolonged tasks (walking, running, cycling). Intra-signal variability of SNR during sEMG recording could be ascribed mainly to the change of noise power, due to the alteration of electrode–skin contact characteristics or to the changes in the ground reference level [7]. This could strongly affect the onset-offset event detection in those portions of the sEMG signal where SNR deteriorates.

Machine/deep learning has proven to be effective in interpreting sEMG signals for different purposes [14], such as to classify gestures [15], to detect muscle fatigue [16], and to investigate human–machine interaction [17]. Different models were adopted: convolutional and recurrent neural networks for muscle force estimation [18], unsupervised competitive learning for assessing muscle recruitment during pregnancy [19], and multi-layer percep-tron to classify neuromuscular disorders [20,21]. Support vector machines were largely applied to the sEMG signal for classification purposes [22,23] and for the detection of physiological patterns and parameters [24,25]. Attempts were made even for characterizing the walking task, with particular focus on classifying gait phases and assessing gait [26–29].

In spite of the presence of a large literature on the machine-learning based interpreta-tion of sEMG signals, this approach is scarcely adopted to face the challenge of assessing the timing of muscle activation. The problem to solve is essentially an sEMG-based pre-diction of a transition between the period when the muscle is silent and the period when the muscle is active, i.e., to discriminate between actual sEMG activity and noise. Given that, the possibility of adopting a machine learning approach that learns to interpret the shape of the sEMG signals for assessing muscle-activation onset and offset seems to be a feasible solution. A very recent study proposed by Ghisleri et al. adopted long short-term memory (LSTM) recurrent neural networks (RNN) for detecting muscle activity [30]. Very encouraging outcomes were achieved in this study by using a very diversified dataset of sEMG signals to train the network, including simulated signals, signals from able-bodied subjects, and signals from patients affected by neurological or orthopedic pathologies. To run this approach, a large dataset of real sEMG signals from many different subjects is needed. However, recruiting an adequate number of subjects to build the dataset could be a challenging task. This is particularly true if patients affected by different pathologies are included, as in this case. Thus, an alternative way that considers a less demanding approach to neural-network training could be valuable. A first preliminary (and at the moment the only) attempt to provide a different approach to the training phase was proposed, based on the idea of including only simulated sEMG signals in the training procedure [31]. This study used the wavelet spectrogram of sEMG signals as the input to the network. Model performances were provided only in terms of absolute latency of onset-timing detection. Validation performed against two literature methods [32] showed promising results in terms of latency, encouraging research to continue along this path. However, the predic-tion of the offset event is not provided, and the model performances are tested only in a single subject, questioning the clinical impact of this approach and the reliability of the validation procedure.

The goal of the present study was to investigate the suitability of a novel machine-learning-based approach in assessing the onset-offset timing of muscle activation, i.e., the Detector of Muscular Activity by Neural Networks (DEMANN). Specifically, the present approach aimed to predict both onset and offset timing using only simulated sEMG signals with a large range of SNR values for neural network training in order to explore a large range of SNRs without deterioration, which is often encountered in clinical environments. This aspect, together with the simple architecture of the neural network (based on a multi layer perceptron), should help to provide fast training and prediction, making this approach very suitable for clinical purposes. Thus, the main contributions that the present study would like to provide could be summarized as follows:

- To develop a novel high-performance approach (DEMANN) that contributes to sup-port the use of machine learning for muscle activity detection;
- To highlight the advantages of the proposed machine-learning approach, such as the possibility of real-time applications, achieved without loss of accuracy and with respect to existing, non-machine-learning-based systems;

- To limit the deterioration of event assessment associated with low SNRs and the large inter-signal variability of SNR, typical of clinical environments, by training the model with simulated sEMG signals with a large range of SNR values;
- To reduce the complexity of the experimental protocol associated with model training, since no signal acquisition is needed to provide real time activation predictions.

2. Materials and Methods

The robustness of DEMANN was evaluated by a test bench of simulated sEMG signals and two datasets of real sEMG signals. Simulated and real sEMG signals underwent the same procedure described in the following sections. DEMANN was validated by a direct comparison with reference approaches on both simulated and real data.

2.1. Simulated sEMG Signals

A simulation study, using a test bench of signals, was carried out for assessing the performance of the DEMANN approach in predicting onset and offset events of muscular activity. sEMG signals acquired during cyclic movements could be modeled as the superimposition of the actual signal produced by muscle contraction and the background noise [33]. In this study, a Gaussian process with zero mean and variance σ^2_{noise} was adopted to model the sEMG-signal where the muscle was silent and only background noise was acknowledged. To simulate the sEMG-signal portion where the muscle is recruited, the background uncorrelated noise was added to a band-limited stochastic process with zero-mean Gaussian distribution of amplitude and a fixed power level [6]. This distribution was achieved by band-pass filtering (80–120 Hz) a Gaussian series of uncorrelated samples, according to [6]. This Gaussian distribution was truncated to simulate the sEMG activity due to muscle activation. Each simulated sEMG signal was generated with a sampling frequency fs = 2000 Hz, a time window = 1 s, and a variable value of the Gaussian-distribution median, μ, ranging from 0 to 1. Different simulated sEMG signals were created varying the standard deviation, σ, and the time support, $2 \times \alpha \times \sigma$, of the Gaussian distribution, in order to simulate the physiological variability associated with the recruitment of different muscles. The variation of σ was achieved according to the desired value of SNR, where:

$$SNR = 10 * \log \frac{\left(\sigma^2_{signal}\right)}{\left(\sigma^2_{noise}\right)} \tag{1}$$

Simulated sEMG signals were generated from all the different combinations of the values adopted for σ (50, 100, and 150 ms), for α (1, 1.5, 2, and 2.4), and SNR values from 1 dB to 30 dB, with step = 1. In [30], Ghisleri et al. trained LSTM recurrent neural networks by means of simulated sEMG signals, with SNR ranging from 3 dB to 30 dB. In the present paper, this SNR range was slightly expanded to consider even worse conditions.

2.2. Real sEMG Signals

Two different datasets of real sEMG signals were considered. The first dataset is available in [3] (https://github.com/TenanATC/EMG, accessed on 23 April 2021), including the ground truth. The experimental protocol consisted of acquiring sEMG signals from 18 participants performing knee extension and elbow flexion. Knee extension was performed in subjects seated in a stationary chair, with a mass (2.3 kg) applied to the right ankle. Elbow flexion was performed with a mass (2.3 kg) applied to the right wrist. sEMG probes were applied over vastus lateralis (VL) for monitoring knee extension and over biceps brachii (BB) for elbow flexion. A total of 103 sEMG signals were acquired with 0 dB < SNR < 13 dB. Three experts visually analyzed the signals and noted down the activation onsets in a randomized and double-blind fashion. Every trial was inspected twice by each expert. The average over the six onset values was the ground truth for the experiments in [3] and it was adopted also here. Further details can be found in [3].

The second dataset consisted of foot–floor contact and the sEMG data collected during 30 healthy adults walking, retrospectively taken from the database built at the Movement Analysis Lab, Università Politecnica delle Marche, Ancona, Italy and used for previous studies [28,29]. Data are freely available, consulting the public repository of medical research data PhysioNet [29,34,35]. Overweight and obese people (body mass index, BMI > 25) and subjects affected by any pathological condition, joint pain, or undergone orthopedic surgery were not considered. Gait data were captured (sampling rate: 2 kHz; resolution: 12 bit) by the multichannel recording system Step32 (Medical Technology, Torino, Italy). sEMG signals were acquired in each leg by single differential probes placed over gastrocnemius lateralis (GL), tibialis anterior (TA), and vastus lateralis (VL). SNR values ranged between 3 dB and 30 dB. SENIAM guidelines for sEMG-sensor positioning were respected [36]. Foot–floor contact signals were measured by three footswitches placed under the heel and the first and the fifth metatarsal heads of the foot. Subjects walked barefoot at a self-selected pace for about 5 min, following an eight-shaped path, which involved natural deceleration, acceleration, and reversing. Further details are reported in [28]. The research was undertaken following the ethical principles of the Helsinki Declaration and was approved by the local ethical committee.

2.3. Signal Pre-Processing

Simulated and real sEMG signals were band-pass filtered (2nd-order Butterworth filter, cut-off frequency 10–500 Hz). Then, signals were pre-processed to extract the linear envelope (LE), the root mean square (RMS), and the wavelet scalogram, which were concomitantly used as input to the neural network. LE was extracted by low-pass filtering of the signal (2nd-order Butterworth filter; cut-off frequency 5 Hz). RMS was extracted by computing the following formula over overlapping sliding 60-sample windows that scan the whole signal:

$$\text{RMS} = \sqrt{\frac{1}{T} \int_0^T |x(t)|^2 \, dt} \tag{2}$$

Continuous wavelet transform (CWT) was used for providing energy localization in the time-frequency domain of sEMG signals in terms of CWT scalogram function, P_{sEMG}, defined as the square of the absolute value of CWT coefficients, W_{sEMG}:

$$P_{sEMG}\,(a,b) = |W_{sEMG}\,(a,b)|^2 \tag{3}$$

Wavelet transform was implemented by adopting Morse of order 4 with 6 levels of decomposition as mother wavelet.

2.4. Data Preparation

To adopt the most suitable input to the neural network, preliminary experiments were performed, evaluating four different alternatives: LE, RMS, CWT scalogram, and their concatenation (LE + RMS + CWT). The concatenation consisted of a min–max normalization of the outputs of the different processing procedures, thus mapping the values in a [0, 1] range, and a concatenation of outputs of the different processing procedures (Figure 1). These choices were motivated by the related literature, where LE and RMS of the sEMG proved to be suitable signals to train the neural network for gait analysis [27–29], even if the prediction tasks were different from the one addressed here. Outputs of time-frequency analysis (spectrograms, scalograms) were also features often used in sEMG analysis, as for example in [31] to predict muscle activations. Before training the classifier, the concatenated vector was segmented in overlapping sliding windows of 10 samples, where each window was shifted of one sample with respect to the previous window. Each window was used to label that single sample, according to the value of the related ground truth in the window. The single sample was labeled as 1 (muscle activity) or 0 (no muscle activity), according to the most frequent ground truth value identified in the window. The size of the processed windows, the simple neural network architecture, and the use of sliding windows provided a very low latency of 3–4 milliseconds, which could be suitable for real-time applications.

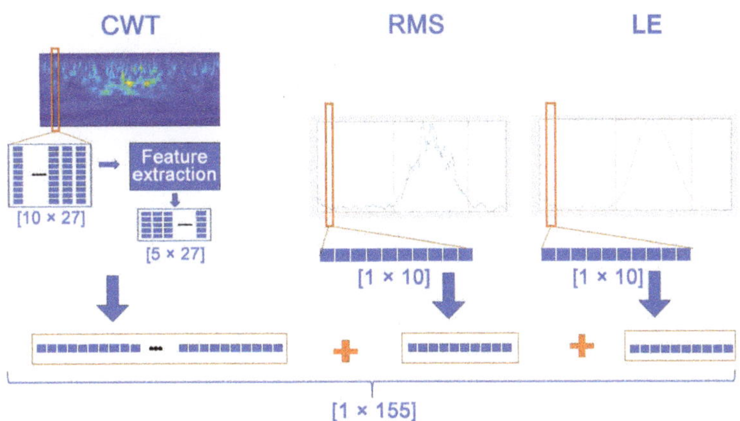

Figure 1. Realization of sEMG vectors used as input to DEMANN model.

2.5. Training the Classifier

The classifier was a hidden single-layer (32 units) fully-connected neural network. A Rectified Linear Unit (ReLU) activation function was used, and a sigmoid function was adopted to map the network output to a 0–1 interval. The binary output was achieved by using a standard threshold of 0.5. The model was trained with a learning rate of 0.001, a batch size of 512 for 40 epochs using the standard stochastic gradient descent (SGD) optimization algorithm, and by adding a L2 regularization penalty set to 0.0001. The training set was composed of only simulated sEMG signals: 8 signals for each combination of σ (50, 100, and 150 ms) and α (1, 1.5, 2, and 2.4) were chosen, for a total of 96 signals for each SNR. Considering 30 SNR values (from 1 dB to 30 dB, step = 1), a total of 2880 simulated signals were included.

The classifier performances were evaluated on three different testing sets. The first one was composed of only simulated sEMG signals. Eight signals were generated for each combination of σ, α, and SNR. Nine different SNRs were considered, specifically 3, 6, 10, 13, 16, 20, 23, 26, and 30 dB, as suggested in [6]. A total of 864 simulated signals were achieved. No overlapping occurred between the training and testing set, i.e., none of the simulated signals generated to train the model were used during testing. The ground truth of muscle activity was the vector composed of the same number of samples of the simulated sEMG signal, where samples can assume only two values: "0" and "1". The ground truth was "1" if the truncated Gaussian distribution assumed values > 0, "0" otherwise. The DEMANN performance was provided in terms of precision, recall, F1-score, and mean absolute error (MAE), assessed in true positives as defined in Section 2.6. MAE was the average time distance between the predicted event and the one of the same kind in the ground truth signal. A comparison of the results achieved in the first test set was reported in Table 1, in terms of the mean F1-score ($\pm$SD) of classification. The overall best F1-score was achieved by LE + RMS + CWT (Table 1). Thus, this input was adopted to feed the neural network.

The second test set was composed of 103 real sEMG signals proposed in [3]. The performance of the DEMANN approach was provided in terms of prediction accuracy and MAE, assessed in all 103 signals of the dataset.

Table 1. Mean classification accuracy in the simulated test dataset associated with different inputs.

Input	F1-Score $\pm$ SD (%)			
	Activity Area	**Silent Area**	**Macro**	**Weighted**
LE	95.0 $\pm$ 0.4	87.9 $\pm$ 0.8	91.4 $\pm$ 0.6	92.8 $\pm$ 0.5
RMS	96.4 $\pm$ 0.3	91.5 $\pm$ 0.6	93.9 $\pm$ 0.4	94.9 $\pm$ 0.4
CWT	98.0 $\pm$ 0.2	95.5 $\pm$ 0.4	96.8 $\pm$ 0.3	97.3 $\pm$ 0.2
LE + RMS + CWT	98.3 $\pm$ 0.1	96.0 $\pm$ 0.3	97.2 $\pm$ 0.2	97.6 $\pm$ 0.2

The third test set included foot–floor contact and sEMG data collected during 30 healthy adults walking, as described in Section 2.2. Sequences of five consecutive gait cycles were selected randomly. Two experts analyzed three different versions of the same signal: raw sEMG signal, rectified band-pass-filtered sEMG signal, and RMS of the sEMG signal. Then, the experts identified onset-offset instants of muscular activity by visual inspection. The mean over the six onset values represented the ground truth for the experiments. A total of 538 events were identified (269 onsets and 269 offsets). The reference chosen for validation was the acknowledged double thresholding algorithm (DT) [5,6]. The performances were reported in terms of precision, recall, and F1-score of the event prediction.

For all the three test sets, model validation and performance were computed in signals never used during the training of the model.

2.6. Identification of sEMG Onset-Offset

To achieve the model output, segmented sEMG signals were provided as input to the trained model. Thus, the model output was composed of sequences of 0 (no muscle activity) alternating with sequences of 1 (muscle activity). This signal was chronologically scanned to identify the transitions between the two conditions: the transition from 0 to 1 identified the onset event and the transition from 1 to 0 detected the offset event. This was achieved by the following procedure: a time tolerance T of 100 ms was adopted, as suggested in [10]. Then, we acknowledged as true positive each predicted event at time t_p if an event of the same kind occurred in the ground-truth signal at time t_g, such that $|t_g - t_p| < T$. Otherwise, the predicted event was acknowledged as false positive. Moreover, a post-processing procedure was performed, consisting of cleaning the signal by discarding those sequences of samples that were too short to be physiologically plausible; it was acknowledged, indeed, that muscle recruitments lasting less than 30 ms had no effect in controlling joint motion [6]. Thus, sequences of 0 (or sequences of 1) shorter than 60 samples were removed.

2.7. Statistics

The Shapiro-Wilk test was adopted to appraise the normality of data distribution. A two-tailed, non-paired Student's t-test was applied to verify the significance of difference between the normally-distributed samples. The Mann-Whitney test was applied to verify the significance of difference between the non-normally-distributed samples. Statistical significance was established at 5%.

3. Results

3.1. Simulated sEMG Signals

The mean classification accuracy computed in the testing set stratified for different SNR is shown in Table 2. The accuracy on the simulated test set increased with increasing SNR from 3 dB (accuracy = 95.3%) to 23 dB (accuracy = 99.2%), and then it remained practically unaltered. Likewise, SD decreased with increasing SNR (from 4.8 to 0.7%).

Table 2. Mean classification accuracy stratified for different SNR.

SNR (dB)	Accuracy (%)
3	95.3 ± 4.8
6	96.2 ± 4.3
10	97.3 ± 3.3
13	98.1 ± 2.1
16	98.4 ± 2.0
20	98.9 ± 1.4
23	99.2 ± 0.9
26	99.1 ± 1.0
30	99.1 ± 0.7
Mean $\pm$ SD	97.8 ± 3.0

Table 3 reports the mean classification performances in the testing set computed separately in the portions of sEMG signals where muscle activity was acknowledged (activity area) and where it was not (silent area). The effect of SNR on the classification performances was preserved.

Table 3. Mean classification performances computed in the test set separately for the activity area and the silent area, stratified for different SNR.

Activity Area			
SNR (dB)	Precision (%)	Recall (%)	F1-Score (%)
3	95.1 ± 7.6	91.6 ± 10.3	92.7 ± 6.1
6	96.0 ± 6.1	93.4 ± 8.2	94.2 ± 4.2
10	97.8 ± 3.7	94.2 ± 7.3	95.7 ± 3.8
13	98.8 ± 2.3	95.2 ± 6.2	96.7 ± 3.2
16	98.8 ± 2.0	96.5 ± 4.5	97.5 ± 2.4
20	98.7 ± 1.5	97.9 ± 3.1	98.2 ± 1.5
23	98.9 ± 1.6	98.3 ± 2.4	98.5 ± 1.3
26	98.5 ± 2.2	98.5 ± 2.1	98.5 ± 1.4
30	98.4 ± 2.0	98.3 ± 4.2	98.3 ± 2.4
Mean ($\pm$SD)	97.9 ± 4.0	96.0 ± 6.4	96.7 ± 3.8
Silent Area			
SNR (dB)	Precision (%)	Recall (%)	F1-Score (%)
3	94.8 ± 8.0	97.8 ± 4.7	96.0 ± 5.0
6	94.8 ± 9.5	98.8 ± 1.8	96.5 ± 5.4
10	96.3 ± 6.5	99.3 ± 1.3	97.6 ± 3.7
13	97.4 ± 3.8	99.4 ± 1.1	98.4 ± 1.9
16	97.7 ± 4.6	99.5 ± 0.9	98.6 ± 2.6
20	98.4 ± 3.8	99.5 ± 0.7	98.9 ± 2.0
23	99.1 ± 2.1	99.6 ± 0.7	99.3 ± 1.1
26	99.0 ± 2.8	99.3 ± 1.0	99.1 ± 1.5
30	99.3 ± 1.4	99.3 ± 0.9	99.3 ± 0.7
Mean ($\pm$SD)	97.4 ± 5.6	99.2 ± 1.9	98.2 ± 3.3

While in the present study, a shallow neural network was used as a classifier, the DEMANN approach can be flexibly modified to embed a different machine-learning model. Support vector machines (SVM) are identified in literature as suitable modeling tools [22–25]. Thus, a direct comparison was performed, with results achieving replacing the neural network with a linear kernel SVM classifier on the same dataset of simulated sEMG signals. The SVM model was trained with the Stocastic Gradient Descent optimizer on a Hinge loss function and by applying a L2 regularization with coefficient 0.0001. The results of this comparison are shown in the following Table 4.

Table 4. Mean (±SD) performances of the onset and offset prediction provided by DEMANN and SVM over all the simulated sEMG signals.

	DEMANN		SVM	
	Onset	**Offset**	**Onset**	**Offset**
MAE (ms)	10.0 ± 17.5 *	10.1 ± 17.3 [§]	20.6 ± 28.2 *	19.3 ± 23.8 [§]
Precision (%)	99.0 ± 9.6	99.4 ± 7.4	97.0 ± 16.9	98.5 ± 11.9
Recall (%)	99.2 ± 9.0	99.5 ± 6.8	97.1 ± 16.8	98.6 ± 11.7
F1-score (%)	99.0 ± 9.2	99.4 ± 9.6	97.0 ± 16.8	98.5 ± 11.8

* means that the difference between the two mean onset values is statistically significant ($p < 0.05$); [§] means that the difference between the two mean offset values is statistically significant ($p < 0.05$).

A significantly lower mean MAE ($p < 0.05$) was provided by the DEMANN approach for both onset and offset timing. No significant differences were detected in precision, recall, or F1-score between the performances of the two models.

Figure 2 reports an example of simulated sEMG signal, where onset and offset events predicted by DEMANN and DT approaches (rectangular lines) are highlighted and compared with the ground truth, i.e., the truncated Gaussian function used to model the simulated signal.

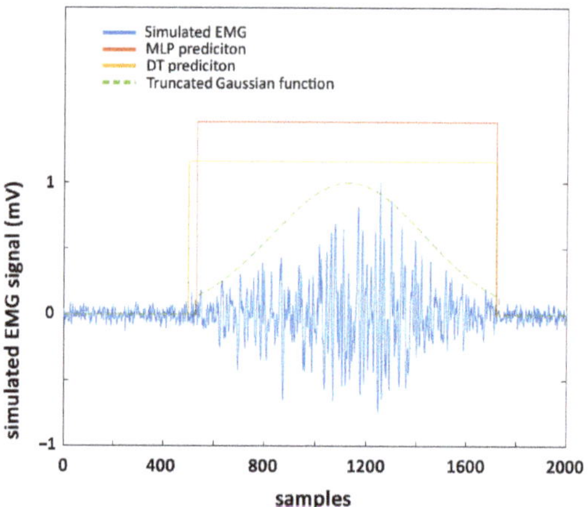

Figure 2. Example of simulated sEMG signal (blue line). The truncated Gaussian function used to model the simulated signal (green dashed line), predictions by DEMANN (red rectangle), and DT (yellow rectangle) of onset and offset events are superimposed.

The average performances of the onset-offset prediction over the simulated-signal dataset provided by the DEMANN and DT approaches are reported in Table 5.

Table 5. Mean (±SD) performances of onset and offset prediction provided by DEMANN and DT over all the simulated signals.

	DEMANN		DT	
	Onset	**Offset**	**Onset**	**Offset**
MAE (ms)	10.0 ± 17.5	10.1 ± 17.3 *	11.5 ± 21.9	16.1 ± 26.9 *
Precision (%)	99.0 ± 9.6	99.4 ± 7.4	98.5 ± 12.1	96.9 ± 17.4
Recall (%)	99.2 ± 9.0	99.5 ± 6.8	98.4 ± 12.3	96.8 ± 17.4
F1-score (%)	99.0 ± 9.2	99.4 ± 9.6	98.5 ± 12.2	96.9 ± 17.4

* means that the difference between the two mean values is statistically significant ($p < 0.05$).

The variability of MAE in the function of α, σ, and SNR is quantified in Table 6. A color-level coded representation was adopted to allow a visual interpretation of results.

Table 6. Variability of MAE in the function of simulated-signal parameters α, σ, and SNR (dB) for onset and offset prediction.

Onset—MAE (ms)											
σ = 50 ms				σ = 100 ms				σ = 150 ms			
SNR	$\alpha = 1$	$\alpha = 1.5$	$\alpha = 2$	$\alpha = 2.4$	$\alpha = 1$	$\alpha = 1.5$	$\alpha = 2$	$\alpha = 2.4$	$\alpha = 1$	$\alpha = 1.5$	$\alpha = 2$	$\alpha = 2.4$
3	7.2	6.9	8.9	21.0	11.4	15.1	18.0	60.8	21.7	10.3	37.3	72.8
6	8.4	11.1	13.5	18.2	6.2	9.6	12.6	52.6	5.2	5.6	33.9	87.4
10	5.1	6.1	7.1	10.0	5.9	12.1	9.4	42.6	4.6	1.6	32.5	62.4
13	3.1	2.7	3.0	11.1	3.3	7.1	8.1	31.6	8.2	4.4	11.8	35.0
16	1.8	2.5	6.2	9.5	2.0	3.9	9.9	28.5	2.0	8.4	4.6	33.1
20	1.2	1.9	4.0	5.4	1.1	3.1	4.7	13.8	2.3	3.8	4.6	20.3
23	1.8	2.8	4.6	4.9	1.3	1.7	3.8	8.7	1.2	3.4	4.5	3.9
26	2.6	1.8	3.3	6.4	2.4	2.3	5.1	7.0	2.0	4.3	2.9	12.0
30	1.1	2.0	5.7	9.8	2.0	3.4	3.8	3.0	2.2	4.8	5.0	11.5

Offset—MAE (ms)											
σ = 50 ms				σ = 100 ms				σ = 150 ms			
SNR	$\alpha = 1$	$\alpha = 1.5$	$\alpha = 2$	$\alpha = 2.4$	$\alpha = 1$	$\alpha = 1.5$	$\alpha = 2$	$\alpha = 2.4$	$\alpha = 1$	$\alpha = 1.5$	$\alpha = 2$	$\alpha = 2.4$
3	5.2	6.6	12.1	29.1	9.8	14.8	34.8	55.9	10.2	13.3	35.1	102.3
6	3.3	10.1	14.6	15.5	8.2	3.0	22.9	45.4	5.8	8.8	36.4	75.7
10	1.5	2.9	6.4	19.6	2.5	5.2	8.5	44.9	6.3	5.0	36.8	66.6
13	4.4	7.4	3.8	17.2	3.4	2.5	10.0	49.3	1.0	5.8	12.3	36.0
16	1.2	9.3	7.4	10.5	2.6	3.6	3.6	23.4	2.7	6.8	4.4	37.5
20	1.6	4.4	4.6	6.8	3.3	3.0	3.9	11.3	1.4	4.6	12.3	31.9
23	2.2	3.6	5.3	6.0	1.3	2.5	5.1	12.7	1.5	2.5	2.7	26.5
26	2.8	3.1	5.3	4.6	1.6	5.2	2.6	4.8	2.6	4.6	6.1	24.5
30	2.0	9.5	1.8	7.7	2.1	3.9	4.9	6.6	1.6	2.4	6.3	6.2

All the areas with different levels of green indicate MAE values < 10 ms. Progressively darker green indicate progressively lower MAE. All the yellow, orange, and red areas indicate MAE values $\geq$ 10 ms. Progressively darker colors indicate progressively higher MAE. The value of 10 ms was chosen since it was the mean MAE value over the whole dataset (Table 5).

The direct comparison of performances achieved by DEMANN and DT is depicted in Figure 3, stratified for different SNR. An improvement of the F1-score of offset prediction was introduced by DEMANN for signals with SNR $\leq$ 6 dB ($p < 0.05$, Figure 3B). No significant differences were detected for SNR > 6 dB. The F1-score was comparable for onset prediction in the whole SNR range ($p > 0.05$, Figure 3A). Lower MAEs in onset-offset prediction were provided by DEMANN. Details of statistical significance are reported in Figure 3C,D.

3.2. Real sEMG signals

A first validation was performed on the sEMG dataset available in [3]. In [13], four onset-detection algorithms and two filtering approaches were tested on this dataset characterized by SNR $\leq$ 8 dB. The same 52 sEMG signals were considered here (first four lines, Table 7).

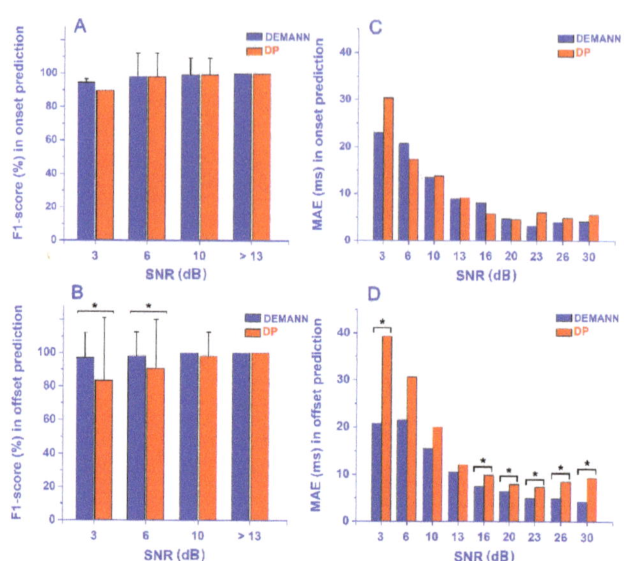

Figure 3. Mean F1-score computed in onset (panel **A**) and offset (panel **B**) prediction and mean MAE computed in onset (panel **C**) and offset (panel **D**) prediction for each SNR value by DEMANN (blue bars) vs. DT algorithm (red bars). * indicates statistically significant difference.

Table 7. Absolute error of onset prediction in the function of SNR ranges in terms of mean, standard deviation (SD), median, 25-percentile, and 75-percentile.

SNR (dB)	Number of Signals	Mean (ms)	SD (ms)	Median (ms)	25-Perc (ms)	75-Perc (ms)
≤ 2	6	209.9	182.0	131.6	66.9	368.8
$2 \div 4$	10	187.5	163.7	116.0	60.4	338.0
$4 \div 6$	15	76.7	53.7	77.6	32.7	107.4
$6 \div 8$	21	24.0	27.7	13.2	6.8	32.2
$8 \div 10$	20	15.8	16.9	11.5	3.9	16.4
$10 \div 12$	6	12.2	2.9	12.9	11.6	14.3
≤ 8	52	92.1	120.3	54.2	13.2	93.9
>8	26	14.9	14.6	12.0	7.1	11.6

As in [13], the 52-signal dataset was split according to four ranges of increasing SNR values (step = 2 dB) to facilitate the comparison of results. The absolute error of the onset prediction provided by DEMANN is reported in Table 7, in terms of mean, standard deviation (SD), median, 25-percentile, and 75-percentile. Validation was performed against the four algorithms tested in [13]: the double-threshold statistical algorithm (DT) [6]; the wavelet-based approach (WLT) [9]; the method grounded on CUSUM logic [37]; and the technique based on profile-likelihood maximization, employing discrete Fibonacci search (PROLIFIC) [38]. DEMANN provided the lowest values of absolute error for all the metrics (Table 8), except for SD (best value = 114.8 ms; DEMANN-value =120.3 ms). Similar consideration could be performed for signals with 6 < SNR < 8 dB. For lower SNR (<6 dB), DEMANN provided performances comparable to the other algorithms (Table 8). The results of signals with 8 < SNR < 12 are also reported in Table 7. Precision, recall, and F1-score were dependent on the choice of the tolerance used to identify true positives. In this case, all the events were detected within the tolerance range, leading to a precision, recall, and F1-score of 100% for DEMANN and for all the algorithms chosen for validation.

Table 8. Comparison among the absolute errors of the onset prediction provided in the same population by DEMANN approach and by the four algorithms introduced in Section 3.2. The best values for each parameter and each SNR are highlighted in bold.

SNR (dB)	DEMANN	DT		WLT		CUSUM		PROLIFIC	
		TKEO	ETKEO	TKEO	ETKEO	TKEO	ETKEO	TKEO	ETKEO
Mean (ms)									
≤2	209.9	733.5	243.7	504.5	139.8	827.1	**126.6**	357.9	303.4
2 ÷ 4	187.5	225.5	154.0	191.5	**145.9**	1143.8	222.8	460.0	185.5
4 ÷ 6	**76.7**	201.3	101.3	248.5	165.2	708.1	93.8	371.4	123.4
6 ÷ 8	**24.0**	182.3	116.9	158.8	92.2	618.0	65.5	229.7	39.4
≤8	**92.1**	259.7	134.2	230.9	129.1	769.2	115.1	410.4	122.2
SD (ms)									
≤2	182.0	456.4	381.0	578.5	**66.9**	584.3	91.9	534.6	519.4
2 ÷ 4	163.7	115.0	**92.0**	254.5	146.3	489.2	170.4	392.2	185.5
4 ÷ 6	**53.7**	266.9	106.2	335.4	272.5	492.2	92.0	453.3	123.4
6 ÷ 8	**27.7**	305.3	229.8	311.4	170.2	579.2	54.8	462.0	39.4
≤8	120.3	330.1	203.6	352.9	192.3	558.7	**114.8**	443.5	122.2
Median (ms)									
≤2	131.6	765.4	**92.5**	208.3	133.3	999.0	149.7	111.1	104.0
2 ÷ 4	116.0	231.2	125.0	121.3	**95**	1134.5	148.7	396.7	136.0
4 ÷ 6	77.6	104.0	58.6	122.6	104.9	793.5	55.7	93.8	**50.8**
6 ÷ 8	**13.2**	69.8	41.5	61.0	35.2	729.0	48.8	135.7	36.6
≤8	**54.2**	109.6	69.3	116.9	78.6	958.0	78.6	137.9	54.9
25-Percentile (ms)									
≤2	66.9	400.9	42.9	127.4	126.5	153.8	90.8	**57.6**	56.6
2 ÷ 4	60.4	121.1	87.9	**30.3**	33.7	883.3	128.9	140.1	124.5
4 ÷ 6	32.7	47.8	27.6	71.2	31.7	257.1	**23.4**	40.4	20.7
6 ÷ 8	**6.8**	28.3	25	19.4	7.8	35.9	32.3	30.5	11.7
≤8	**13.2**	45.2	31.5	33.0	25.2	100.3	40.3	46.6	24.9
75-Percentile (ms)									
≤2	368.8	1042.9	**192.2**	861.3	197.7	1182.1	239.7	409.2	157.7
2 ÷ 4	338.0	309.1	245.6	191.9	**188.5**	1483.9	298.3	746.1	173.3
4 ÷ 6	107.4	231.9	146.6	223.4	128.5	1124.8	131.2	680.9	**88.6**
6 ÷ 8	**32.2**	107.5	94.2	122.4	130.5	1115.2	80.9	884.7	54.8
≤8	**93.9**	304.7	146.5	192.9	146.5	1181.4	152.6	756.6	125.5

A second validation was performed on the sEMG dataset acquired during walking (Section 2.2), with a direct comparison to the DT algorithm. Outcomes are reported in Figure 4. A significant mean increase over the whole population ($p < 0.05$) of recall and F1-score was provided by DEMANN, for onset and offset prediction. This improvement ($p < 0.05$) was preserved also considering signals from a single muscle, for both TA and for GL. No significant differences ($p > 0.05$) were identified in the VL signals and for all the prediction parameters.

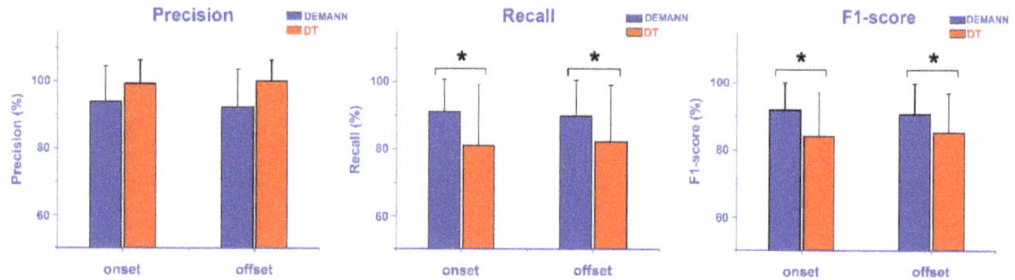

Figure 4. Mean (±SD) precision, recall, and F1-score computed in the onset and offset prediction by the DEMANN approach (blue bars) vs. the DT algorithm (red bars) achieved in real sEMG data during able-bodied walking. * indicates statistically significant difference.

4. Discussion

The present study was designed to test the capability of a novel machine-learning-based approach of estimating onset and offset timing of muscle activation. One of the main advantages of the present DEMANN approach is that the neural network was trained by means of only simulated sEMG signals (no real signal was needed to train the neural network), thus avoiding all the possible complications and costs associated with a typical experimental procedure. A further advantage was the running time. Without considering the processing time, which depends on the processing capability of the running device (in the case of the present neural network, it was less than 1 ms on an i-7 processor), once the model was trained, the maximum delay of activation prediction was 10 ms (the size of the windows). Although this paper did not explicitly target real-time applications, such a delay can be acceptable even under real-time constraints [26], making DEMANN suitable for the detection of muscle activity in sEMG-driven assistive devices, such as orthoses and exoskeletons. Otherwise, this could be an issue for the algorithmic (non-machine-learning) approaches. For example, the recent literature proposed a novel algorithm for detecting muscle activation in a time-frequency domain, based on Continuous Wavelet Transform (CWT) [11]. This study focused on quantifying the frequency content of the muscle activations and needed to detect muscle activation in the time domain in order to properly compute the frequency range (maximum and minimum). This approach could be very useful for specific aims and could open a new way to deepen the knowledge of neuromotor disorders. However, as most of the algorithm-based approaches, it was based on the computation of a threshold value in order to identify the activation onset and offset [5–11]. Thus, a portion of the sEMG signals must be processed to compute the threshold. This introduces a time-delay of at least the duration of the chosen portion, increasing the running time. In cyclic tasks such as walking, such a portion corresponds to a complete gait cycle. This would introduce a delay of at least 1 s, limiting the application of the approach to environments where real-time application is requested, such as in sEMG-driven exoskeletons. This is not needed in the DEMANN approach, where activations are predicted on subsequent 10 ms windows. Moreover, to identify each single gait cycle, kinematic or dynamic data are needed, such as signals from foot-switch sensors, pressure mats, stereo-photogrammetric systems, and inertial measurements units. This introduces a further complexity in experimental settings, potentially raising the costs, the time consumption, and the intrusiveness on patients. DEMANN does not suffer of these limitations, as it is based on a "blind" segmentation in short time segments.

In the present study, DEMANN proved to provide high performances in three different datasets: (1) a test bench of 864 simulated sEMG signals; (2) 103 real sEMG signals acquired in vastus lateralis during knee extension and in biceps brachii during elbow flexion; and (3) real sEMG signals from gastrocnemius lateralis, tibialis anterior, and vastus lateralis collected during 30 subjects walking. Details are reported in the following two sections.

4.1. Simulated sEMG Signals

DEMANN provided a high classification performance, quantified by a mean accuracy ($\pm$SD) of 97.8 $\pm$ 3.0% and supported by the accuracy = 95.3% in the worst-case scenario (SNR = 3 dB, Table 2). Differences due to increasing SNR values were very small (<4%), suggesting a good robustness to SNR variability. The classification performances of activity vs. silent area confirmed these findings (Table 3).

The effective classification capability and the efficient post-processing of model output provided mean prediction very close to 100% (Table 5). The variability of MAE in the function of α, σ, and SNR is reported in Table 6. Independently from the SNR effect, MAE increased where α and σ assumed the highest values. This means that the quality of prediction worsened, enlarging the activation time-duration, being the time support (i.e., the duration of a single activation) defined as $2 \times \alpha \times \sigma$. However, for activations lasting up to 45% of the simulated-signal duration (450 ms), MAE was <15 ms for both onset and offset predictions, except for sporadic low-SNR situations (<6 dB). MAE > 50 ms was reported mainly for those activations characterized by the concomitant conditions of time durations > 60% of the simulated-signal duration (600 ms) and SNR < 10 dB (red areas, Table 6). It is worth noticing that, in cyclic tasks such as walking, a single muscle activation longer than 50% of signal period (gait cycle, for walking) is rare. Continuous muscular recruitment longer than 60% of the gait cycle is practically not realistic during walking. Muscle groups such as ankle plantar flexors (gastrocnemius, soleus, peroneus) and knee extensors and flexors (vastii, rectus femoris, biceps femoris) are typically recruited for short periods, covering up to 35% of the gait cycle [39]. Only ankle dorsi flexors (tibialis anterior, extensor digitorum longus) may rarely present activations that last up to 50% of the gait cycle. Thus, for most practical applications, DEMANN can provide onset-offset estimation affected by MAE < 20 ms for a wide SNR range (3–30 dB), confirming a good classifier robustness for SNR variability.

The efficiency of the DEMANN approach was firstly proved versus a different machine-learning model. The support vector machine (SVM) was chosen among the models proposed in the literature as a suitable tool for this purpose [22–25]. A comparison, in the whole dataset of 864 simulated sEMG signals, specifically generated for the current experiments, showed DEMANN outperforming SVM, in terms of both onset and offset MAE (Table 4). Moreover, the DEMANN robustness was supported by comparison with the DT algorithm on the same simulated data (Table 5). DEMANN predicted offset values with better accuracy for the lowest SNR values (SNR < 6; Figure 3B). Moreover, DEMANN provided F1-score = 100% in offset prediction for SNR $\geq$ 10 dB; DT only for SNR $\geq$ 13. Likewise, mean offset MAE over the whole dataset was reduced in the DEMANN prediction, compared to DT (Table 5, $p < 0.05$). This was true also considering each single SNR value (Figure 3D); the reduction was significant ($p < 0.05$) for SNR = 3 and for SNR $\geq$ 16. An absence of statistical significance for $6 \leq$ SNR ≤ 13 was likely due to the very large mean SD (28.8 ms) associated with the mean MAE computed over DT predictions in this range. Particularly relevant was the 47% reduction of MAE for SNR = 3 dB, suggesting that DEMANN improved DT performances especially in the lowest SNR values. Although an overall reduction of onset-MAE was visible in the DEMANN prediction (Figure 3C), no significant difference was detected.

One of the most reliable sEMG timing detectors reported in the literature is the wavelet-based approach described in [10]. In that study, the robustness of algorithm performances was also tested on simulated sEMG signals. However, a suitable comparison of the results of the current study with those reported in [10] was hard to accomplish because of the many differences in the generation of the simulated signals (different values of α, σ, and SNR) and in the metrics used to evaluate the algorithm performances (MAE in the present study and bias in [10]). Nevertheless, in the attempt of giving the readers further tools to evaluate the robustness of the present approach, the bias has been computed also in the present data as the relative (with sign) value of the time distance between the predicted and the ground-truth value. Results computed in the signals characterized by SNR = 20 dB (the only value in common between the present study and the one reported in [10]) were compared with those reported in [10]: mean bias was 1.7 ms for DEMANN vs. 7.1 ms in [10] for the onset and −2.8 ms for DEMANN vs. 4.1 ms in [10] for the offset. Signs "−" and "+" were adopted to indicate that the predicted event occurred earlier and later than the corresponding value in the ground-truth signal, respectively.

4.2. Real sEMG Signals

The dataset introduced in [3] was mainly chosen for the specific characteristics of the motor tasks (knee extension and elbow flexion), which allow for achieving a reliable detection of the onset event and consequently a trustworthy ground truth. Only onset events were tested, because the ground truth for offset events was not available in [3]. Outcomes of the application of DEMANN to this dataset are shown in Table 7. At first glance, it seems that a substantial difference exists between MAE values obtained for the simulated (Table 5) and real sEMG signals when using DEMANN. However, considering the same SNR range (3 dB $\leq$ SNR $\leq$ 12 dB), the distance between the two MAE values was strongly reduced (MAE-simulated = 19.1 $\pm$ 25.5 ms vs. MAE-real = 38.5 $\pm$ 56.4 ms); MAE and SD are about twice as many in real signals. This difference may be mainly due to a couple of reasons: (1) the neural network was trained with only simulated signals; (2) the larger variability of real sEMG signals due to the eight-shaped path followed by subjects during the experimental procedure that introduced further sEMG variability (caused by curves, reversing, deceleration, and acceleration [40]) and thus affected the performance of classification and prediction.

Table 8 highlights that the DEMANN approach globally outperformed the performance of the algorithms tested in [13], providing: (1) the lowest absolute error values over the whole 52-signal dataset (SNR $\leq$ 8 dB) for all considered metrics; (2) a relevant reduction of mean and median values over the whole 52-signal dataset of absolute error compared to the best value (ETKEO) reported for DT (mean 31.4%; median 21.8%), WLT (mean 28.7%; median 31.0%), CUSUM (mean 20.3%; median 31.0%), and PROLIFIC (mean 24.6%; median comparable); (3) the same result also for the signals with 6 dB < SNR < 8 dB; and 4) performances comparable with those achieved by the four algorithms, for SNR < 4 dB. As conducted in [13], this dataset was adopted to evaluate the performance of the proposed approach on sEMG signals characterized by a range of low SNR ($\leq$12 dB). For 6 dB $\leq$ SNR $\leq$ 12 dB, absolute error was practically not affected by SNR variability (Table 7). It was reported that, in limb movement studies, time differences from stimulus to sEMG onset with neurological diseases, aging, and postural sets may be as low as 20 ms [41]. The performances of DEMANN in the SNR range from 6 dB to 12 dB complied with these requirements. For lower SNR values (<6 dB), the absolute error was proportionally increasing with decreasing SNR, up to 200 ms for SNR < 2 dB. For this SNR range, and for these specific motor tasks (knee extension and elbow flexion), all the algorithms considered in Table 8 reported high values of absolute error, not complying with the abovementioned clinical needs. However, for these very low SNR values, the identification of onset timing by visual inspection could be very hard also when performed by actual experts, as shown in [13]. Thus, onset prediction is affected not only by the reduction of algorithm performances but also by the uncertainty associated with ground truth identification. In our opinion, this consideration may contribute to explain the high values of absolute error, especially for SNR < 4 dB. This would contribute to also explain the fact that, for similar SNR (=3 dB), the mean MAE provided by DEMANN in the simulated signals was around 20 ms (Figure 3C).

Since walking is one of the most useful tasks to obtain insights on human movement, DEMANN was tested also on a dataset of sEMG data collected during 30 healthy adults walking. Despite the high data variability due to curves, reversing, deceleration, and acceleration during the eight-shaped path, prediction performances were >90% for both the onset and offset prediction (Figure 4). Performances provided by DEMANN were validated vs. the DT algorithm. Significantly higher values ($p < 0.05$, Figure 4) of recall and F1-score for onset and offset prediction showed that DEMANN outperformed the DT algorithm in correctly identifying these events. This was true ($p < 0.05$) also considering the mean values over the signals from the same muscle, in the case of TA and GL. Otherwise, for VL, no significant difference was detected between the two approaches. TA and GL are mainly ankle flexor muscles and VL is a knee extensor; it is acknowledged that ankle muscles are typically more involved in the walking task [39]. Given that differences between DEMANN and DT were significant for TA and GL but not VL, one interesting direction to follow in the future studies could be the analysis of possible muscle specificity of the present approach.

5. Conclusions

The present outcomes suggest the feasibility of predicting onset-offset timing of muscular recruitment of the proposed machine-learning-based method, which was able to provide high performances also in condition of large variability of the sEMG signal. The adoption of DEMANN introduced several further advantages, such as a running time compatible with real time applications, a small deterioration of event detection due to low SNR values and to a large within-signal variability of SNR, and reduced complexity of the experimental protocol associated with model training, since no real signal is needed. All these advantages make this approach suitable for clinical practice and for being included in the procedure for controlling sEMG-driven assistive devices, such as orthoses and exoskeletons.

The DEMANN approach was validated in simulated sEMG signals and in real sEMG signals acquired in young able-bodied subjects, but not in elderly and pathological populations. This is acknowledged as a limitation of the present study. Future studies will be focused on assessing the reliability of the DEMANN approach to provide a robust prediction of activation events also in these populations and on the possible improvements to implement for adapting the model to different conditions and environments. While the present study showed that relatively simple supervised methods, such as shallow neural networks, can be suitable for muscle activation detection, further experiments should be made to determine an optimal classifier to embed in the detecting system.

Author Contributions: Conceptualization, F.D.N. and C.M.; methodology, F.D.N. and C.M.; software, C.M. and A.N.; validation, A.N., C.M. and F.D.N.; investigation, C.M. and F.D.N.; resources, F.D.N. and S.F.; data curation, F.D.N., A.N. and C.M.; writing—original draft preparation, F.D.N.; writing—review and editing, F.D.N., C.M., A.C., A.N. and S.F.; supervision, A.C. and S.F. All authors have read and agreed to the published version of the manuscript.

Funding: This research received no external funding.

Institutional Review Board Statement: The study was conducted in accordance with the Declaration of Helsinki. Ethical review and approval were waived for this study because only simulated data and data available in the literature and already employed in previous studies were used.

Informed Consent Statement: Informed consent was obtained from all subjects involved in the study. Written informed consent has been obtained from the patients to publish this paper.

Data Availability Statement: Data supporting reported results can be found by contacting the corresponding author.

Conflicts of Interest: The authors declare no conflict of interest.

References

1. Sutherland, D.H. The evolution of clinical gait analysis part l: Kinesiological EMG. *Gait Posture* **2001**, *14*, 61–70. [CrossRef]
2. Rosati, S.; Ghislieri, M.; Dotti, G.; Fortunato, D.; Agostini, V.; Knaflitz, M.; Balestra, G. Evaluation of Muscle Function by Means of a Muscle-Specific and a Global Index. *Sensors* **2021**, *21*, 7186. [CrossRef] [PubMed]
3. Tenan, M.S.; Tweedell, A.J.; Haynes, C.A. Analysis of statistical and standard algorithms for detecting muscle onset with surface electromyography. *PLoS ONE* **2017**, *12*, e0177312. [CrossRef] [PubMed]
4. Selvan, S.E.; Allexandre, D.; Amato, U.; Yue, G.H. Unsupervised Stochastic Strategies for Robust Detection of Muscle Activation Onsets in Surface Electromyogram. *IEEE Trans. Neural Syst. Rehabil. Eng.* **2018**, *26*, 1279–1291. [CrossRef]
5. Staude, G.; Flachenecker, C.; Daumer, M.; Wolf, W. Onset detection in surface electromyographic signals: A systematic comparison of methods. *EURASIP J. Appl. Signal Process.* **2001**, *2001*, 867853. [CrossRef]

6. Bonato, P.; D'Alessio, T.; Knaflitz, M. A statistical method for the measurement of muscle activation intervals from surface myoelectric signal during gait. *IEEE Trans. Biomed. Eng.* **1998**, *45*, 287–299. [CrossRef]

7. Severini, G.; Conforto, S.; Schmid, M.; D'Alessio, T. Novel formulation of a double threshold algorithm for the estimation of muscle activation intervals designed for variable SNR environments. *J. Electromyogr. Kinesiol.* **2012**, *22*, 878–885. [CrossRef]

8. Olmo, G.; Laterza, F.; Lo Presti, L. Matched wavelet approach in stretching analysis of electrically evoked surface EMG signal. *Signal Process.* **2000**, *80*, 671–684. [CrossRef]

9. Merlo, A.; Farina, D.; Merletti, R. A fast and reliable technique for muscle activity detection from surface EMG signals. *IEEE Trans. Biomed. Eng.* **2003**, *50*, 316–323. [CrossRef]

10. Vannozzi, G.; Conforto, S.; D'Alessio, T. Automatic detection of surface EMG activation timing using a wavelet transform based method. *J. Electromyogr. Kinesiol.* **2010**, *10*, 767–772. [CrossRef]

11. Di Nardo, F.; Basili, T.; Meletani, S.; Scaradozzi, D. Wavelet-Based Assessment of the Muscle-Activation Frequency Range by EMG Analysis. *IEEE Access* **2022**, *10*, 9793–9805. [CrossRef]

12. Li, X.; Zhou, P.; Aruin, A.S. Teager–Kaiser energy operation of surface EMG improves muscle activity onset detection. *Ann. Biomed. Eng.* **2007**, *35*, 1532–1538. [CrossRef] [PubMed]

13. Tigrini, A.; Mengarelli, A.; Cardarelli, S.; Fioretti, S.; Verdini, F. Improving EMG Signal Change Point Detection for Low SNR by Using Extended Teager-Kaiser Energy Operator. *IEEE Trans. Med. Robot. Bionics* **2020**, *2*, 661–669. [CrossRef]

14. Faust, O.; Hagiwara, Y.; Hong, T.J.; Lih, O.S.; Acharya, U.R. Deep learning for healthcare applications based on physiological signals: A review. *Comput. Methods Programs Biomed.* **2018**, *161*, 1–13. [CrossRef] [PubMed]

15. Lee, K.H.; Min, J.Y.; Byun, S. Electromyogram-Based Classification of Hand and Finger Gestures Using Artificial Neural Networks. *Sensors* **2021**, *22*, 225. [CrossRef]

16. Wang, J.; Sun, S.; Sun, Y. A Muscle Fatigue Classification Model Based on LSTM and Improved Wavelet Packet Threshold. *Sensors* **2021**, *21*, 6369. [CrossRef]

17. Xiong, D.; Zhang, D.; Zhao, X.; Zhao, Y. Deep Learning for EMG-based Human-Machine Interaction: A Review. *IEEE/CAA J. Autom. Sin.* **2021**, *8*, 512–533. [CrossRef]

18. Xu, L.; Chen, X.; Cao, S.; Zhang, X.; Chen, X. Feasibility Study of Advanced Neural Networks Applied to sEMG-Based Force Estimation. *Sensors* **2018**, *18*, 3226. [CrossRef]

19. Moslem, B.; Diab, M.; Khalil, M.; Marque, C. Classification of multichannel uterine EMG signals by using unsupervised competitive learning. In Proceedings of the 2011 IEEE Workshop on Signal Processing Systems (SiPS), Beirut, Lebanon, 4–7 October 2011; pp. 267–272. [CrossRef]

20. Subasi, A.; Yilmaz, M.; Ozcalik, H.R. Classification of EMG signals using wavelet neural network. *J. Neurosci. Methods* **2006**, *156*, 360–367. [CrossRef]

21. Elamvazuthi, I.; Duy, N.; Ali, Z.; Su, S.; Khan, M.; Parasuraman, S. Electromyography (EMG) based Classification of Neuromuscular Disorders using Multi-Layer Perceptron. *Procedia Comput. Sci.* **2015**, *76*, 223–228. [CrossRef]

22. Subasi, A. Classification of EMG signals using PSO optimized SVM for diagnosis of neuromuscular disorders. *Comput. Biol. Med.* **2013**, *43*, 576–586. [CrossRef]

23. Cai, S.; Chen, Y.; Huang, S.; Wu, Y.; Zheng, H.; Li, X.; Xie, L. SVM-Based Classification of sEMG Signals for Upper-Limb Self-Rehabilitation Training. *Front. Neurorobot.* **2019**, *13*, 31. [CrossRef] [PubMed]

24. Li, S.S.W.; Chu, C.C.F.; Chow, D.H.K. EMG-based lumbosacral joint compression force prediction using a support vector machine. *Med. Eng. Phys.* **2019**, *74*, 15–120. [CrossRef] [PubMed]

25. Quitadamo, L.R.; Cavrini, F.; Sbernini, L.; Riillo, F.; Bianchi, L.; Seri, S.; Saggio, G. Support vector machines to detect physiological patterns for EEG and EMG-based human-computer interaction: A review. *J. Neural Eng.* **2017**, *14*, 011001. [CrossRef] [PubMed]

26. Kidziński, Ł.; Delp, S.; Schwartz, M. Automatic real-time gait event detection in children using deep neural networks. *PLoS ONE* **2019**, *14*, e0211466. [CrossRef]

27. Nazmi, N.; Abdul Rahman, M.; Yamamoto, S.I.; Ahmad, S. Walking gait event detection based on electromyography signals using artificial neural network. *Biomed. Signal Process. Control* **2019**, *47*, 334–343. [CrossRef]

28. Di Nardo, F.; Morbidoni, C.; Mascia, G.; Verdini, F.; Fioretti, S. Intra-subject approach for gait-event prediction by neural network interpretation of EMG signals. *Biomed. Eng. Online* **2020**, *19*, 58. [CrossRef]

29. Di Nardo, F.; Morbidoni, C.; Cucchiarelli, A.; Fioretti, S. Influence of EMG-Signal Processing and Experimental Set-up on Prediction of Gait Events by Neural Network. *Biomed. Signal Process. Control* **2021**, *63*, 102232. [CrossRef]

30. Ghislieri, M.; Cerone, G.L.; Knaflitz, M.; Agostini, V. Long short-term memory (LSTM) recurrent neural network for muscle activity detection. *J. Neuroeng. Rehabil.* **2021**, *18*, 153. [CrossRef]

31. Khowailed, I.A.; Abotabl, A. Neural muscle activation detection: A deep learning approach using surface electromyography. *J. Biomech.* **2019**, *95*, 109322. [CrossRef]

32. Staude, G.H. Precise onset detection of human motor responses using a whitening filter and the log-likelihood-ratio test. *IEEE Trans. Biomed. Eng.* **2001**, *48*, 1292–1305. [CrossRef]

33. Agostini, V.; Knaflitz, M. An algorithm for the estimation of the signal-to-noise ratio in surface myoelectric signals generated during cyclic movements. *IEEE Trans. Biomed. Eng.* **2012**, *59*, 219–225. [CrossRef] [PubMed]

34. Di Nardo, F.; Morbidoni, C.; Fioretti, S. Surface Electromyographic Signals Collected during Long-Lasting Ground Walking of Young Able-Bodied Subjects (Version 1.0.0). *PhysioNet*. Available online: https://physionet.org/content/semg/1.0.0/ (accessed on 17 March 2022).
35. Goldberger, A.; Amaral, L.; Glass, L.; Hausdorff, J.; Ivanov, P.C.; Mark, R.; Mietus, J.E.; Moody, G.B.; Peng, C.K.; Stanley, H.E. PhysioBank, PhysioToolkit, and PhysioNet: Components of a new research resource for complex physiologic signals. *Circulation* **2000**, *101*, e215–e220. [CrossRef] [PubMed]
36. Hermens, H.; Freriks, B.; Disselhorst-Klug, C.; Rau, G. Development of recommendations for SEMG sensors and sensor placement procedures. *J. Electromyogr. Kinesiol.* **2000**, *10*, 361–374. [CrossRef]
37. Basseville, M.; Nikiforov, I. *Detection of Abrupt Changes: Theory and Application*; Prentice-Hall: Englewood Cliffs, NJ, USA, 1993; Volume 104.
38. Suviseshamuthu, E.S.; Allexandre, D.; Amato, U.; Vecchia, B.D.; Guang, H.Y. Prolific: A fast and robust profile-likelihood-based muscle onset detection in electromyogram using discrete Fibonacci search. *IEEE Access* **2020**, *8*, 105362–105375. [CrossRef]
39. Perry, J. *Gait Analysis—Normal and Pathological Function*, 2nd ed.; Slack Inc.: West Deptford, NJ, USA, 1992.
40. Winter, D.A.; Yack, H.J. EMG profiles during normal human walking: Stride-to-stride and inter-subject variability. *Electroencephalogr. Clin. Neurophysiol.* **1987**, *7*, 402–411. [CrossRef]
41. Hodges, P.W.; Bui, B.H. A comparison of computer-based methods for the determination of onset of muscle contraction using electromyography. *Electroencephalogr. Clin. Neurophysiol.* **1996**, *101*, 511–519.

Article

Trunk Muscle Coactivation in People with and without Low Back Pain during Fatiguing Frequency-Dependent Lifting Activities

Tiwana Varrecchia [1,2], Silvia Conforto [2,*], Alessandro Marco De Nunzio [3,4], Francesco Draicchio [1], Deborah Falla [5] and Alberto Ranavolo [1]

[1] Department of Occupational and Environmental Medicine, Epidemiology and Hygiene, INAIL, 00078 Rome, Italy; t.varrecchia@inail.it (T.V.); f.draicchio@inail.it (F.D.); a.ranavolo@inail.it (A.R.)
[2] Department of Industrial, Electronic and Mechanical Engineering, Roma Tre University, 00146 Rome, Italy
[3] Department of Sport and Exercise Science, LUNEX International University of Health, Exercise and Sports, 4671 Luxembourg, Luxembourg; alessandro.denunzio@lunex-university.net
[4] Luxembourg Health & Sport Sciences Research Institute A.s.b.l., 4671 Luxembourg, Luxembourg
[5] Centre of Precision Rehabilitation for Spinal Pain (CPR Spine), School of Sport, Exercise and Rehabilitation Sciences, University of Birmingham, Birmingham B15 2TT, UK; D.Falla@bham.ac.uk
* Correspondence: silvia.conforto@uniroma3.it

Citation: Varrecchia, T.; Conforto, S.; De Nunzio, A.M.; Draicchio, F.; Falla, D.; Ranavolo, A. Trunk Muscle Coactivation in People with and without Low Back Pain during Fatiguing Frequency-Dependent Lifting Activities. *Sensors* **2022**, *22*, 1417. https://doi.org/10.3390/s22041417

Academic Editor: Dimitrios A. Patikas

Received: 25 January 2022
Accepted: 9 February 2022
Published: 12 February 2022

Publisher's Note: MDPI stays neutral with regard to jurisdictional claims in published maps and institutional affiliations.

Abstract: Lifting tasks are manual material-handling activities and are commonly associated with work-related low back disorders. Instrument-based assessment tools are used to quantitatively assess the biomechanical risk associated with lifting activities. This study aims at highlighting different motor strategies in people with and without low back pain (LBP) during fatiguing frequency-dependent lifting tasks by using parameters of muscle coactivation. A total of 15 healthy controls (HC) and eight people with LBP performed three lifting tasks with a progressively increasing lifting index (LI), each lasting 15 min. Bilaterally erector spinae longissimus (ESL) activity and rectus abdominis superior (RAS) were recorded using bipolar surface electromyography systems (sEMG), and the time-varying multi-muscle coactivation function (TMCf) was computed. The TMCf can significantly discriminate each pair of LI and it is higher in LBP than HC. Collectively, our findings suggest that it is possible to identify different motor strategies between people with and without LBP. The main finding shows that LBP, to counteract pain, coactivates the trunk muscles more than HC, thereby adopting a strategy that is stiffer and more fatiguing.

Keywords: fatiguing frequency-dependent lifting; low back pain; trunk muscle coactivation; sEMG

1. Introduction

Lifting tasks are manual material-handling activities and are commonly associated with work-related low back disorders (WLBDs) [1–3], which include both low back pain (LBP) and low back injuries. An accurate and precise biomechanical-risk assessment allows not only for the prevention of the onset of WLBDs but also an evaluation of the effectiveness of ergonomic interventions [3–9], i.e., redesign the working environment or work station [10] used to reduce WLBDs occurrences and costs [11].

In recent years, to integrate the Revised National Institute for Occupational Safety and Health (NIOSH) Lifting Equation (RNLE), which is the most widely used approach for the biomechanical risk assessment of lifting heavy loads [2,12–14], instrument-based tools have been designed and developed [15]. These quantitative approaches, which have been further optimized with machine-learning techniques [16,17], rely on kinematic, kinetic and surface electromyography (sEMG) indexes (i.e., such as mechanical energy consumption, compression and shear forces on the spine and trunk muscle coactivation) associated with different lifting risk conditions which are positively correlated to compressive and shear forces at the lumbosacral region of the spine [15,18–20].

The above-mentioned indexes have significant advantages, as they can be used in scenarios in which RNLE cannot, their calculation has a very low computational cost, and the sensors used to record the signals from the human body are unobtrusive, wireless, wearable, miniaturized, and have low power consumption [15]. However, they have never been tested in workers with LBP. When LBP occurs, many workers continue to work despite pain, exposing themselves to an unknown risk [21]. The presence of LBP commonly implies the adoption of different motor strategies (e.g., stiffening the spine, avoiding motion and increasing trunk reflex gains [22]) typically aimed at reducing the pain [23]. A common strategy adopted is to increase trunk stiffness, most likely due to augmented trunk muscle activity and changes in the reflex control of trunk muscles. This mechanism, which intends to protect the spinal structures, could have long-term consequences for spinal health and pain recurrence due to decreased damping compromising trunk dynamics [24]. Trunk stiffness is increased by increasing antagonist trunk-muscle coactivation [18,20,25,26], which is a common adaptation in people with LBP, seen in various conditions, even in standing [27]. Several approaches were proposed to estimate muscle coactivation [28], and they have been applied in different experimental studies. Studies have revealed that antagonist muscle activity counteracts the agonist actions producing functionally unfavorable moments that do not contribute to the required net trunk moment [29–36]. Furthermore, increased muscle coactivation generates increased compressive and shear forces across the spine [37–40] and an increased risk of WLBDs [25,41–43]. On the other hand, other studies have shown that during lifting tasks, the coactivation of the trunk muscles increases as the level of risk increases to improve spine stability and prevent the development of LBDs [15,20,25].

The biomechanical risk has been studied mainly in single frequency-independent lifting tasks with no adjustments for the influence of muscle fatigue. Just recently, a study carried out by our group provided the first risk assessment for fatiguing frequency-dependent lifting tasks [44] based on bipolar and high-density surface electromyography parameters.

The current study aimed to highlight motor strategies by comparing trunk muscle coactivation in people with and without LBP during the execution of fatiguing frequency-dependent lifting tasks at three increasing levels of risks. The time-varying multi-muscle coactivation function (TMCf) [18] was selected as the method to compute muscle coactivation. We hypothesized that people with LBP will show a higher level of muscle coactivation than asymptomatic participants and will develop muscle fatigue at a faster rate.

2. Materials and Methods

2.1. Participants

Fifteen healthy control (HC) participants (nine females and six males; age: 27.87 ± 3.98 years; body mass index [BMI]: 25.26 ± 3.21 kg/m^2) and eight (four females and four males; age: 25.15 ± 6.5 years; BMI: 23.51 ± 4.59 kg/m^2) people with LBP were enrolled. All of the participants with LBP reported pain in the low lumbar region. Research brochures were distributed and people contacted us if they were interested in taking part in the study. Before enrolling, we confirmed that the study's inclusion requirements were met using a standardized questionnaire. The following eligibility criteria were used:

- capacity to give informed written consent;
- no concurrent systemic, rheumatic or neuro-musculoskeletal disorders, which may confound testing, or on high doses of opioids (>30 mg of morphine equivalent dose);
- no current pregnancy;
- HC did not have a relevant history, over the last three years, of back and lower-limb pain or injury that limited their daily activities and/or required treatment from a health professional;
- LBP participants presented with chronic pain for at least 3 months during the past 6 months, which was not attributed to a specific pathology.
- LBP participants had not received treatment from a therapist in the last three months before the date of enrolment.

Before taking part in the study, all participants provided written informed consent, which was carried out in accordance with the Declaration of Helsinki at the Centre of Precision Rehabilitation for Spinal Pain (CPR Spine), the University of Birmingham, United Kingdom, and approved by the School of Sport, Exercise & Rehabilitation Sciences Ethics Committee (protocol number MCR260319-1). To eliminate expectation bias, no information about the expected results was provided to the participants.

2.2. Experimental Procedure

The experimental procedure presented in Varrecchia et al. 2021 was performed [44]. Briefly, the participants performed lifting tasks in three different lifting conditions (see Table 1) selected to obtain Lifting Index (LI) values of 1, 2, and 3 [12]. LI was calculated as follows:

$$\text{LI} = \frac{\text{L}}{\text{RWL}} = \frac{\text{L}}{\text{LC} \times \text{HM} \times \text{VM} \times \text{DM} \times \text{AM} \times \text{FM} \times \text{CM}} \tag{1}$$

where:

- L is the actual weight of the lifted load;
- RWL is the recommended weight limit that provides an estimate of the level of physical demand associated with the lifting task [12];
- LC is the constant load of 23 kg [12];
- HM, VM, DM and AM are the horizontal distance, vertical location, vertical displacement and asymmetry multipliers calculated by using equations or derived by tables by measuring the following parameters (see Figure 1A): horizontal distance (H); vertical location (V); vertical displacement (D); angle of asymmetry (A);
- CM is the coupling multiplier for the quality of gripping;
- FM is the frequency multiplier depending on lifting frequency (F), lifting duration and vertical location [12].

The three conditions differed only in the values attributed to F and FM in order to study the effect of frequency in this frequency-dependent task, while keeping the other NIOSH parameters constant for each risk condition (see Table 1). Notably, hand-to-object coupling was defined as "good" for all three lifting tasks [12].

Standing in a neutral body position [12,45] with the feet positioned parallel at a natural standing distance, the participants lifted the load (L = 10 kg, Table 1) represented by a plastic crate (34 × 29 × 13 cm) filled with a weight, using both hands in three distinct sessions, one for each LI, performed three different days. The different lifting sessions were randomized across the three sessions to avoid any confounding influence from a predefined order of the sequence of risk conditions. Each session was 72 h apart and was conducted at the same time of the day for each participant to avoid confounding effects due to fatigue or daily habits [46].

For a total lifting-task length of 15 min, the number of repetitions was determined by the frequency parameter utilized to obtain the specific LI for each session. Specifically, during the LI = 1, 2 and 3, 4, 11 and 15 lifts per minute were performed, respectively (Table 1).

Participants with LBP were asked to complete the lifting repetition until exhaustion if lasting less than 15 min. A timer and acoustic feedback were used to monitor the duration of lifting and the frequency of tasks, respectively. Specifically, each time the acoustic signal was heard, the participants raised the load to the defined height (V + D = 75 cm + 40 cm = 115 cm, Table 1), and then immediately lowered it again. Then, they released it by standing upright while waiting for the next acoustic signal. The task was performed with a self-selected strategy and no instructions were provided for the technique of task execution (e.g., bending of the trunk or limbs). In each of the three sessions, before the lifting tasks were performed, isometric maximum voluntary contractions (iMVCs) were performed for the trunk flexor and extensor muscles [47].

Table 1. For each task (A, B, and C), the values of the load constant (LC), the load weight (L), the horizontal (H) and vertical (V) locations, the vertical travel distance (D), the asymmetry angle (A), the lifting frequency (F) and the hand-to-object coupling (C) and the corresponding values of the multipliers (HM, VM, DM, AM, FM, CM), the recommended weight limit (RWL) and the lifting index (LI).

Task	LC (Kg)	H (Cm)	HM	V (Cm)	VM	D (Cm)	DM	A (°)	AM	F (Lift/Min)	FM	C	CM	L (KG)	RWL	LI
A	23	44	0.57	75	0.99	40	0.93	0	1	4	0.83	good	1	10	10	1
B	23	44	0.57	75	0.99	40	0.93	0	1	11	0.41	good	1	10	5	2
C	23	44	0.57	75	0.99	40	0.93	0	1	15	0.28	good	1	10	3.33	3

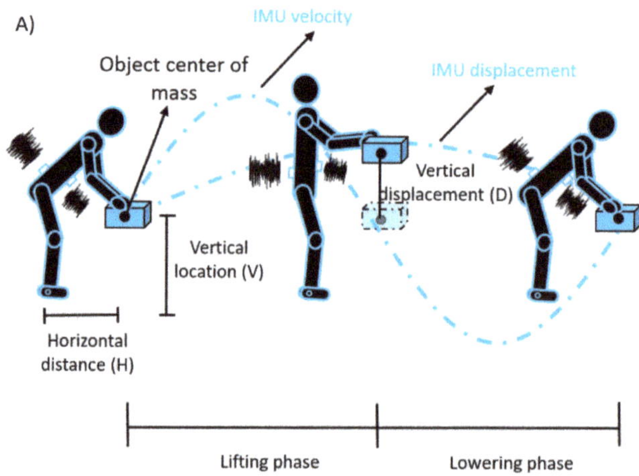

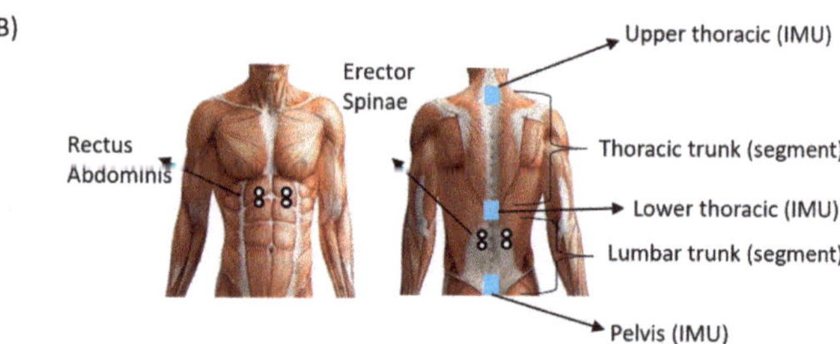

Figure 1. (**A**) description of the experimental setup and of the lifting cycle: displacement and velocity of the IMU placed on the load were used to define the lifting and lowering phases (see Section 2.5.1 for further details). (**B**) Locations of the IMU (blue squares) and sEMG (white circles) electrodes.

2.3. Electromyographic and Inertial Measurement Unit Recordings

Data from the bipolar sEMG and Inertial Measurement Unit (IMU) were acquired simultaneously. All devices were synchronized via a synching device (MyoSync, Noraxon, USA Inc., Scottsdale, AZ, USA).

2.3.1. Bipolar sEMG

Four wireless bipolar sEMG sensors (Ultimium EMG system, Noraxon, USA Inc., Scottsdale, AZ, USA) were placed over the right and left erector spinae longissimus (RESL and LESL, see Figure 1B) and the right and left rectus abdominis superior (RRAS and

LRAS, see Figure 1B) following the electrode placement guidelines [48] and the atlas of muscle innervation zones [49]. Before applying the sensors, the skin was prepared by shaving the area if needed, applying an abrasive paste (SPES Medica, Genova, Italy), and finally washing and drying the region. Then, the sensors were placed using bipolar disposable, wet-gel, self-adhesive Ag/AgCl snap electrodes (2 cm diameter; Dual EMG wet gel electrodes, Noraxon, USA Inc., Scottsdale, AZ, USA). The bipolar sEMG sampling frequency was set to 2000 Hz.

2.3.2. Inertial Measurement Unit

Three inertial sensors (myoMotion Research PRO IMU, Noraxon) were used to acquire movements of the following body segments (Figure 1B): upper thoracic (T2), lower thoracic (over the spine at L1/T12) and pelvis (bony area of the sacrum at L5-S1 level). An additional IMU was placed on the plastic crate (z-axis in the vertical direction). Calibration was carried out with the participant in an upright standing position. The inertial sensor-sampling frequency was set at 2000 Hz, so as to be consistent with sEMG recordings.

2.4. Questionnaires Data

At the end of each session, participants completed the Borg scale to rate fatigue (with 6–20 as anchor points for extremely light and extremely hard perceived exertion, respectively [50]). The pain level in the low back region was measured using a visual analogue scale (VAS, [51]) (with 0–100 as anchor points for no pain and the worst pain imaginable, respectively). In both groups, pain ratings were recorded before and after the session and every minute of the lifting exercise for individuals with LBP.

2.5. Data Analysis

Data were processed using Matlab (version 2018b 9.5.0.1178774, MathWorks, Natick, MA, USA). The IMU and sEMG data during the lifting task were time-normalized to the duration of the lifting and lowering phases. A linear interpolation procedure was used to obtain 200 samples per phase to compare different lifting tasks with different durations [45].

2.5.1. Lifting Cycles Detection

The vertical displacement and velocity of the IMU placed over the load were calculated by integrating the acceleration of the IMU (3rd order low-pass Butterworth filtered by applying a 10 Hz cut-off frequency, [44,52]) once and then twice, respectively, with the drift correction considering a null vertical acceleration and speed before and after the lifting action. Each whole-lifting cycle was subdivided into lifting and lowering phases. The onset and termination of the lifting phase were defined as the time point at which the IMU vertical velocity exceeded a threshold of 0.025 m/s and the peak of the IMU vertical displacement, respectively. The same threshold was used to set the termination of the lowering phase (see Figure 1A). After selecting the whole-lifting cycles, a Dynamic Time Warping approach [53] was used to align the curves that were shifted if wrong events were detected [44].

2.5.2. Bipolar sEMG Preprocessing

To decrease low-frequency artefacts and high-frequency noise, the sEMG signals recorded for both iMVC and tasks were band-pass filtered using a 3rd order Butterworth filter of 25–400 Hz [54,55]. Full-wave rectification and low-pass filtering using a 4th order Butterworth filter at 5 Hz were used to extract the envelope of sEMG signals of each lifting task [33]. The sEMG envelope was amplitude-normalized to the average iMVC peak value for each muscle [56–59].

2.5.3. Time-Varying Multi-Muscle Coactivation Function (TMCf)

The time-varying multi-muscle coactivation function (TMCf) [18,20] was computed to estimate the coactivation of the four trunk muscles during the lifting task using the following formula:

$$TMCf(d(k), k) = \left(1 - \frac{1}{1 + e^{-12(d(k) - 0.5)}}\right) \cdot \frac{\left(\sum_{m=1}^{M} sEMG_m(k)/M\right)^2}{max_{m=1...M}[sEMG_m(k)]} \tag{2}$$

where:

- $d(k)$ is the mean of the differences between the kth samples of each pair of sEMG signals:

$$d(k) = \frac{\sum_{m=1}^{M-1} \sum_{n=m+1}^{M} |sEMG_m(k) - sEMG_n(k)|}{J(M!/(2!(M-2)!))} \tag{3}$$

- J is the length of the signal;
- M is the number of considered muscles;
- $sEMG_m(k)$ and $sEMG_n(k)$ are the kth sample value of the envelope of the sEMG signals of the mth and nth muscles, respectively.

As coactivation indices, the mean (TMCf$_{Mean}$) and the maximum (TMCf$_{Max}$) values within the cycles were calculated. This function and indices were calculated within the lifting and lowering phase.

TMCf$_{Mean}$ and TMCf$_{Max}$ in all the conditions (LI = 1, 2 and 3) of all of the lifting tasks were time-averaged across all the cycles and over one-minute consecutive windows to compare data with a different number of repetitions of the lifting cycles. The first five cycles of each lifting condition were averaged to compare the two groups in a non-fatigued condition.

2.5.4. Range of Motion and Trunk Stability Parameters

The flexion-extension range of motion (RoM) of thoracic (angle between upper thoracic and lower thoracic IMU) and lumbar (between lower thoracic and pelvis) trunk were extracted from the IMU system by calculating the difference between the maximum and minimum angle values within the lifting and lowering phase (RoM$_{Thoracic}$ and RoM$_{Lumbar}$). The stability parameters were extracted via the Root Mean Square (RMS) of the acceleration of upper (RMS$_{upper}$) and lower (RMS$_{lower}$) thoracic IMUs (see Figure 1B), an increase in which indicates a decrease in stability [52].

2.6. Statistical Analysis

The statistical analysis was performed using Matlab (version 2018b 9.5.0.1178774, MathWorks, Natick, MA, USA) to verify the difference between HC and LBP groups and the effect of the risk levels via the TMCf parameters across the total number of lifting repetitions, for the repetitions on the consecutive one-minute windows and separately for the lifting and lowering phases. The group effect was measured via the statistical analysis of the TMCf parameters for the first five cycles, as non-fatigued lifting cycles of each lifting condition. For each parameter, the normality of data distribution was checked using the Shapiro–Wilks test. Then, in each group (HC and LBP), one-way repeated-measures analysis of variance (ANOVA) or a corresponding Friedman t-test (if data not normally distributed) was performed to determine whether LI levels determine significant changes in each parameter. Post hoc analyses were performed using a paired t-test with Bonferroni's corrections when significant differences were observed. For each LI, the unpaired two-sample t-test or Mann–Whitney (MW) test was used to evaluate differences in TMCf parameter between LBP and HC. The same statistical approach was used to verify the difference between HC and LBP groups, and the effect of the risk levels on and VAS and Borg scales. Statistical significance level was set as p-value < 0.05.

3. Results

3.1. TMCf

Figure 2 shows, for the HC (panel A) and LBP (panel B) groups, the mean envelopes of the LESL, RESL, LRAS and RRAS and the mean envelopes (±standard deviation, SD) of TMCf function among all subjects for each lifting condition, considering all the cycles of the task.

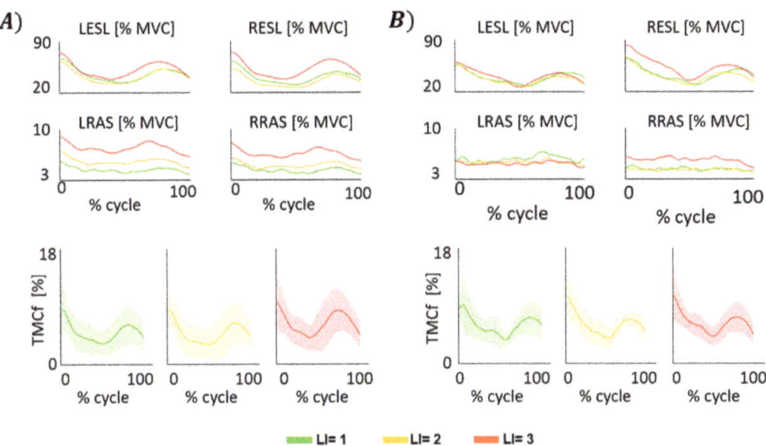

Figure 2. Mean envelopes among all subjects for each lifting condition of the left (LESL) and right (RESL) erector spinae longissimus and the left (LRAS) and right (RRAS) rectus abdominis, considering all the cycles of the task and mean envelopes (±SD) among all subjects for each lifting condition of TMCf function in both groups: healthy controls (**A**) and people with Low Back Pain (**B**). LI: Lifting index.

Figure 3 shows the mean and standard deviation of $TMCf_{Max}$ and $TMCf_{Mean}$ for both groups considering the cycles in the total duration of the task for each lifting condition (panel A) and only the first five cycles of all conditions (panel B). For HC, statistically significant effects of LI for $TMCf_{Max}$ in both lifting (F = 3.73, df = 2, p = 0.0367) and lowering (F = 3.49, df = 2, p = 0.044) phases were found while there was no significant effect for $TMCf_{Mean}$ in both lifting (Chi = 2.13, df = 2, p = 0.344) and lowering (F = 1.9, df = 2, p = 0.169) phases (see Figure 3A). A post hoc analysis showed significant differences (p < 0.05) between LI = 1 and LI = 3 for $TMCf_{Max}$ in both phases (Figure 3A).

For the LBP group, statistically significant effects of LI for $TMCf_{Max}$ in both lifting (Chi = 0.25, df = 2, p = 0.883) and lowering (Chi = 0.75, df = 2, p = 0.687) as well as for $TMCf_{Mean}$ in both lifting (F = 0.05, df = 2, p = 0.956) and lowering (F = 0.84, df = 2, p = 0.452) were found. Moreover, considering all the cycles in the total duration of the task (Figure 3A), statistically significant effects of group for $TMCf_{Max}$ at LI = 1 in both lifting (p = 0.042) and lowering phase (p = 0.026), at LI = 2 in lowering phase (p < 0.001) and for $TMCf_{Mean}$ at LI = 2 (p = 0.001) were found. Furthermore, considering only the first cycles of all conditions (Figure 3B), statistically significant effects of group for $TMCf_{Max}$ in both lifting (p = 0.049) and lowering (p = 0.002) phases and for $TMCf_{Mean}$ in both lifting (p = 0.036) and lowering phases (p = 0.012) were found.

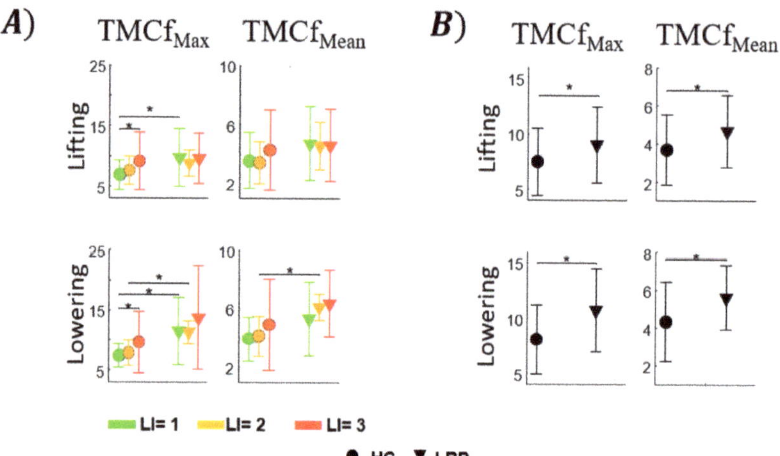

Figure 3. Mean ± SD for each risk level in both groups for the mean (TMCf$_{Mean}$) and the maximum (TMCf$_{Max}$) values of TMCf function considering all repetitions within the entire session, in lifting lowering phases (**A**) and considering the first 5 cycles for each lifting conditions (**B**). TMCf: Time-varying multi-muscle coactivation function. [* statistical significance ($p < 0.05$)].

Figure 4 shows the mean values and standard deviation of TMCf$_{Max}$ and TMCf$_{Mean}$ of all the participants for both groups and each LI during each minute of the task. For each period, statistically significant effects of LI ($p < 0.05$) for TMCf$_{Max}$ and the TMCf$_{Mean}$ were found. The statistical significances for the post hoc analysis are reported in Figure 4. Statistical differences were found between groups, as shown in Figure 4.

3.2. Trunk Motion

Figure 5 shows the mean (±SD) of RoM$_{Thoracic}$, RoM$_{Lumbar}$, RMS$_{upper}$ and RMS$_{lower}$ for both groups, considering all the cycles in the total duration of the task for each lifting condition.

For the HC, the statistically significant effects of LI for RoM$_{Thoracic}$ (F = 4.75, df = 2, $p = 0.017$) in the lifting phase, for RMS$_{upper}$ in both lifting (F = 20.17, df = 2, $p < 0.001$) and lowering (F = 21.25, df = 2, $p < 0.001$) phases and for RMS$_{lower}$ in both lifting (Chi = 17.73, df = 2, $p < 0.001$) and lowering phases (Chi = 12.13, df = 2, $p = 0.002$) were found.

For the LBP group, statistically significant effects of LI for RMS$_{upper}$ in both lifting (Chi = 7, df = 2, $p = 0.030$) and lowering (Chi = 9.75, df = 2, $p = 0.008$) phases and for RMS$_{lower}$ in both lifting (Chi = 7.75, df = 2, $p = 0.021$) and lowering (Chi = 9, df = 2, $p = 0.011$) phases were found.

The post hoc analysis showed significant differences ($p < 0.05$) for HC between LI = 1 and LI = 2 and LI = 1 and LI = 3 for RoM$_{Thoracic}$; for HC between each pair of LI (1 vs. 2, 2 vs. 3 and 1 vs. 3) for RMS$_{upper}$ in both lifting and lowering phases; for HC between LI = 1 and LI = 3 for RMS$_{lower}$ in both lifting and lowering phases and between LI = 1 and LI = 2 for RMS$_{lower}$ in the lifting phase; for those with LBP between LI = 1 and LI = 3 for RMS$_{upper}$ and RMS$_{lower}$ in both the lifting and lowering phase (Figure 5).

Statistical significant differences between the groups ($p < 0.05$) for RMS$_{upper}$ in the lifting phase at LI = 1 and for RMS$_{lower}$ in lifting and lowering phases at LI = 1, LI = 2 and LI = 3 were found (Figure 5).

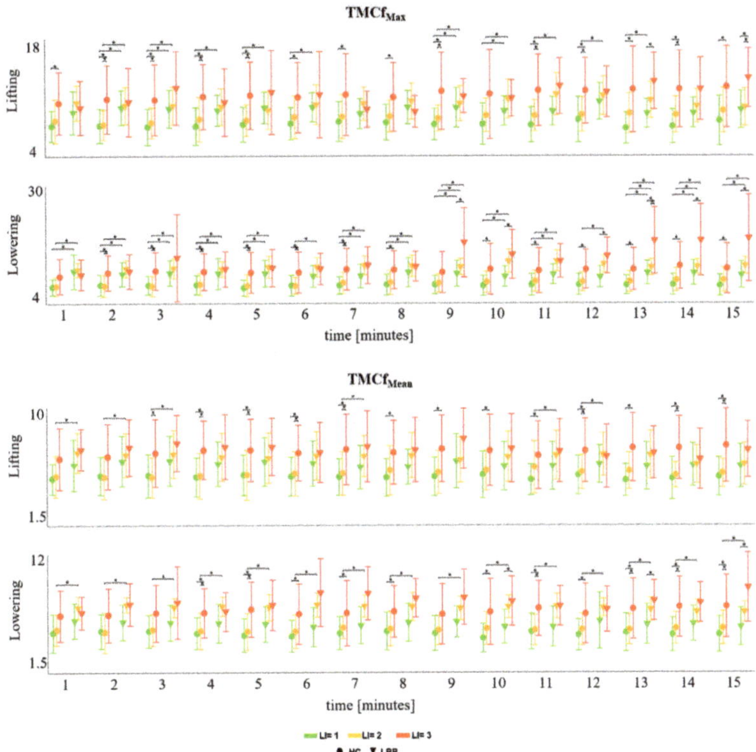

Figure 4. Mean $\pm$ SD for each risk level in both groups for the mean (TMCf$_{Mean}$) and the maximum (TMCf$_{Max}$) values of TMCf function considering all repetitions within each minute of lifting and lowering cycles. [* statistical significance ($p < 0.05$)].

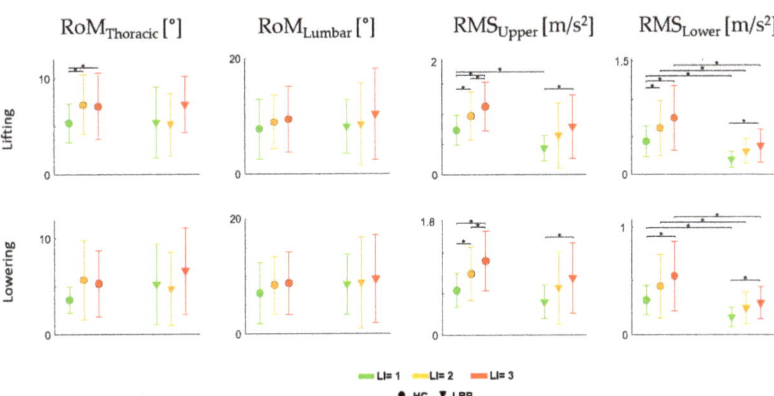

Figure 5. Mean $\pm$ SD for each risk level in both groups for the Range of Motion (RoM) of Thoracic (RoM$_{Thoracic}$) and Lumbar (RoM$_{Lumbar}$) regions and the Root mean square of the acceleration of upper (RMS$_{upper}$) and lower trunk (RMS$_{lower}$) values considering all repetitions within the entire session, in lifting and lowering phases. [* statistical significance ($p < 0.05$)].

3.3. TMCf and Trunk Motion

Figure 6 shows the mean values of $TMCf_{Max}$ (Figure 6A,B) or $TMCf_{Mean}$ (Figure 6C,D), with RoM of the trunk ($RoM_{Thoracic}$) and the RMS of the trunk acceleration (RMS_{upper} in Figure 6A,C or RMS_{lower} in Figure 6B,D), considering all repetitions within each minute of lifting and lowering cycles for both groups.

3.4. Questionnaires

VAS and Borg scale average values at the end of each session are reported in Table 2 for both groups. For the HC, no significant effect of the LI was observed on perceived pain ($p = 0.114$), but there was a significant increase in perceived fatigue ($p < 0.001$). The post hoc analysis showed significant differences for perceived fatigue between LI = 1 and LI = 2 ($p = 0.04$) and between LI = 1 and LI = 3 ($p = 0.002$). For the LBP group, there was no significant effect of the LI on either pain intensity or perceived fatigue ($p > 0.05$). Statistically significant effects of LI = 1, LI = 2 ($p < 0.01$) and LI = 3 ($p < 0.01$) for pain intensity and of LI = 1 and LI = 3 ($p = 0.04$) for fatigue were found between the groups, while no significant effect of LI = 2 ($p = 0.09$) was observed for fatigue.

Table 2. Pain and fatigue scores measured at the end of each lifting task. VAS, visual analogue scale (0–100); HC: healthy controls; LBP: Low Back Pain participants; LI, Lifting Index. Values are presented as mean ± SD.

Scale	LI	HC (Mean ± SD)	LBP (Mean ± SD)
	1	1.4 ± 3.22	42.25 ± 28.48
VAS Pain Intensity (0–100)	2	4.73 ± 10.81	45.71 ± 17.7
	3	11.6 ± 22.78	45.4 ± 17.02
	1	7.53 ± 1.55	10.13 ± 2.47
Borg Scale (6–20)	2	9.2 ± 2.62	13.13 ± 1.96
	3	10.1 ± 2.65	13.5 ± 2.78

Figure 7A shows the mean and standard deviation values in each minute for the VAS score for pain intensity normalized to the values before starting the lifting for those with LBP. Figure 7B shows the mean values of VAS values with $TMCf_{Max}$ (first row) or $TMCf_{Mean}$ (second row) and RMS_{lower} considering all repetitions within each minute of the lifting and lowering cycles for both groups. For each considered period, the statistical analysis revealed significant effects for VAS considering LI ($p < 0.05$). The statistical significances for the post hoc analysis are reported in Figure 7A.

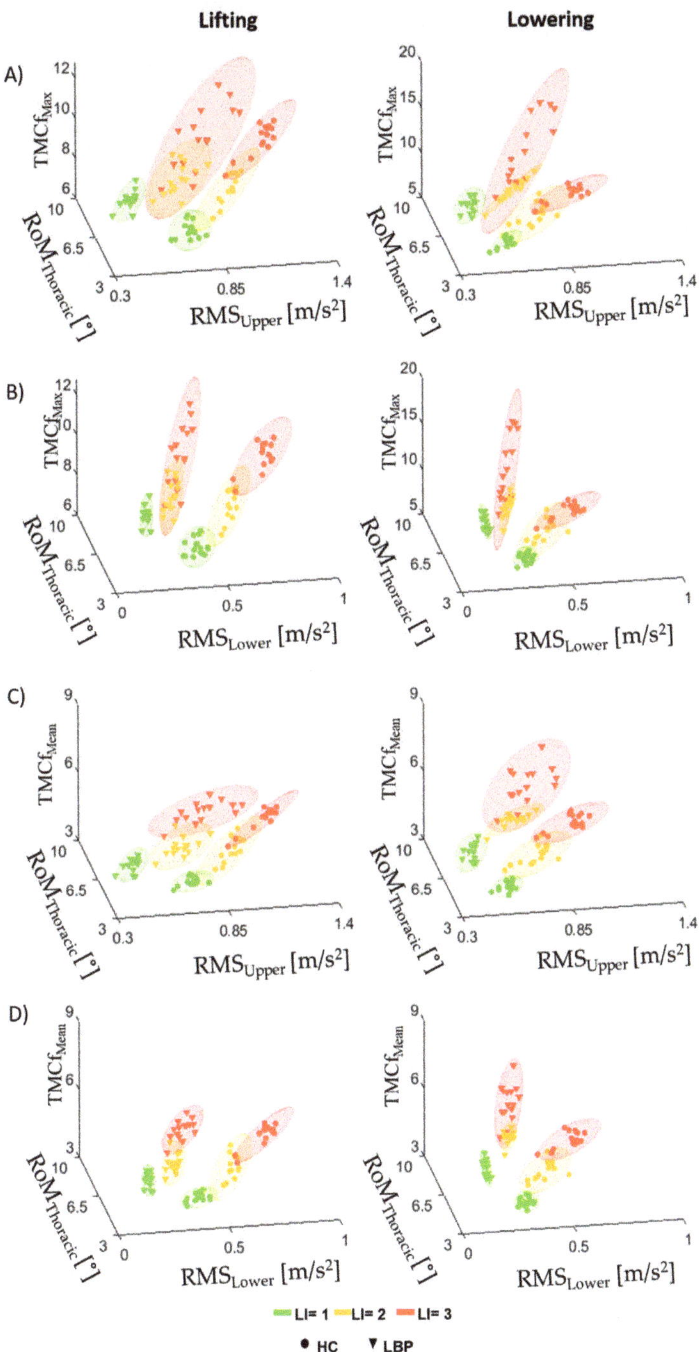

Figure 6. Plot 3D with mean for each risk level in both groups for the max (TMCf$_{Max}$, (**A**,**B**)) and mean (TMCf$_{Mean}$, (**C**,**D**)) values of TMCf function, the RoM of the flexion-extension of the Thoracic region (RoM$_{Thoracic}$) and the RMS of the acceleration of the upper (RMS$_{upper}$, (**A**,**C**)) and lower trunk (RMS$_{lower}$, (**B**,**D**)) considering all repetitions within each minute of lifting and lowering cycles.

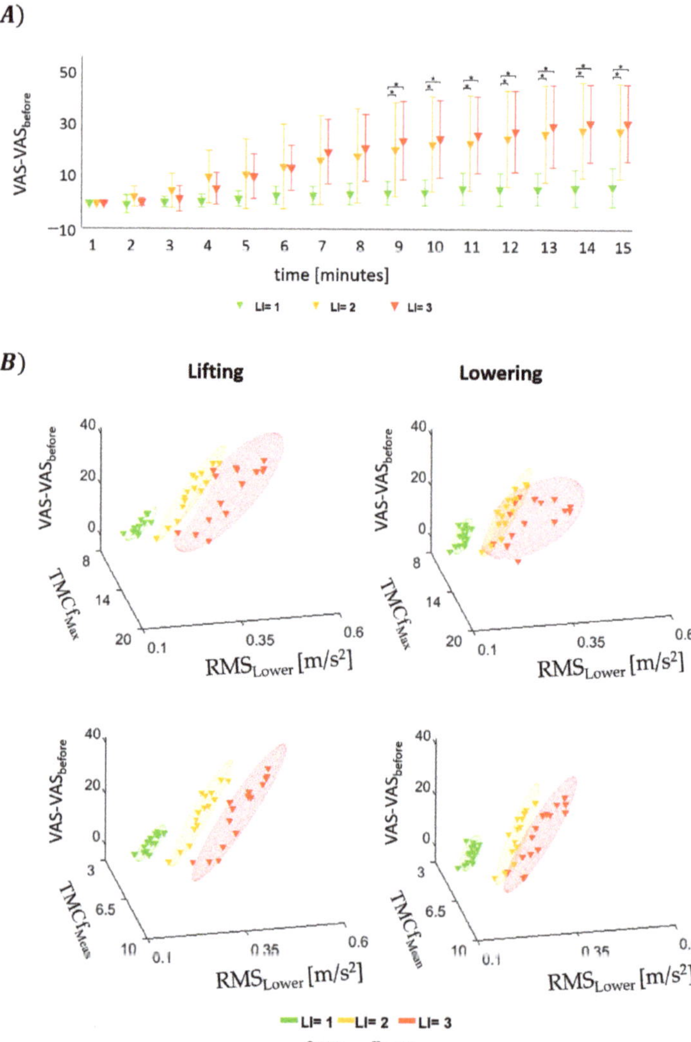

Figure 7. Mean values $\pm$ SD in each minute of VAS for pain intensity normalized to the values before starting the lifting (**A**). Plot 3D (**B**) with mean for each risk level in for the VAS, the max (TMCf$_{Max}$) and mean (TMCf$_{Mean}$) values of TMCf function and the RMS of the acceleration of lower trunk (RMS$_{lower}$) considering all repetitions within each minute of lifting and lowering cycles. [* statistical significance ($p < 0.05$)].

4. Discussion

This study investigated trunk muscle coactivation and trunk movement strategies adopted by people with and without LBP during the execution of fatiguing frequency-dependent lifting tasks characterized by three different levels of risks.

At the beginning of both the lifting and lowering phases, the TMCf showed high values with a reduction until the upright position was reached (end of lifting phase) and the load was released (end of lowering phase) (Figure 2). In addition, from a qualitative point of view, the curves representing the envelopes of LESL and RESL muscle activity and

TMCf (Figure 2) are slightly different, and they show an earlier activation when the risk level increases.

When we analyse the first 5 cycles of the task, we notice that the coactivation of the trunk muscles of those with LBP is significantly higher than that of HC, independently from both the lifting index and phase of the task (lifting and lowering phases, Figure 3B). The increased values of both the mean and maximum of TMCf rely on increased muscle activity during the entire task duration. This can be interpreted as those with LBP, regardless of the fatiguing conditions, are likely exposed to more significant stresses at the L5-S1 segment (increased TMCf corresponds to an increased load at the L5-S1 joint [20]), and they undergo a greater risk for increased pain and injury [37–43].

Additional differences between people with and without LBP (Figure 3A) were found within each risk level (Figure 3A). Specifically, there was an increase in the maximum value of coactivation for those with LBP for LI = 1 in lifting and for LI = 1 and LI = 2 during the lowering phase. The results suggest that the lowering phase is more challenging for those with LBP, especially at low and medium risk levels, likely due to the eccentric nature of antagonist-muscle activation [60].

Both groups did not show statistically significant differences between the different levels of risk, except for differences between LI = 1 and LI = 3 for $TMCf_{Max}$. This result reveals that the coactivation, calculated as an average across the entire task duration, does not vary across the risk levels as its time-varying nature is hidden by the averaging approach. On the contrary, the differences between the levels of risks for both groups and between groups were revealed by analysing the data on a minute-by-minute basis. These differences are particularly evident in the lowering phase, and they appear in both the maximum and the mean values (Figure 4), with the level of the maximum coactivation being constantly higher in those with LBP compared to the HC, especially during lowering.

Considering trunk motion ($RoM_{Thoracic}$ and RoM_{Lumbar} in Figure 5), the two groups have similar kinematic strategies and range of motion in each risk level (the only statistically significant difference was present for LI = 1 and both LI = 2 and 3 for HC during the lifting phase). This is widely justified by the geometry of the task, and was expected because the risk levels were different only for the lifting frequency. However, with the same kinematic strategy, the movement of those with LBP is much less efficient with a greater activation of the antagonist muscles [19], especially in the lowering phase.

The evidence that the RRMS of IMU acceleration is significantly lower for those with LBP than for HC (Figure 5) can be justified by the consideration that the lower back, which is the main body segment involved during lifting and lowering tasks [61], is controlled by a strategy to minimize perturbations and so the reduction in this parameter implies an increase in stability [53].

A clustering approach (Figure 6) applied to the analyzed parameters (both muscular and kinematics) showed a significant difference between the two groups for each level of risk. Such a clustering approach will be applied and extended in future studies that aim to detect the onset of LBP based on the analysis of the trunk muscle and kinematic strategies adopted to reduce the biomechanical effort in lifting tasks.

Differences in pain intensity (VAS) [51] and fatigue [50] between groups are consistent with the indexes of movement and muscular activity and allow different clustering results between the two groups.

The limitations of this study are the small sample size and the case–control study design. Future studies could consider larger sample sizes, other age groups, evaluate men and women separately and could also test other lifting conditions with the same LI values but with different multiplier values.

Collectively, our findings suggest that it is possible to identify different muscular and kinematics strategies between people with and without LBP: the main result shows that people with LBP coactivate their trunk muscles more than HC, by adopting a fatiguing trunk-stiffening strategy, possibly to avoid/minimize pain. This strategy implies an increased risk level that can be quantitively assessed using the trunk muscles coactivation indexes.

The findings of this study provide a preliminary basis for future studies aiming to detect the onset of LBP via the TMCf assessment using a time-varying approach.

Author Contributions: Conceptualization, T.V., A.R., S.C.; methodology, T.V., A.R.; software, T.V., A.R.; validation, T.V., A.R.; formal analysis, T.V., A.R., A.M.D.N., S.C.; investigation, T.V., D.F., A.M.D.N.; resources, A.M.D.N.; data curation, T.V., A.M.D.N.; writing—original draft preparation, T.V., A.R., S.C.; writing—review and editing, T.V., A.R., S.C., F.D., D.F., A.M.D.N.; visualization, T.V.; supervision, A.R., S.C., D.F.; project administration, S.C., A.R., A.M.D.N., D.F.; funding acquisition S.C., A.R. All authors have read and agreed to the published version of the manuscript.

Funding: The research presented in this article was carried out as part of the program BRIC 2016-ID10 funded by INAIL and as part of the SOPHIA project, which has received funding from the European Union's Horizon 2020 research and innovation programme under Grant Agreement No. 871237.

Institutional Review Board Statement: The study was conducted in accordance with the Declaration of Helsinki, and approved by the School of Sport, Exercise & Rehabilitation Sciences Ethics Committee (protocol number MCR260319-1).

Informed Consent Statement: Informed consent was obtained from all subjects involved in the study.

Data Availability Statement: The data presented in this study are available on request from the corresponding author.

Acknowledgments: We are extremely grateful to Amal M Alsubaie; Nadège Haouidji-Javaux; David Jiménez-Grande; Michail Arvanitidis of the Centre of Precision Rehabilitation for Spinal Pain (CPR Spine), School of Sport, Exercise and Rehabilitation Sciences, University of Birmingham, Edgbaston, B152TT, United Kingdom for their help with the experimental procedure.

Conflicts of Interest: The authors declare no conflict of interest.

References

1. Kuijer, P.P.; Verbeek, J.H.; Visser, B.; Elders, L.A.; Van Roden, N.; Van den Wittenboer, M.E.; Lebbink, M.; Burdorf, A.; Hulshof, C.T. An evidence based multidisciplinary practice guideline to reduce the workload due to lifting for preventing work-related low back pain. *Ann. Occup. Environ. Med.* **2014**, *26*, 16. [CrossRef] [PubMed]
2. Lu, M.L.; Waters, T.R.; Krieg, E.; Werren, D. Efficacy of the revised NIOSH lifting equation to predict risk of low-back pain associated with manual lifting: A one-year prospective study. *Hum. Factors* **2014**, *56*, 73–85. [CrossRef] [PubMed]
3. Garg, A.; Boda, S.; Hegmann, K.T.; Moore, J.S.; Kapellusch, J.M.; Bhoyar, P.; Thiese, M.S.; Merryweather, A.; Deckow-Schaefer, G.; Bloswick, D.; et al. The NIOSH lifting equation and low-back pain, Part 1: Association with low-back pain in the backworks prospective cohort study. *Hum. Factors* **2014**, *56*, 6–28. [CrossRef] [PubMed]
4. Garg, A.; Kapellusch, J.M.; Hegmann, K.T.; Moore, J.S.; Boda, S.; Bhoyar, P.; Thiese, M.S.; Merryweather, A.; Deckow-Schaefer, G.; Bloswick, D.; et al. The NIOSH lifting equation and low-back pain, Part 2: Association with seeking care in the backworks prospective cohort study. *Hum. Factors* **2014**, *56*, 44–57. [CrossRef]
5. Griffith, L.E.; Shannon, H.S.; Wells, R.P.; Walter, S.D.; Cole, D.C.; Côté, P.; Frank, J.; Hogg-Johnson, S.; Langlois, L.E. Individual participant data meta-analysis of mechanical workplace risk factors and low back pain. *Am. J. Public Health* **2012**, *102*, 309–318. [CrossRef]
6. Kwon, B.K.; Roffey, D.M.; Bishop, P.B.; Dagenais, S.; Wai, E.K. Systematic review: Occupational physical activity and low back pain. *Occup. Med.* **2011**, *61*, 541–548. [CrossRef]
7. Wai, E.K.; Roffey, D.M.; Bishop, P.; Kwon, B.K.; Dagenais, S. Causal assessment of occupational lifting and low back pain: Results of a systematic review. *Spine J.* **2010**, *10*, 554–566. [CrossRef]
8. Bakker, E.W.; Verhagen, A.P.; van Trijffel, E.; Lucas, C.; Koes, B.W. Spinal mechanical load as a risk factor for low back pain: A systematic review of prospective cohort studies. *Spine* **2009**, *34*, E281–E293. [CrossRef]
9. Hoogendoorn, W.E.; Bongers, P.M.; de Vet, H.C.W.; Ariëns, G.A.; van Mechelen, W.; Bouter, L.M. High physical work load and low job satisfaction increase the risk of sickness absence due to low back pain: Results of a prospective cohort study. *Occup. Environ. Med.* **2002**, *59*, 323–328. [CrossRef]
10. Westgaard, R.H.; Winkel, J. Ergonomic intervention research for improved musculoskeletal health: A critical review. *Int. J. Ind. Ergon.* **1997**, *20*, 463–500. [CrossRef]
11. Burgess-Limerick, R. Participatory ergonomics: Evidence and implementation lessons. *Appl. Ergon.* **2018**, *68*, 289–293. [CrossRef]
12. Waters, T.R.; Putz-Anderson, V.; Garg, A. *Applications Manual for the Revised NIOSH Lifting Equation*; U.S. Department of Health and Human Services: Cincinnati, OH, USA, 1994.

13. Waters, T.R.; Putz-Anderson, V.; Garg, A.; Fine, L.J. Revised NIOSH equation for the design and evaluation of manual lifting tasks. *Ergonomics* **1993**, *36*, 749–776. [CrossRef] [PubMed]

14. Waters, T.R.; Lu, M.L.; Piacitelli, L.A.; Werren, D.; Deddens, J.A. Efficacy of the revised NIOSH lifting equation to predict risk of low back pain due to manual lifting: Expanded cross-sectional analysis. *J. Occup. Environ. Med.* **2011**, *53*, 1061–1067. [CrossRef] [PubMed]

15. Ranavolo, A.; Draicchio, F.; Varrecchia, T.; Silvetti, A.; Iavicoli, S. Wearable Monitoring Devices for Biomechanical Risk Assessment at Work: Current Status and Future Challenges—A Systematic Review. *Int. J. Environ. Res. Public Health* **2018**, *15*, 2001. [CrossRef] [PubMed]

16. Varrecchia, T.; De Marchis, C.; Draicchio, F.; Schmid, M.; Conforto, S.; Ranavolo, A. Lifting activity assessment using kinematic features and neural networks. *Appl. Sci.* **2020**, *10*, 1989. [CrossRef]

17. Varrecchia, T.; De Marchis, C.; Rinaldi, M.; Draicchio, F.; Serrao, M.; Schmid, M.; Conforto, S.; Ranavolo, A. Lifting activity assessment using surface electromyographic features and neural networks. *Int. J. Ind. Ergon.* **2018**, *66*, 1–9. [CrossRef]

18. Ranavolo, A.; Mari, S.; Conte, C.; Serrao, M.; Silvetti, A.; Iavicoli, S.; Draicchio, F. A new muscle coactivation index for biomechanical load evaluation in work activities. *Ergonomics* **2015**, *58*, 966–979. [CrossRef]

19. Ranavolo, A.; Varrecchia, T.; Rinaldi, M.; Silvetti, A.; Serrao, M.; Conforto, S.; Draicchio, F. Mechanical lifting energy consumption in work activities designed by means of the "revised NIOSH lifting equation". *Ind. Health* **2017**, *55*, 444–454. [CrossRef]

20. Ranavolo, A.; Varrecchia, T.; Iavicoli, S.; Marchesi, A.; Rinaldi, M.; Serrao, M.; Conforto, S.; Cesarelli, M.; Draicchio, F. Surface electromyography for risk assessment in work activities designed using the "revised NIOSH lifting equation". *Int. J. Ind. Ergon.* **2018**, *68*, 34–45. [CrossRef]

21. Dean, S.G.; Hudson, S.; Hay-Smith, E.J.C.; Milosavljevic, S. Rural workers' experience of low back pain: Exploring why they continue to work. *J. Occup. Rehabil.* **2011**, *21*, 395–409. [CrossRef]

22. Trost, Z.; France, C.R.; Sullivan, M.J.; Thomas, J.S. Pain-related fear predicts reduced spinal motion following experimental back injury. *Pain* **2012**, *153*, 1015–1021. [CrossRef] [PubMed]

23. Griffioen, M.; van Drunen, P.; Maaswinkel, E.; Perez, R.S.G.M.; Happee, R.; van Dieën, J.H. Identification of intrinsic and reflexive contributions to trunk stabilization in patients with low back pain: A case-control study. *Eur. Spine J.* **2020**, *29*, 1900–1908. [CrossRef] [PubMed]

24. Hodges, P.; van den Hoorn, W.; Dawson, A.; Cholewicki, J. Changes in the mechanical properties of the trunk in low back pain may be associated with recurrence. *J. Biomech.* **2009**, *42*, 61–66. [CrossRef] [PubMed]

25. Granata, K.P.; Marras, W.S. Cost-benefit of muscle cocontraction in protecting against spinal instability. *Spine* **2000**, *25*, 1398–1404. [CrossRef] [PubMed]

26. Marras, W.S.; Mirka, G.A. Electromyographic studies of the lumbar trunk musculature during the generation of lowlevel trunk acceleration. *J. Orthop. Res.* **1993**, *11*, 811–817. [CrossRef]

27. Kiers, H.; van Dieën, J.H.; Brumagne, S.; Vanhees, L. Postural sway and integration of proprioceptive signals in subjects with LBP. *Hum. Mov. Sci.* **2015**, *39*, 109–120. [CrossRef]

28. Rinaldi, M.; D'Anna, C.; Schmid, M.; Conforto, S. Assessing the influence of SNR and pre-processing filter bandwidth on the extraction of different muscle co-activation indexes from surface EMG data. *J. Electromyogr. Kinesiol.* **2018**, *43*, 184–192. [CrossRef]

29. Le, P.; Aurand, A.; Dufour, J.S.; Knapik, G.G.; Best, T.M.; Khan, S.N.; Mendel, E.; Marras, W.S. Development and testing of a moment-based coactivation index to assess complex dynamic tasks for the lumbar spine. *Clin. Biomech.* **2017**, *46*, 23–32. [CrossRef]

30. Rosa, M.C.; Marques, A.; Demain, S.; Metcalf, C.D.; Rodrigues, J. Methodologies to assess muscle co-contraction during gait in people with neurological impairment—A systematic literature review. *J. Electromyogr. Kinesiol.* **2014**, *24*, 179–191. [CrossRef]

31. Olney, S.J. *Quanitative Evaluation of Cocontraction of Knee and Ankle Muscles in Normal Walking*; Human Kinetics Publishers: Champaign, IL, USA, 1985.

32. Frost, G.; Dowling, J.; Dyson, K.; Bar-Or, O. Cocontraction in Three Age Groups of Children during Treadmill Locomotion. *J. Electromyogr. Kinesiol.* **1997**, *7*, 179–186. [CrossRef]

33. Winter, D.A. *Biomechanics and Motor Control of Human Movement*, 4th ed.; John Wiley & Sons, Inc.: Hoboken, NJ, USA; University of Waterloo: Waterloo, ON, Canada, 2009.

34. Falconer, F.K.; Winter, D. Quantitative Assessment of Cocontraction at the Ankle Joint during Walking. *Electromyogr. Clin. Neurophysiol.* **1985**, *25*, 135–149. [PubMed]

35. Macaluso, A.; Nimmo, M.A.; Foster, J.E.; Cockburn, M.; McMillan, N.C.; De Vito, G. Contractile Muscle Volume and Agonist–Antagonist Coactivation Account for Differences in Torque Between Young and Older Women. *Muscle Nerve* **2002**, *25*, 858–863. [CrossRef] [PubMed]

36. Brookham, R.L.; Middlebrook, E.E.; Grewal, T.J.; Dickerson, C.R. The Utility of an Empirically Derived Coactivation Ratio for Muscle Force Prediction Through Optimization. *J. Biomech.* **2011**, *44*, 1582–1587. [CrossRef] [PubMed]

37. Lewek, M.D.; Rudolph, K.S.; Snyder-Mackler, L. Control of Frontal Plane Knee Laxity during Gait in Patients with Medial Compartment Knee Osteoarthritis. *Osteoarthr. Cartil.* **2004**, *12*, 745–751. [CrossRef] [PubMed]

38. Childs, J.D.; Sparto, P.J.; Fitzgerald, G.K.; Bizzini, M.; Irrgang, J.J. Alterations in Lower Extremity Movement and Muscle Activation Patterns in Individuals with Knee Osteoarthritis. *Clin. Biomech.* **2004**, *19*, 44–49. [CrossRef] [PubMed]

39. Griffin, T.; Guilak, F. The Role of Mechanical Loading in the Onset and Progression of Osteoarthritis. *Exerc. Sport Sci. Rev.* **2005**, *33*, 195–200. [CrossRef]

40. Collins, A.; Blackburn, J.T.; Olcott, C.; Yu, B.; Weinhold, P. The Impact of Stochastic Resonance Electrical Stimulation and Knee Sleeve on Impulsive Loading and Muscle Co-contraction during Gait in Knee Osteoarthritis. *Clin. Biomech.* **2011**, *26*, 853–858. [CrossRef]
41. Granata, K.P.; Marras, W.S. An EMG-assisted Model of Loads on the Lumbar Spine during Asymmetric Trunk Extensions. *J. Biomech.* **1993**, *26*, 1429–1438. [CrossRef]
42. Granata, K.P.; Marras, W.S. The Influence of Trunk Muscle Coactivity on Dynamic Spinal Loads. *Spine* **1995**, *20*, 913–919. [CrossRef]
43. Marras, W.S.; Granata, K.P. Spine Loading During Trunk Lateral Bending Motions. *J. Biomech.* **1996**, *30*, 697–703. [CrossRef]
44. Varrecchia, T.; Ranavolo, A.; Conforto, S.; De Nunzio, A.M.; Arvanitidis, M.; Draicchio, F.; Falla, D. Bipolar versus high-density surface electromyography for evaluating risk in fatiguing frequency-dependent lifting activities. *Appl. Ergon.* **2021**, *95*, 103456. [CrossRef]
45. Moore, S.M.; Torma-Krajewski, J.; Steiner, L.J. *Practical Demonstrations of Ergonomic Principles*; Technical Report of investigations for National Institute for Occupational Safety and Health, 9684; DHHS Publication: Washington, DC, USA, 2011.
46. Filho, J.C.; Gobbi, L.T.; Gurjão, A.L.; Gonçalves, R.; Prado, A.K.; Gobbi, S. Effect of different rest intervals, between sets, on muscle performance during leg press exercise, in trained older women. *J. Sports Sci. Med.* **2013**, *12*, 138–143. [PubMed]
47. Vera-Garcia, F.J.; Moreside, J.M.; McGill, S.M. MVC techniques to normalize trunk muscle EMG in healthy women. *J. Electromyogr. Kinesiol.* **2010**, *20*, 10–16. [CrossRef] [PubMed]
48. Seniam. Available online: www.seniam.org (accessed on 9 October 2020).
49. Barbero, M.; Merletti, R.; Rainoldi, A. *Atlas of Muscle Innervation Zones: Understanding Surface Electromyography and Its Applications*; Springer: New York, NY, USA, 2012. [CrossRef]
50. Borg, G.A. Psychophysical bases of perceived exertion. *Med. Sci. Sports Exerc.* **1982**, *14*, 377–381. [CrossRef] [PubMed]
51. Faiz, K.W. VAS—visuell analog skala [VAS—visual analog scale]. *Tidsskr Nor Laegeforen* **2014**, *134*, 323. [CrossRef] [PubMed]
52. Fazio, P.; Granieri, G.; Casetta, I.; Cesnik, E.; Mazzacane, S.; Caliandro, P.; Pedrielli, F.; Granieri, E. Gait measures with a triaxial accelerometer among patients with neurological impairment. *J. Neurol. Sci.* **2013**, *34*, 435–440. [CrossRef] [PubMed]
53. Muscillo, R.; Conforto, S.; Schmid, M.; Caselli, P.; D'Alessio, T. Classification of motor activities through derivative dynamic time warping applied on accelerometer data. In Proceedings of the 2007 29th Annual International Conference of the IEEE Engineering in Medicine and Biology Society, Lyon, France, 22–26 August 2007; pp. 4930–4933.
54. Butler, H.L.; Newell, R.; Hubley-Kozey, C.L.; Kozey, J.W. The Interpretation of Abdominal Wall Muscle Recruitment Strategies Change when the Electrocardiogram (ECG) is Removed from the Electromyogram (EMG). *J. Electromyogr. Kinesiol.* **2009**, *19*, 102–113. [CrossRef]
55. Drake, J.D.; Callaghan, J.P. Elimination of electrocardiogram contamination from electromyogram signals: An evaluation of currently used removal techniques. *J. Electromyogr. Kinesiol.* **2006**, *16*, 175–187. [CrossRef] [PubMed]
56. Ranavolo. *Principi di Elettromiografia di Superficie. Dal Potenziale D'azione alle Applicazioni nei Diversi Settori Della Medicina e Dell'ingegneria*; Edizioni Universitarie Romane: Rome, Italy, 2021.
57. Burden, A. How should we normalize electromyograms obtained from healthy participants? What we have learned from over 25 years of research. *J. Electromyogr. Kinesiol.* **2010**, *20*, 1023–1035. [CrossRef]
58. Hermens, H.J.; Freriks, B.; Disselhorst-Klug, C.; Rau, G. Development of recommendations for SEMG sensors and sensor placement procedures. *J. Electromyogr. Kinesiol.* **2000**, *10*, 361–374. [CrossRef]
59. Marras, W.S.; Davis, K.G. A non-MVC EMG normalization technique for the trunk musculature: Part 1. Method development. *J. Electromyogr. Kinesiol.* **2001**, *11*, 1–9. [CrossRef]
60. Shirado, O.; Ito, T.; Kaneda, K.; Strax, T.E. Concentric and eccentric strength of trunk muscles: Influence of test postures on strength and characteristics of patients with chronic low-back pain. *Arch. Phys. Med. Rehabil.* **1995**, *76*, 604–611. [CrossRef]
61. Cole, M.H.; Grimshaw, P.N. Low back pain and lifting: A review of epidemiology and aetiology. *Work* **2003**, *21*, 173–184. [PubMed]

Article

Modified Functional Reach Test: Upper-Body Kinematics and Muscular Activity in Chronic Stroke Survivors

Giorgia Marchesi [1,*,†], Giulia Ballardini [1,†], Laura Barone [2], Psiche Giannoni [1], Carmelo Lentino [2], Alice De Luca [1,3,‡] and Maura Casadio [1,‡]

[1] Department of Informatics, Bioengineering, Robotics and Systems Engineering, University of Genoa, 16145 Genoa, Italy; giulia.ballardini@edu.unige.it (G.B.); psichegi@tin.it (P.G.); alice.deluca@movendo.technology (A.D.L.); maura.casadio@unige.it (M.C.)

[2] Recovery and Functional Reeducation Unit, Rehabilitation Department, Santa Corona Hospital, 17027 Pietra Ligure, Italy; roboticarrf.pietra@asl2.liguria.it (L.B.); c.lentino@asl2.liguria.it (C.L.)

[3] Movendo Technology s.r.l., 16128 Genoa, Italy

* Correspondence: giorgia.marchesi@edu.unige.it; Tel.: +39-0103536550

† These authors contributed equally to this work.

‡ These authors contributed equally to this work.

Abstract: Effective control of trunk muscles is fundamental to perform most daily activities. Stroke affects this ability also when sitting, and the Modified Functional Reach Test is a simple clinical method to evaluate sitting balance. We characterize the upper body kinematics and muscular activity during this test. Fifteen chronic stroke survivors performed twice, in separate sessions, three repetitions of the test in forward and lateral directions with their ipsilesional arm. We focused our analysis on muscles of the trunk and of the contralesional, not moving, arm. The bilateral activations of latissimi dorsi, trapezii transversalis and oblique externus abdominis were left/right asymmetric, for both test directions, except for the obliquus externus abdominis in the frontal reaching. Stroke survivors had difficulty deactivating the contralesional muscles at the end of each trial, especially the trapezii trasversalis in the lateral direction. The contralesional, non-moving arm had muscular activations modulated according to the movement phases of the moving arm. Repeating the task led to better performance in terms of reaching distance, supported by an increased activation of the trunk muscles. The reaching distance correlated negatively with the time-up-and-go test score.

Keywords: sitting balance; trunk control; ipsilesional arm; MFRT; sEMG

Citation: Marchesi, G.; Ballardini, G.; Barone, L.; Giannoni, P.; Lentino, C.; De Luca, A.; Casadio, M. Modified Functional Reach Test: Upper-Body Kinematics and Muscular Activity in Chronic Stroke Survivors. *Sensors* **2022**, 22, 230. https://doi.org/10.3390/s22010230

Academic Editors: Francesco Di Nardo, Valentina Agostini and Silvia Conforto

Received: 1 November 2021
Accepted: 24 December 2021
Published: 29 December 2021

Publisher's Note: MDPI stays neutral with regard to jurisdictional claims in published maps and institutional affiliations.

1. Introduction

Core stability and proper trunk muscle control are fundamental in most daily living activities, such as standing up, sitting down, walking and stabilizing distal limbs [1]. Both are necessary for sitting balance, to maintain stable posture and to shift the body weight inside the base of support while performing a variety of self-initiated actions, such as eating or taking a glass from the table [2].

Following a stroke, the upper motor neuron syndrome induces abnormal muscular activations and motor patterns, with phenomena categorized as "positive" or "negative" in relation, respectively, to the presence of overt behaviors due to muscle overactivity or to the loss of overt behaviors, indicating muscle and motor impairments [3]. For example, spasticity and increased cutaneous reflexes are considered "positive" phenomena, while weakness, impaired control and fatigue are considered to be "negative" phenomena [3].

The synchronized activity of several trunk muscles is necessary for maintaining stability in momentary postures, executing movements and shifting body weight [4]. However, after stroke, the phenomena described above together with the weakness of the trunk flexor and extensor muscles [5], often determine the delayed onset of muscular activations and poor synchronization of muscle pairs [1]. Thus, stroke affects postural and dynamic stability [6–8], leading to incorrect distribution of body weight and inability to shift it according to the task requirements [7,8]. When sitting, stroke survivors tend to have asymmetrical weight-bearing and reduced ability to shift the center of pressure, both in the anteroposterior and medio-lateral directions [9,10].

Moreover, the deficits in trunk control and sitting balance are predictors of functional mobility [11–14], i.e., 45% to 71% of the variance reported for functional recovery can be explained by different trunk control abilities of stroke survivors [13,15]. For these reasons, they are primary goals in rehabilitation and targets for early interventions [14], and their quantitative and standardized evaluation is crucial.

In current clinical day-to-day practice, for assessing performance, therapists mostly use clinical scales—based on ordinal scores—that are standardized and validated, but they are qualitative, subjective and often with low resolution [16]. Those limitations can be overcome by using technological solutions to characterize and quantify performance, allowing for a more complete, functional and objective assessment [17]. Among assessment techniques, surface electromyography (sEMG) is an easy-to-use tool to characterize the muscular activation patterns and to investigate the neural control mechanisms underlying the kinematic measures [18]. More specifically, sEMG is a fundamental tool to investigate the activity of trunk muscles, which have a key-role in maintaining upright trunk posture in standing and sitting and in controlling movements to counteract gravity [2].

In this work, we aimed at investigating, in depth, the activations of trunk muscles during a widely used clinical assessment, the Modified Functional Reach Test (MFRT). It is an adaptation of the Functional Reach Test developed by Duncan [19], where participants must reach forward with one arm, maintaining 90° of shoulder flexion, while standing. The modified version is performed while sitting, being suitable for a larger population of people with motor impairments [20]. The MFRT is a functional clinical assessment for evaluating the risk of fall and determining the limits of stability while sitting, focusing on the ability to shift the body weight maintaining the equilibrium in a self-initiated movement [19–23]. It is widely used in clinical practice for both neurotypical individuals and heterogenous populations with sensorimotor deficits [19–26]. In this work, we considered reaching movements both in forward and lateral directions because these reveal different components of trunk stability [20], and there could be no strong relation between their results [27], i.e., knowledge of performance in one direction may not be predictive for performance in the other direction [28]. MFRT is of broad interest, for both its simplicity and its functional implications. In stroke survivors, this test has been mainly characterized in terms of reaching distance [19–23].

To the best of our knowledge, a kinematic and muscular characterization of MFRT, focusing on the upper body, is still missing, despite the fact that trunk muscles are fundamental for postural control during reaching movements. Our work fills this gap, having the primary goal of characterizing the trunk and upper-body muscular activations as well as the kinematic performance of chronic stroke survivors engaging in the forward and lateral MFRT. Specifically, we focused on the bilateral trunk muscles' activations; on the muscular activity of the contralesional arm, not actively involved in the reaching movement; on the effects of the reaching movement repetition.

2. Materials and Methods

2.1. Participants

Fifteen chronic stroke survivors (eight females, age range: 48 to 78 years old, see Table 1 more details) participated in the study.

Table 1. Demographic data and clinical evaluation.

	Age (ys)	Gender	TSS (ys)	Etiology	PS	FMA-AD (0–66)	FMA-H (0–12)	TIS (0–23)	WMFT (0–85)	TUG (s)
ID01	66	F	11	I	R	39	12	13	49	15.3
ID02	48	F	6	H	L	23	10	14	32	11.3
ID03	65	F	2	H	R	14	12	19	9	15.5
ID04	52	F	5	I	R	37	0	16	44	15.4
ID05	68	M	16	I	R	55	12	11	75	9.4
ID06	68	M	1.5	H	R	33	12	12	57	16.7
ID07	60	F	8	I	R	9	10	12	14	18.5
ID08	62	F	4	I	L	56	11	16	78	36.2
ID09	60	M	4	I	R	57	9	17	78	10.9
ID10	69	M	4	I	L	50	12	16	74	13.3
ID11	68	F	1	I	L	14	12	16	14	35.7
ID12	70	F	7	I	R	44	3	12	62	26.8
ID13	78	M	1	I	L	52	12	13	73	9.5
ID14	72	M	12	I	R	11	3	13	1	25.3
ID15	60	M	10	I	R	14	3	14	30	19.2
					All *					
	64.4 ± 7.4	8 F 7 M	6.2 ± 4.3	12 I 3 H	10 R 5L	33.9 ± 17.6	8.9 ± 4.2	14.2 ± 2.2	46.0 ± 26.7	18.60 ± 8.39

Abbreviations: FMA: Upper Extremity portion of the Fugl-Meyer Assessment; AD: motor sections max 66; H: sensory section max 12; TIS: Trunk Impairment Scale max 23, WMFT: Wolf Motor Function Test max 85; TUG: Time Up and Go; ID01–ID15: Participant identifiers; F: female; M: male; TSS: time since stroke; I: ischemic; H: hemorrhagic; PS: paretic side; R: right; L: left. * population results are reported in mean ± std.

The inclusion criteria were: (i) chronic post-stroke stage, i.e., more than one year after the stroke-event; (ii) Mini-Mental State Examination above 24; (iii) no botulinum toxin injection within the past four months; (iv) no functional surgery in the previous six months; (v) absence of neglect; (vi) no changes in the clinical scores—stable clinical condition—for at least three months. All participants declared to be right-handed before the stroke event.

All participants were enrolled among the outpatient population of the Recovery and Functional Re-education Unit of the Santa Corona Hospital (Pietra Ligure, SV, Italy). All study procedures and consent forms conformed to the ethical standards of the 1964 Declaration of Helsinki and were approved by the institutional review board of the hospital (55/2012/CE2). The participants provided informed consent to participate in the study and to the publication of the results.

Before the experiment, a qualified physiotherapist evaluated the motor, functional and proprioceptive status of each participant using a series of clinical assessments (Table 1): the Upper Extremity portion of the Fugl-Meyer Assessment (FMA-UE), which includes tests of motor impairment (sections A–D, max score 66) and somatosensation (section H, max score 12) in the contralesional arm [29] (higher FMA-UE scores indicate less impairment); (ii) the Trunk Impairment Scale (TIS max score 23), which assesses static and dynamic sitting balance and trunk coordination in sitting position [30] (higher scores indicate less impairment); (iii) the Wolf Motor Function Test (WMFT max score 85), which is a quantitative measure of upper extremity motor ability through timed and functional daily living activities [31] (higher scores indicate less impairment); (iv) the Time Up and Go (TUG), which is used to determine fall risk and measure the balance, sit to stand and walking ability [32] (higher time indicates worse performance).

2.2. Experimental Set-Up and Protocol

Participants completed two separated sessions within six weeks. In each session, the participants were seated on a stool without back support and were asked to perform the modified functional reach test in two specific directions: forward (RF) and lateral (RL) with their ipsilesional arm (Figure 1). The RF required an anterior shoulder flexion of 90 degrees, while the RL a shoulder abduction of 90°. In each session, participants performed two reaching blocks; each block was characterized by three trials (T1–T3) toward the same direction for a total of six trials.

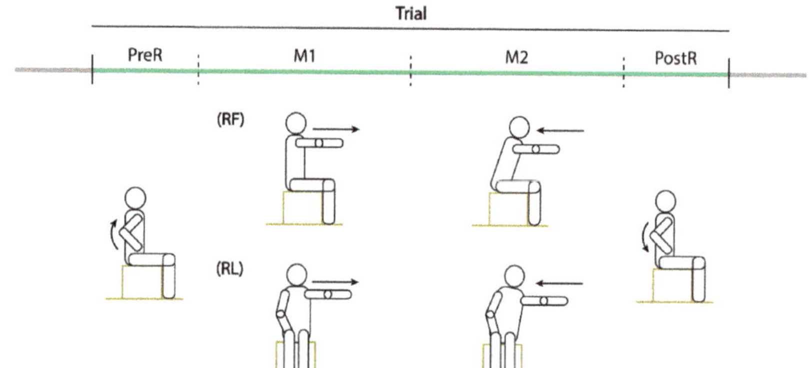

Figure 1. Experimental tasks. Drawings of one trial of the MFRT in the forward (RF) and lateral (RL) directions. Participants were seated on a stool without back support inside the acquisition volume defined by the motion capture system and were asked to maintain the feet on the stool support with shoulder width apart and the contralesional arm along the side with the hand on the thigh. Then, starting from the sitting posture with the trunk as straight as possible, they had to perform the reaching movement with their ipsilesional arm at their comfortable speed. Each trial was divided in four phases: PreR (raising the ipsilesional arm up), M1 (reaching movement to the maximum distance), M2 (reaching movement back to the initial position) and PostR (lowering the arm).

The seat height was adjusted for each participant anthropometric measures, i.e., the height of the feet support was set depending on the length of the participant's legs, so that the feet were always on the stool support, shoulder width apart. The knee, ankle and hip flexion were 90°. Instructions were to maintain the sitting posture with the trunk as straight as possible, with their Contralesional arm (C) resting along the side of the body with the hand on their thigh and to reach forward (or laterally) as far as possible with their Ipsilesional arm (I), without falling, at their comfortable speed.

2.3. Data Acquisition

The body motion was recorded at 100 Hz with a motion capture system (SMART DX, BTS Bioengineering, Milan, Italy) consisting of eight SMART-DX 5000 infrared cameras and two RGB cameras, placed frontally and laterally with respect to the participant. A set of 18 reflective spherical markers with a diameter of 15 mm was used, each marker was placed on the skin in correspondence of the following anatomical landmarks (Figure 2): forehead's center, sternum, spinal process of sacrum, C7; bilaterally on tip of the index, head of the index metacarpus, wrist (styloid process of the ulna), elbow (lateral epicondyle of the elbow), acromion, anterior superior iliac spine (ASIS), posterior superior iliac spine (PSIS).

We recorded the activity of ten selected muscles (Figure 2), six on the torso and four on the contralesional arm, with surface electromyography (sample frequency 1000 Hz, recording device: POCKETEMG, BTSBioengineering, Milan, Italy). The surface electrodes were placed according to SENIAM guidelines [33] to bilaterally record the trunk muscles trapezius trasversalis (TrapT), latissimus dorsi (LD) and obliquus externus abdominis (OEA), and unilaterally record the contralesional arm muscles deltoideus posterior (Delt), sternal head of pectoral major (Pect), caput lungus of triceps brachii (Tric) and of biceps brachii (Bic).

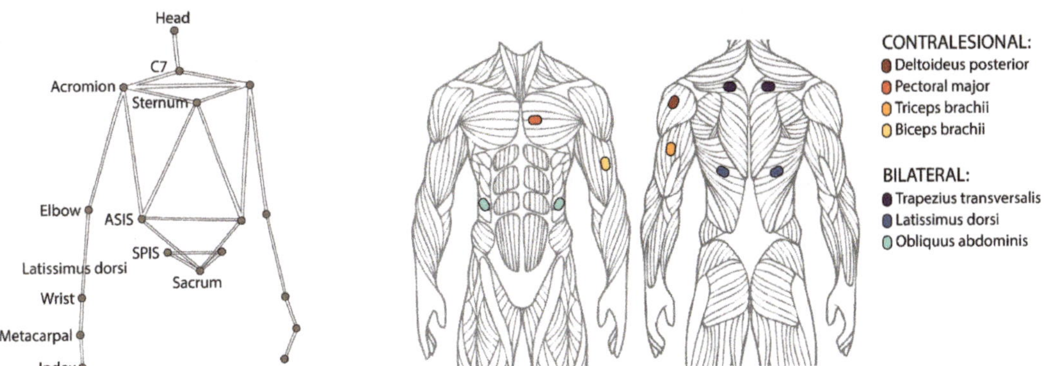

Figure 2. (left) Positions of the 18 reflective markers placed on the skin in correspondence of anatomical landmarks and used for recording the body motion with a motion capture system. (right) Muscles selected for recording the muscular activation patterns, four on the contralesional arm and three bilaterally on the torso.

2.4. Data Analysis

All the pre-processing and analysis were performed in MATLAB (Mathworks Inc.). We filtered the marker data by a fourth order Butterworth filter with a 12 Hz cut-off frequency. Then, we computed the marker velocity for the hand (H, i.e., the marker placed on the metacarpus landmark) and the shoulder (S, i.e., the marker placed on the acromion landmark), then we divided each trial in four sequential phases (Figure 1) based on the speed of the ipsilesional arm.

- PreR. The participants raised up the ipsilesional arm. This phase started when the hand speed along the y-axis reached 10% of its peak speed (Hy) and ended when the shoulder speed along the movement direction (x or z for RL and RF, respectively) was higher than 10% of its peak speed (Sx or Sz).
- Reaching movement (M1). The participant performed the reaching movement to the maximum distance. This phase started at the end of the PreR phase and ended when the maximum distance was reached (D).
- Return movement (M2). The participant moved back to the starting position. Specifically, this phase started at the end of the M1 phase and ended when the shoulder speed in the movement direction was lower than 10% of its peak speed (Sx or Sz).
- PostR. The participant lowered the ipsilesional arm. This phase started at the end of the M2 phase and ended when the hand speed along the y-axis was lower than 10% of its peak speed (Hy).

To characterize the kinematic performance, we computed the following quantitative metrics for both the MFRT in both directions (RF and RL):

- Normalized reaching distance, i.e., the maximum distance reached by the acromion marker in the movement direction, normalized by the arm length. It is computed as follows:

$$\text{Normalized reaching distance} = \frac{|l(TD) - l(T0)|}{L} \qquad (1)$$

where l is the coordinate of the marker acromion along the reaching direction (x or z for RL and RF, respectively) at the time instants T0 (starting of the reaching trial with the arm extended) and TD (time when the maximum distance is reached in the movement direction), and L is the length of the ipsilesional arm;

- Movement time, i.e., the time to complete the M1 and M2 phases of the movement;

- Δpelvis, computed as the mean value, over the M1 and M2 phases, of the absolute difference between coordinates along the vertical y-axis of the contralesional (Y_C) and ipsilesional (Y_I) superior iliac spine markers:

$$\Delta\text{pelvis} = \frac{\sum_{i=1}^{N}|Y_C(i) - Y_I(i)|}{N} \tag{2}$$

where N is the number of samples of M1 and M2 in each trial.

The sEMG signals recorded at 1 kHz were pre-processed using a fourth order bandpass Butterworth filter between 40 and 450 Hz. The filtered data were rectified, and then a fourth order lowpass Butterworth filter with 4 Hz cut-off frequency was applied to obtain the envelopes. The EMG envelopes were segmented according to the kinematic phases described above. Then, to make each phase comparable across participants independent of their duration, we interpolated the EMG envelopes over a time base with 50 points for the PreR and PostR phases and 100 points for the M1 and M2 phases. Moreover, to directly compare the modulation in amplitude and to average the EMG envelopes across participants, we normalized them for their baseline activations (i.e., the signals before the first PreR phase for each task).

We also verified that a different normalization (using the median value and or its maximum) did not change the main results that we obtained.

Statistical Analysis

In this work, we characterized the frontal and lateral MFRT, by investigating:

- the trunk muscles' activations in the contra and ipsilesional side;
- the muscles' activations in the contralesional arm, not actively engaged in the reaching movement;
- the effects of the reaching movement repetition on the muscles' activations and kinematic performance.

To test if there was an effect of repetition of the tasks between the two sessions, we ran a repeated measure ANOVA on both the kinematic and the muscular activity. Since there was no statistical difference between the two sessions, we averaged the results of these two. For our characterization, on the kinematic parameters and the unilateral muscular activity (arm muscles), we ran a repeated measure ANOVA with the trials as the only within-subjects factor (three levels: T1–T3). For the trunk muscles, which are bilaterally recorded, we run a repeated measure ANOVA with two within-subject factors: 'trials' (three levels: T1–T3) and 'body sides' (two levels: C, I). A main significant effect of the trial factor would support the hypothesis that a repetition of the tasks is inducing a change in the muscular activation pattern or on kinematic performance. A main effect of the body side in the trunk muscles' activation would support the hypothesis that the trunk muscles are activated in a different manner in the two sides.

Moreover, to compare the modulation of muscular activity during the reaching trials, we used the Statistical Parametric Mapping (SPM) approach (spm1d.org, accessed on 30 October 2021 [34]), which allows analyzing statistical differences among continuous signals, such as the EMG envelopes.

Before running the ANOVAs, we checked the normality of the kinematic data by Anderson–Darling test [35]. When the null hypothesis was rejected (only for the Δpelvis), the data were corrected applying the Box Cox transformation [36]. We also tested for sphericity using Mauchly's test [37] (which was rejected only for the Δpelvis and the reaching distance in RF), and we used the Greenhouse–Geisser correction. Statistical significance was set for all statistics at the family-wise error rate of $\alpha = 0.05$.

Lastly, since we expected the kinematic performance to be related to the participant movement ability, we investigated whether there was a correlation between kinematic performance and clinical evaluation. Furthermore, we expected a correlation between the kinematic performance of the MFRT and the clinical scores directly related to trunk control ability, such as the TIS and the TUG. To this end, we computed the correlation coefficient between the normalized reaching distance averaged across the repetitions and the scores of TIS (Spearman's coefficient) and TUG (Pearson's coefficient) separately. For the sake of completeness, we also computed the correlation (Spearman's coefficient) between the normalized reaching distance averaged across the repetitions and the other clinical scores used to characterize our population (FMA-UE and WMFT). Correlation coefficients ranging from 0.20 to 0.39 were considered as moderate, from 0.40 to 0.59 as relatively strong, from 0.60 to 0.79 as strong, and higher as very strong correlation [38]. We also reported the probability p that the observed correlation was due to chance, i.e., lower *p*-value indicates that the observed correlation is unlikely to be due to chance.

3. Results

All our participants completed the two evaluation sessions without problems or discomfort.

Since the two directions of the test highlight different aspects of trunk control [20] and induce different kinematic and muscular strategies, in the following we report the results of the RF and RL separately. Unless otherwise stated, all descriptive data in the text, tables and figures are mean $\pm$ SE.

3.1. Frontal Reaching

Kinematic performance. The reaching distance (Table 2) significantly increased with the repetitions of the reaching movement ($F_{2,17.8} = 10.31$, $p = 0.003$), without changes in the movement time ($F_{2,28} = 2.98$, $p = 0.067$). The Δpelvis, when executing the frontal reaching, was small (mean: 1.1 cm) and did not change with the trials' repetition ($F_{1.3,18.6} = 1.58$, $p = 0.224$).

Table 2. Results for the kinematic parameters, mean and standard error of the whole population.

Parameters	T1	T2	T3
Normalized reaching distance **	0.58 ± 0.03	0.63 ± 0.03	0.67 ± 0.03
Movement time (s)	5.66 ± 0.49	5.17 ± 0.43	5.20 ± 0.44
Δpelvis (cm)	1.01 ± 0.10	1.02 ± 0.10	1.11 ± 0.12

** $p < 0.001$.

Correlation between reaching kinematics and clinical tests. To evaluate the relationship between the kinematic performance and the clinical scores, we computed the correlation coefficients between the mean normalized reaching distance and the clinical scales. For the TUG, we found a significant and relatively strong negative correlation ($r = -0.56$, $p = 0.033$). Instead, we did not find a correlation with the TIS ($\rho = 0.03$, $p = 0.918$), the FMA-UE ($\rho = -0.12$, $p = 0.684$) and the WMFT ($\rho = -0.03$, $p = 0.914$).

Muscle Activity

The muscular activations in the different phases of the trial are shown in Figure 3A (for each trunk muscle the effect of repetition is reported on the left and the difference between the two sides on the right) and Figure 3B (contralesional arm muscles). Note that the EMG envelopes are normalized with respect to their baseline values, i.e., values equal to 1 correspond to the at rest (baseline) activations, while higher values indicated an increased muscle activation with respect to baseline.

Trunk muscles. As expected, in our stroke survivors, the OEAs were active in the M1 phase without significant differences between the two sides of the body (Figure 3A, right panel). The OAEs were not active during the PreR and PostR phases. Conversely, the LDs had different activation in two sides of the body during the forward and the beginning of the backward reaching movement, with a higher activation for the contralesional side. Both sides were equally involved during the arm movement in the PreR and PostR phases.

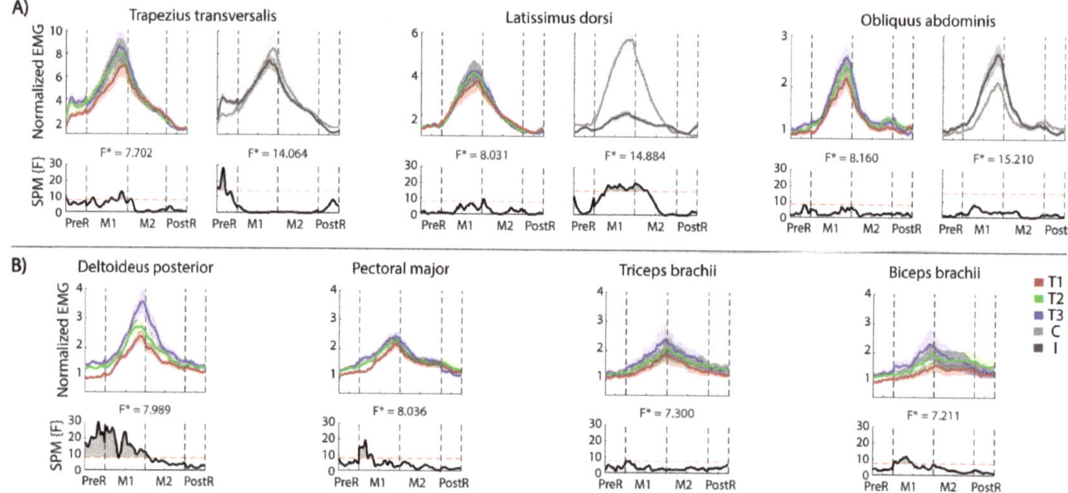

Figure 3. MFRT in the forward direction. (**A**) Bilateral EMG activity of the trunk muscles: for each pair of muscles we reported their envelopes (top) and the statistical analysis (down, the significance difference is the part of the curve above the dashed red line) separated for the trials' repetition (left, the mean activity of the muscles pair is reported in red, green and blue for T1, T2 and T3, respectively) and the body sides (right, the mean activity of each side is reported in lighter and darker gray for the contralesional and the ipsilesional side, respectively); (**B**) EMG activity of the contralesional arm muscles separated for the trials' repetition and the statistical analysis (bottom). F* is the F* value from the statistical analysis.

The TrapTs became active in the PreR phase, with a significantly more pronounced activation on the ipsilesional side. There was not a significant trend in the PostR phase, despite most of the participants appeared to reduce the activation in the ipsilesional TrapT, maintaining the activation longer on the contralesional side.

The trial repetition (Figure 3A, left panels) had a significant effect only on the TrapTs at the end of the forward reaching, which could be related to an increased muscle activation required to reach a farther distance.

Contralesional (non moving) arm. We evaluated the muscle activations (Figure 3B), despite that this arm was not actively engaged in the movement. We found that all the muscle activations followed the timing of the reaching movement, with the higher activations in the M1 and M2 phases.

The activation of Delt was affected by the trial repetitions, i.e., its activation significantly increased in the PreR and M1 phases. The activation started from a baseline level in the first trial and increased across the trials. The biceps and the pectoralis had a significantly higher activation only in the M1 phase across repetitions, while the triceps did not change with the repetitions.

3.2. Lateral Reaching

Kinematic performance. The distance reached (Table 3) by the participants significantly increased from the first to the last movement ($F_{2,28} = 19.66$, $p < 0.001$), without change in movement time ($F_{2,28} = 0.57$, $p = 0.575$). The increased reaching distance, instead, was followed by an increased Δpelvis ($F_{2,28} = 4.55$, $p = 0.037$).

Table 3. Results for the kinematic parameters, mean and standard error of the whole population.

Parameters	T1	T2	T3
Normalized reaching distance	0.29 ± 0.02	0.34 ± 0.02	0.36 ± 0.03
Movement time (s)	4.81 ± 0.33	4.82 ± 0.33	5.06 ± 0.42
Δpelvis (cm) *	1.47 ± 0.30	2.89 ± 0.62	2.70 ± 0.65

* $p < 0.05$.

Correlation between reaching kinematics and clinical tests. To evaluate the relation between the kinematic performance and the clinical scores, we computed the correlation coefficient between the normalized reaching distance and the scores of all the clinical scales. For the TUG, we found a relatively strong negative correlation (r = −0.46), close to the significance threshold ($p = 0.083$). In contrast, we did not find a correlation between the mean normalized reaching distance and the scores of the other scales: TIS (ρ = −0.09, $p = 0.743$), FMA-UE (ρ = −0.39, $p = 0.154$) and WMFT (ρ = −0.37, $p = 0.172$).

Muscle Activity

The involvement of muscles in the lateral reaching is shown in Figure 4A (for each trunk muscle the effect of the repetition is reported on the left and the difference between the two sides on the right) and B (contralesional arm muscles).

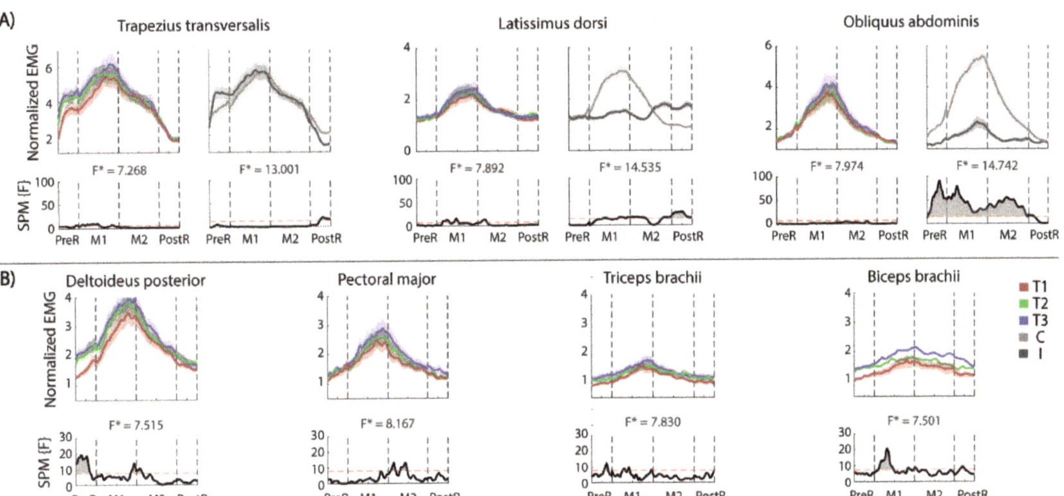

Figure 4. MFRT in the lateral direction (**A**) bilateral EMG activity of the trunk muscles: for each pair of muscles we reported the envelops (top) and the statistical analysis (down, the significance difference is the part of the curve above the dashed red line) separated for the trials' repetition (left, the mean activity of the muscles pair is reported in red, green and blue for T1, T2 and T3, respectively) and the side (right, the mean activity of each side is reported in lighter and darker gray for the contralesional and the ipsilesional side, respectively); (**B**) EMG activity of the contralesional arm muscles separated for the trials repetition and the statistical analysis (bottom). F* is the F* value from the statistical analysis.

Trunk. As expected, the OEAs were differently activated on the two sides. The activation of the contralesional OEA started when the trunk began to move and shifted laterally, and the pelvis tilted in the frontal plane (x-y). The activation of the OEAs stopped in the PostR phase. The activation of ipsilesional OEA followed the same pattern, but with negligible activations compared to the contralesional side. The LDs had different timing and level of activation between the two sides of the body. In fact, the contralesional LD was significantly more active than the ipsilesional in the M1 phase with a maximum peak when the maximal distance was reached, while the ipsilesional LD was significantly more active than the contralesional LD at the end of the M2 and in the PostR phases. In the PostR phase, only the ipsi-LD was active to perform the adduction of the shoulder and lower the arm.

In this task–as in RF–the ipsilesional TrapT started its activity in the PreR phase to stabilize the shoulder girdle. The activation of contralesional TrapT in the PostR phase was maintained significantly more active than the ipsilesional TRapT.

The repetition of the reaching movements (Figure 4A, left panels) increased the activation of LDs and TrapTs mainly in the M1 phase. We did not find any significant effect of the trials' repetitions in OEAs.

Contralesional (non-moving) arm. As in RF, the muscle activation of this arm followed the timing of the movement (Figure 4B), with the highest activations in M1 and M2 phases. A significant repetition effect was observed for the Delt. This was mainly due to the PreR phase; in fact, the activation level in the first trial was significantly lower than in the others, as highlighted by the statistical analysis (Figure 4B, lower panels). Moreover, in the M1 phase, this muscle had a higher (maximum) activation in the third repetition, when the ipsilesional arm reached the highest maximum distance. This trend is also evident in the M2 phase and almost absent in the PostR phase. The pectoral and biceps at the beginning of the M1 phase increase their activity while repeating the same gesture, while there were not significant repetition effects for the triceps.

4. Discussion

In this work, we extensively characterized the frontal and lateral modified functional reach test, in terms of both kinematic performance and muscular activity of the upper body after stroke, with a specific focus on trunk control and on the non-moving contralesional arm. The MFRT is a clinical tool, fundamental to assess the risk of falls and to evaluate the sitting balance abilities [12].

The trunk muscles had activations that differed between ipsi- and contralesional sides of the body, except for the oblique externus abdominis in the forward direction. Our stroke survivors had difficulty deactivating the muscles in the contralesional side at the end of each trial, especially the trapezii trasversalis in the lateral direction. The contralesional arm, despite not being actively involved in the movement, had muscular activations modulated depending on the movement phases of the moving arm. We also found that the repetition of the same movement improved performance in terms of reaching distance that was associated with an increased activation of the trunk muscles, with no changes in movement time. Lastly, the reaching distance negatively correlated with the time up-and-go test score but not with the other clinal scores. The implications, novelty and limitations of these results are discussed in detail below.

4.1. Trunk Muscles Activity

The MFRT required different kinematic strategies and muscle activations when performed in the forward or in the lateral direction. Stroke survivors increased the Δpelvis in the lateral but not in the forward direction.

As for the corresponding trunk muscle activations, in the RF, we observed a greater engagement of the contralesional LD and a symmetrical and synchronized activation of both OEAs. This suggests that OEAs operate as rotator muscles and cooperate in the trunk's forward flexion together with the activity of the rectus abdominis [39].

In the RL, the contralesional muscles should increase their activity when supporting the weight-shift in the lateral direction [40]. Each phase of the movement requires different activation levels of the trunk muscles. Indeed, both LDs and OEAs had different roles in different phases of the trial depending on the side: in PreR and M1 those contralesional muscles had a strong concentric activity to maintain the center of mass inside the base of support [41], whereas in M2 and PostR, the ipsi-LD had a concentric activation to bring the trunk back to the starting position and to lower the arm [42].

In addition, stroke survivors had difficulty deactivating the contralesional muscles at the end of each trial, especially the TrapT in the lateral direction. In fact, this muscle has high activation for its major role as adductor [43]. The prolonged muscle activations of this muscle have been previously observed also in different tasks, as in [44].

4.2. Muscles' Activations in the Contralesional Arm, Not Actively Engaged in the Movement

During the MFRT, the muscular activity of the non-moving (contralesional) arm, to the best of our knowledge, has not been investigated yet. Interestingly, we found that the muscular activation of the contralesional arm was modulated depending on the phases of the movement of the ipsilesional arm. Specifically, contralesional arm muscles were more active during M1 and M2 phases. Given the existing literature, this activation could depend on an interplay of one or more factors, such as a cross educational effect [45–49] or on deficits in the selective activation of muscles following stroke [50].

Moreover, the muscles activations at the end of the trial did not revert completely to baseline levels. Delay on deactivating the muscles was more evident in Bic muscle both in RF and RL. This could be due to an increased muscle activity to maintain balance while reaching more distant targets, to the atypical muscular over-activation [3] or to the prolonged activation induced by the stroke-event [44].

4.3. Repetitions of Reaching Movement

The three repetitions of the same gesture led to better performance in terms of reaching distance in both frontal and lateral MFR, without effects on the movement duration. Considering the limited number of repetitions in our tasks, it is unlikely that the improvement can be attributed to learning [51]. More likely, our stroke survivors could have increased the confidence in their ability and/or have become more familiar with the tasks, allowing themselves to operate closer to their stability limits. This trend has been previously observed in (standing) functional reach tests, performed by neurotypical individuals of different age [51,52] and by stroke survivors [52].

As for the muscular activations, the effect of repetition was evident in the increased activation of TrapTs and Delt in the first half of the reaching movement in both directions. The activation of TrapTs in both body sides, could be an answer to the need of maintaining the stability of the shoulder during the ipsilesional arm movement. In contrast, the activation of the contralesional Delt could result from the difficulty of stroke survivors to deactivate this muscle at the end of the trial, as already observed in other tasks [44].

4.4. Correlation between Clinical Test and Distance Reached in Frontal and Lateral MFRT

We found that the participants who employed less time to perform the TUG test reached also farther in the MFRT. Less difficulties on the TUG corresponded to a more efficient trunk control, which allowed them also to maintain balance while reaching farther with the ipsilesional arm. On one hand, this finding was expected, since the MFRT is already used in the clinical assessments for evaluating the risk of falls [19–26], and it evaluates balance in a dynamic task, as the TUG test [52,53]. On the other hand, the strong or relatively strong correlation we found was not a given, since the TUG requires to stand-up against gravity and walk, while MFRT is performed while sitting.

Conversely, we did not find significant correlation between the score of the TIS scale and the normalized reaching distance. This may be the case because the TIS evaluates only trunk control abilities in static conditions [30,54], while, as mentioned above, the TUG requires dynamic trunk control [32] as the MFRT. Moreover, the TIS is a more subjective scale, based on the evaluation of the operator and on a discrete rating score [16]. Instead, the TUG outcome is an objective, continuous parameter (as it measures the time) [32]. Nevertheless, [55] found a correlation between TIS and the reaching distance in the affected side in stroke survivors, but while performing the (standing) functional reach test.

Other studies found that the TUG score negatively correlates with other metrics that assess balance ability and trunk control [52,53], as the Berg Balance Scale score in neurotypical individuals [52] and with the center of pressure metrics during the functional reach test in standing for both neurotypical individuals of different age and for stroke survivors [53]. This work provides additional new evidence of a relatively strong correlation of TUG with MFRT in chronic stroke survivors.

In summary, this work supports the evidence [11–14,56] suggesting that the functional reach test score is strongly related with standing up and walking abilities and is a predictor for functional mobility, suggesting that these conclusions could be extended also to the MFRT, performed on sitting.

4.5. Limitations and Future Directions

The small number of muscles considered in this study limited the understanding of complex coordinated muscular patterns and synergic activations. Future investigations will record the electromyographic activity also of other muscles, such as the quadratus lumborum and the acromial part of the deltoid. Furthermore, it would be interesting to study both trunk and leg muscles.

A larger number of stroke survivors should be tested to further understand differences due to the level of impairment and to generalize this characterization of the frontal and lateral modified functional reach test in chronic stroke survivors.

In our experimental design, participants performed three repetitions of each reaching movement, consistent with protocols widely adopted in clinical practice. However, increasing the number of repetitions could be of value for investigating the effects of both learning and fatigue.

Hwang and colleagues [55] suggested that, in stroke survivors, the kinematic results of the (standing) functional reach test along the contra and ipsilesional directions are correlated. In this study, we focused on the modified functional reach test in two directions. Future investigations should also involve other directions, providing a more complete characterization and comparing the muscular activations in the two lateral directions.

Lastly in future investigations, our methodology can be used for a complete characterization of the modified functional reach test in an easy way. It could also be transposed or coupled with marker-less algorithms for movement analysis as performed in [57–61], which are taking place in research and in the clinical practice to facilitate the clinical assessments [58]. Currently, given the pandemic situation, those could be even more important, since they can facilitate clinical assessments from home and/or telerehabilitation, also it may be coupled with haptic feedback to improve its outcome [62].

Author Contributions: Conceptualization, A.D.L., M.C., C.I. and P.G.; methodology, all authors; software, A.D.L., G.B., G.M. and M.C.; formal analysis, A.D.L., G.B., G.M. and M.C.; investigation, A.D.L., L.B., M.C. and P.G.; writing—original draft preparation, A.D.L., G.B., G.M., M.C. and P.G.; writing—review and editing, all authors; supervision, M.C. and P.G. All authors have read and agreed to the published version of the manuscript.

Funding: This research was founded by Ministry of Science and Technology, Israel (Joint Israel-Italy lab in Biorobotics "Artificial somatosensation for humans and humanoids"). G.M. was supported by Regione Liguria.

Institutional Review Board Statement: All procedures were approved by the institutional review board of the ASL 2 Savonese (55/2012/CE2) and the participants gave informed consent in accordance with the ethical standards of the 1964 Declaration of Helsinki.

Informed Consent Statement: Informed consent was obtained from all participants involved in the study.

Data Availability Statement: The data presented in this study are available on request from the corresponding author.

Conflicts of Interest: The authors declare no conflict of interest.

References

1. Karthikbabu, S.; Chakrapani, M.; Ganeshan, S.; Rakshith, K.C.; Nafeez, S.; Prem, V. A review on assessment and treatment of the trunk in stroke: A need or luxury. *Neural Regen. Res.* **2012**, *7*, 1974.
2. Dean, C.; Shepherd, R.; Adams, R. Sitting balance I: Trunk–arm coordination and the contribution of the lower limbs during self-paced reaching in sitting. *Gait Posture* **1999**, *10*, 135–146. [CrossRef]
3. Segal, M. Muscle overactivity in the upper motor neuron syndrome: Pathophysiology. *Phys. Med. Rehabil. Clin. N. Am.* **2018**, *29*, 427–436. [CrossRef]
4. McGill, S.M.; Grenier, S.; Kavcic, N.; Cholewicki, J. Coordination of muscle activity to assure stability of the lumbar spine. *J. Electromyogr. Kinesiol.* **2003**, *13*, 353–359. [CrossRef]
5. Meyer, S.; Karttunen, A.H.; Thijs, V.; Feys, H.; Verheyden, G. How do somatosensory deficits in the arm and hand relate to upper limb impairment, activity, and participation problems after stroke? A systematic review. *Phys. Ther.* **2014**, *94*, 1220–1231. [CrossRef] [PubMed]
6. Campbell, F.M.; Ashburn, A.M.; Pickering, R.M.; Burnett, M. Head and pelvic movements during a dynamic reaching task in sitting: Implications for physical therapists. *Arch. Phys. Med. Rehabil.* **2001**, *82*, 1655–1660. [CrossRef] [PubMed]
7. Geiger, R.A.; Allen, J.B.; O'Keefe, J.; Hicks, R.R. Balance and mobility following stroke: Effects of physical therapy interventions with and without biofeedback/forceplate training. *Phys. Ther.* **2001**, *81*, 995–1005. [CrossRef] [PubMed]
8. Cabanas-Valdés, R.; Cuchi, G.U.; Bagur-Calafat, C. Trunk training exercises approaches for improving trunk performance and functional sitting balance in patients with stroke: A systematic review. *NeuroRehabilitation* **2013**, *33*, 575–592. [CrossRef]
9. Tessem, S.; Hagstrøm, N.; Fallang, B. Weight distribution in standing and sitting positions, and weight transfer during reaching tasks, in seated stroke subjects and healthy subjects. *Physiother. Res. Int.* **2007**, *12*, 82–94. [CrossRef]
10. Wiskerke, E.; van Dijk, M.; Thuwis, R.; Vandekerckhove, C.; Myny, C.; Kool, J.; Dejaeger, E.; Beyens, H.; Verheyden, G. Maximum weight-shifts in sitting in non-ambulatory people with stroke are related to trunk control and balance: A cross-sectional study. *Gait Posture* **2021**, *83*, 121–126. [CrossRef] [PubMed]
11. Feigin, L.; Sharon, B.; Czaczkes, B.; Rosin, A.J. Sitting equilibrium 2 weeks after a stroke can predict the walking ability after 6 months. *Gerontology* **1996**, *42*, 348–353. [CrossRef] [PubMed]
12. Lee, K.; Lee, D.; Hong, S.; Shin, D.; Jeong, S.; Shin, H.; Choi, W.; An, S.; Lee, G. The relationship between sitting balance, trunk control and mobility with predictive for current mobility level in survivors of sub-acute stroke. *PLoS ONE* **2021**, *16*, e0251977. [CrossRef]
13. Duarte, E.; Marco, E.; Muniesa, J.M.; Belmonte, R.; Diaz, P.; Tejero, M.; Escalada, F. Trunk control test as a functional predictor in stroke patients. *J. Rehabil. Med.* **2002**, *34*, 267–272. [PubMed]
14. Verheyden, G.; Nieuwboer, A.; De Wit, L.; Feys, H.; Schuback, B.; Baert, I.; Jenni, W.; Schupp, W.; Thijs, V.; De Weerdt, W. Trunk performance after stroke: An eye catching predictor of functional outcome. *J. Neurol. Neurosurg. Psychiatry* **2007**, *78*, 694–698. [CrossRef] [PubMed]
15. Hsieh, C.-L.; Sheu, C.-F.; Hsueh, I.-P.; Wang, C.-H. Trunk control as an early predictor of comprehensive activities of daily living function in stroke patients. *Stroke* **2002**, *33*, 2626–2630. [CrossRef] [PubMed]
16. Blum, L.; Korner-Bitensky, N. Usefulness of the Berg Balance Scale in stroke rehabilitation: A systematic review. *Phys. Ther.* **2008**, *88*, 559–566. [CrossRef] [PubMed]
17. Dukelow, S.P.; Herter, T.M.; Moore, K.D.; Demers, M.J.; Glasgow, J.I.; Bagg, S.D.; Norman, K.E.; Scott, S.H. Quantitative assessment of limb postion sense following stroke. *Neurorehabil. Neural Repair* **2010**, *24*, 178–187. [CrossRef] [PubMed]
18. Reaz, M.B.I.; Hussain, M.S.; Mohd-Yasin, F. Techniques of EMG signal analysis: Detection, processing, classification and applications. *Biol. Proced. Online* **2006**, *8*, 11–35. [CrossRef]
19. Duncan, P.W.; Weiner, D.K.; Chandler, J.; Studenski, S. Functional reach: A new clinical measure of balance. *J. Gerontol.* **1990**, *45*, M192–M197. [CrossRef]
20. Katz-Leurer, M.; Fisher, I.; Neeb, M.; Schwartz, I.; Carmeli, E. Reliability and validity of the modified functional reach test at the sub-acute stage post-stroke. *Disabil. Rehabil.* **2009**, *31*, 243–248. [CrossRef]
21. Hill, K.; Ellis, P.; Bernhardt, J.; Maggs, P.; Hull, S. Balance and mobility outcomes for stroke patients: A comprehensive audit. *Aust. J. Physiother.* **1997**, *43*, 173–180. [CrossRef]
22. Frzovic, D.; Morris, M.E.; Vowels, L. Clinical tests of standing balance: Performance of persons with multiple sclerosis. *Arch. Phys. Med. Rehabil.* **2000**, *81*, 215–221. [CrossRef]
23. Holbein-Jenny, M.A.; Billek-Sawhney, B.; Beckman, E.; Smith, T. Balance in personal care home residents: A comparison of the Berg Balance Scale, the Multi-Directional Reach Test, and the Activities-specific Balance Confidence Scale. *J. Geriatr. Phys. Ther.* **2005**, *28*, 48–53. [CrossRef]
24. Duncan, P.W.; Studenski, S.; Chandler, J.; Prescott, B. Functional reach: Predictive validity in a sample of elderly male veterans. *J. Gerontol.* **1992**, *47*, M93–M98. [CrossRef]
25. Fishman, M.N.; Colby, L.A.; Sachs, L.A.; Nichols, D.S. Comparison of upper-extremity balance tasks and force platform testing in persons with hemiparesis. *Phys. Ther.* **1997**, *77*, 1052–1062. [CrossRef] [PubMed]
26. Daubney, M.E.; Culham, E.G. Lower-extremity muscle force and balance performance in adults aged 65 years and older. *Phys. Ther.* **1999**, *79*, 1177–1185. [CrossRef] [PubMed]

27. Takahashi, T.; Ishida, K.; Yamamoto, H.; Takata, J.; Nishinaga, M.; Doi, Y.; Yamamoto, H. Modification of the functional reach test: Analysis of lateral and anterior functional reach in community-dwelling older people. *Arch. Gerontol. Geriatr.* **2006**, *42*, 167–173. [CrossRef] [PubMed]

28. Newton, R.A. Validity of the multi-directional reach test: A practical measure for limits of stability in older adults. *J. Gerontol. Ser. A Biol. Sci. Med. Sci.* **2001**, *56*, M248–M252. [CrossRef] [PubMed]

29. Fugl-Meyer, A.R.; Jaasko, L.; Leyman, I.; Olsson, S.; Steglind, S. The post-stroke hemiplegic patient. *Scand. J. Rehabil. Med.* **1975**, *7*, 13–31.

30. Verheyden, G.; Nieuwboer, A.; Mertin, J.; Preger, R.; Kiekens, C.; De Weerdt, W. The Trunk Impairment Scale: A new tool to measure motor impairment of the trunk after stroke. *Clin. Rehabil.* **2004**, *18*, 326–334. [CrossRef] [PubMed]

31. Wolf, S.L.; Catlin, P.A.; Ellis, M.; Archer, A.L.; Morgan, B.; Piacentino, A. Assessing Wolf Motor Function Test as outcome measure for research in patients after stroke. *Stroke* **2001**, *32*, 1635–1639. [CrossRef] [PubMed]

32. Podsiadlo, D.; Richardson, S. The Timed Up and Go: A Test of Basic Functional Mobility for Frail Elderly Persons. *J. Am. Geriatr. Soc.* **1991**, *39*, 142–148. [CrossRef]

33. Hermens, H.J.; Freriks, B.; Disselhorst-Klug, C.; Rau, G. Development of recommendations for SEMG sensors and sensor placement procedures. *J. Electromyogr. Kinesiol.* **2000**, *10*, 361–374. [CrossRef]

34. Pataky, T.C. Generalized n-dimensional biomechanical field analysis using statistical parametric mapping. *J. Biomech.* **2010**, *43*, 1976–1982. [CrossRef]

35. Anderson, T.W.; Darling, D.A. A test of goodness of fit. *J. Am. Stat. Assoc.* **1954**, *49*, 765–769. [CrossRef]

36. Sakia, R.M. The Box-Cox transformation technique: A review. *J. R. Stat. Soc. Ser. D* **1992**, *41*, 169–178. [CrossRef]

37. Mauchly, J.W. Significance test for sphericity of a normal n-variate distribution. *Ann. Math. Stat.* **1940**, *11*, 204–209. [CrossRef]

38. Rea, L.M.; Parker, R.A. *Designing and Conducting Survey Research: A Comprehensive Guide*; John Wiley & Sons: San Francisco, CA, USA, 2014; ISBN 1118767039.

39. Kapandji, I.A. *The Physiology of the Joints: The Spinal Column, Pelvic Girdle and Head*, 6th Revised Edition; Churchill Livingstone: Edinburgh, UK, 2007; Volume 3, ISBN 0702029599.

40. Rabuffetti, M.; Carpinella, I.; Ferrarin, M.; Cattaneo, D.; Bonora, G.; Nardone, A.; Bowman, T. Counteracting Postural Perturbations Through Body Weight Shift: A Pilot Study Using a Robotic Platform in Subjects With Parkinson's Disease. *IEEE Trans. Neural Syst. Rehabil. Eng.* **2018**, *26*, 1794–1802. [CrossRef]

41. Morris, J.M.; Lucas, D.B.; Bresler, B. Role of the trunk in stability of the spine. *JBJS* **1961**, *43*, 327–351. [CrossRef]

42. Anderson, K.G.; Behm, D.G. Maintenance of EMG activity and loss of force output with instability. *J. Strength Cond. Res.* **2004**, *18*, 637–640. [PubMed]

43. Phadke, V.; Camargo, P.R.; Ludewig, P.M. Scapular and rotator cuff muscle activity during arm elevation: A review of normal function and alterations with shoulder impingement. *Braz. J. Phys. Ther.* **2009**, *13*, 1–9. [CrossRef] [PubMed]

44. Pellegrino, L.; Coscia, M.; Pierella, C.; Giannoni, P.; Cherif, A.; Mugnosso, M.; Marinelli, L.; Casadio, M. Effects of hemispheric stroke localization on the reorganization of arm movements within different mechanical environments. *Life* **2021**, *11*, 383. [CrossRef]

45. Sakamoto, K.; Nakamura, T.; Uenishi, H.; Umemoto, Y.; Arakawa, H.; Abo, M.; Saura, R.; Fujiwara, H.; Kubo, T.; Tajima, F. Immediate effects of unaffected arm exercise in poststroke patients with spastic upper limb hemiparesis. *Cerebrovasc. Dis.* **2014**, *37*, 123–127. [CrossRef]

46. Pandian, S.; Arya, K.N.; Kumar, D. Effect of motor training involving the less-affected side (MTLA) in post-stroke subjects: A pilot randomized controlled trial. *Top. Stroke Rehabil.* **2015**, *22*, 357–367. [CrossRef] [PubMed]

47. Pandian, S.; Arya, K.N.; Kumar, D. Does motor training of the nonparetic side influences balance and function in chronic stroke? A pilot RCT. *Sci. World J.* **2014**, *2014*, 769726. [CrossRef] [PubMed]

48. Cernacek, J. Contralateral motor irradiation-cerebral dominance: Its changes in hemiparesis. *Arch. Neurol.* **1961**, *4*, 165–172. [CrossRef] [PubMed]

49. Ehrensberger, M.; Simpson, D.; Broderick, P.; Monaghan, K. Cross-education of strength has a positive impact on post-stroke rehabilitation: A systematic literature review. *Top. Stroke Rehabil.* **2016**, *23*, 126–135. [CrossRef] [PubMed]

50. van der Krogt, H.; Kouwijzer, I.; Klomp, A.; Meskers, C.G.M.; Arendzen, J.H.; de Groot, J.H. Loss of selective wrist muscle activation in post-stroke patients. *Disabil. Rehabil.* **2020**, *42*, 779–787. [CrossRef] [PubMed]

51. Billek-Sawhney, B.; Gay, J. The Functional Reach Test: Are 3 trials necessary? *Top. Geriatr. Rehabil.* **2005**, *21*, 144–148. [CrossRef]

52. Bennie, S.; Bruner, K.; Dizon, A.; Fritz, H.; Goodman, B.; Peterson, S. Measurements of balance: Comparison of the Timed" Up and Go" test and Functional Reach test with the Berg Balance Scale. *J. Phys. Ther. Sci.* **2003**, *15*, 93–97. [CrossRef]

53. Portnoy, S.; Reif, S.; Mendelboim, T.; Rand, D. Postural control of individuals with chronic stroke compared to healthy participants: Timed-Up-and-Go, Functional Reach Test and center of pressure movement. *Eur. J. Phys. Rehabil. Med.* **2017**, *53*, 685–693. [CrossRef]

54. Monticone, M.; Ambrosini, E.; Verheyden, G.; Brivio, F.; Brunati, R.; Longoni, L.; Mauri, G.; Molteni, A.; Nava, C.; Rocca, B. Development of the Italian version of the trunk impairment scale in subjects with acute and chronic stroke. Cross-cultural adaptation, reliability, validity and responsiveness. *Disabil. Rehabil.* **2019**, *41*, 66–73. [CrossRef]

55. Hwang, W.-J.; Kim, J.-H.; Jeon, S.-H.; Chung, Y. Maximal lateral reaching distance on the affected side using the multi directional reach test in persons with stroke. *J. Phys. Ther. Sci.* **2015**, *27*, 2713–2715. [CrossRef] [PubMed]

56. Verheyden, G.; Vereeck, L.; Truijen, S.; Troch, M.; Herregodts, I.; Lafosse, C.; Nieuwboer, A.; De Weerdt, W. Trunk performance after stroke and the relationship with balance, gait and functional ability. *Clin. Rehabil.* **2006**, *20*, 451–458. [CrossRef]
57. Mier, C.M. Accuracy and feasibility of video analysis for assessing hamstring flexibility and validity of the sit-and-reach test. *Res. Q. Exerc. Sport* **2011**, *82*, 617–623. [CrossRef] [PubMed]
58. Matsushita, Y.; Tran, D.T.; Yamazoe, H.; Lee, J.-H. Recent use of deep learning techniques in clinical applications based on gait: A survey. *J. Comput. Des. Eng.* **2021**, *8*, 1499–1532. [CrossRef]
59. Capecci, M.; Ceravolo, M.G.; Ferracuti, F.; Grugnetti, M.; Iarlori, S.; Longhi, S.; Romeo, L.; Verdini, F. An instrumental approach for monitoring physical exercises in a visual markerless scenario: A proof of concept. *J. Biomech.* **2018**, *69*, 70–80. [CrossRef] [PubMed]
60. Kanko, R.M.; Laende, E.K.; Davis, E.M.; Selbie, W.S.; Deluzio, K.J. Concurrent assessment of gait kinematics using marker-based and markerless motion capture. *J. Biomech.* **2021**, *127*, 110665. [CrossRef]
61. Moro, M.; Marchesi, G.; Odone, F.; Casadio, M. Markerless gait analysis in stroke survivors based on computer vision and deep learning: A pilot study. In Proceedings of the Proceedings of the 35th Annual ACM Symposium on Applied Computing, Brno, Czech Republic, 30 March–3 April 2020; Association for Computing Machinery: New York, NY, USA, 2020; pp. 2097–2104.
62. Handelzalts, S.; Ballardini, G.; Avraham, C.; Pagano, M.; Casadio, M.; Nisky, I. Integrating Tactile Feedback Technologies into Home-Based Telerehabilitation: Opportunities and Challenges in Light of COVID-19 Pandemic. *Front. Neurorobot.* **2021**, *15*, 4. [CrossRef]

Article

Validation of Visually Identified Muscle Potentials during Human Sleep Using High Frequency/Low Frequency Spectral Power Ratios

Mo H. Modarres [1], **Jonathan E. Elliott** [2,3], **Kristianna B. Weymann** [4], **Dennis Pleshakov** [5], **Donald L. Bliwise** [6] and **Miranda M. Lim** [2,3,7,8,9,*]

1   Mental Illness Research, Education and Clinical Center (MIRECC-VISN1), VA Bedford Health Care System, Bedford, MA 01730, USA; mo.modarres@va.gov
2   VA Portland Health Care System, Portland, OR 97239, USA; elliojon@ohsu.edu
3   Department of Neurology, Oregon Health & Science University, Portland, OR 97239, USA
4   School of Nursing, Oregon Health & Science University, Portland, OR 97239, USA; weymannk@ohsu.edu
5   School of Medicine, Oregon Health & Science University, Portland, OR 97239, USA; pleshako@ohsu.edu
6   Department of Neurology, Emory University, Atlanta, GA 30322, USA; dbliwis@emory.edu
7   National Center for Rehabilitative Auditory Research, VA Portland Health Care System, Portland, OR 97239, USA
8   Department of Behavioral Neuroscience, Oregon Health & Science University, Portland, OR 97239, USA
9   Oregon Institute of Occupational Health Sciences, Oregon Health & Science University, Portland, OR 97239, USA
*   Correspondence: lmir@ohsu.edu; Tel.: +1-503-220-8262 x57404

Citation: Modarres, M.H.; Elliott, J.E.; Weymann, K.B.; Pleshakov, D.; Bliwise, D.L.; Lim, M.M. Validation of Visually Identified Muscle Potentials during Human Sleep Using High Frequency/Low Frequency Spectral Power Ratios. *Sensors* **2022**, *22*, 55. https://doi.org/10.3390/s22010055

Academic Editors: Francesco Di Nardo, Valentina Agostini and Silvia Conforto

Received: 1 November 2021
Accepted: 20 December 2021
Published: 22 December 2021

Publisher's Note: MDPI stays neutral with regard to jurisdictional claims in published maps and institutional affiliations.

Abstract: Surface electromyography (EMG), typically recorded from muscle groups such as the mentalis (chin/mentum) and anterior tibialis (lower leg/crus), is often performed in human subjects undergoing overnight polysomnography. Such signals have great importance, not only in aiding in the definitions of normal sleep stages, but also in defining certain disease states with abnormal EMG activity during rapid eye movement (REM) sleep, e.g., REM sleep behavior disorder and parkinsonism. Gold standard approaches to evaluation of such EMG signals in the clinical realm are typically qualitative, and therefore burdensome and subject to individual interpretation. We originally developed a digitized, signal processing method using the ratio of high frequency to low frequency spectral power and validated this method against expert human scorer interpretation of transient muscle activation of the EMG signal. Herein, we further refine and validate our initial approach, applying this to EMG activity across 1,618,842 s of polysomnography recorded REM sleep acquired from 461 human participants. These data demonstrate a significant association between visual interpretation and the spectrally processed signals, indicating a highly accurate approach to detecting and quantifying abnormally high levels of EMG activity during REM sleep. Accordingly, our automated approach to EMG quantification during human sleep recording is practical, feasible, and may provide a much-needed clinical tool for the screening of REM sleep behavior disorder and parkinsonism.

Keywords: electromyography; EMG; polysomnography; REM sleep without atonia; REM sleep behavior disorder; RBD; parkinsonism; Parkinson's disease; spectral power

1. Introduction

The unambiguous determination of rapid eye movement (REM) sleep relies on the simultaneous collection of electroencephalography, electrooculography, and electromyography (EMG; conventionally via mentalis/submentalis activity) [1]. One hallmark neurophysiologic feature of REM sleep is skeletal muscle paralysis (outside of specific ventilatory musculature, e.g., the diaphragm) reflected by relatively low EMG voltage at or near detectable noise levels [2,3]. Early methods of EMG quantification relied on analog-to-digital hardware and were limited to comparatively basic functions composed of summating signal voltage in an effort to objectively discriminate between REM and NREM [4].

Renewed efforts in the field of EMG signal analysis and quantification came after the first description of REM sleep behavior disorder (RBD), characterized by abnormally elevated REM sleep EMG (i.e., REM sleep without atonia) and dream mentation [5]. It became critical to accurately identify REM sleep without atonia once it became recognized that RBD is one of the earliest clinical manifestations of Parkinson's disease and other related synucleinopathies (e.g., dementia with Lewy bodies and multiple systems atrophy) [6–11]. Previous longitudinal studies have reported that 50–70% of individuals with RBD eventually develop an overt synucleinopathy within 5–10 years of RBD diagnosis [12–14].

Efforts to quantify EMG in RBD have been carried out over the years. Lapierre and Montplaisir were first to quantify phasic mentalis EMG activity during REM sleep in RBD [15]. Bliwise et al. extended this initial report by Lapierre and Montplaisir describing a phasic electromyographic metric in patients with Parkinson's disease [16]. Subsequently, more sophisticated computer algorithms have been developed across multiple groups. In a prior report by Fairley and Bliwise et al., 15 different EMG signal features were compiled after an exhaustive review of prior literature in the field [17]. These various automated quantitative approaches were compared relative to manual/visual review, and the accuracy of specific features tested. These features ranged from descriptive (e.g., skewness, kurtosis, and variance) to more complex algorithms involving nonlinear energy and spectral entropy. However, an intuitively appealing, fundamental approach was a simple ratio of high frequency to low frequency power density—an approach already used extensively in the analyses of electroencephalography (EEG) but not yet applied to EMG in this context [18]. Since EMG is composed of relatively high frequency signals, we elected to adopt this approach in the current work. Novel approaches to processing sleep signals generated from overnight sleep recordings have been introduced in recent years [18–30], many involving machine learning classification and/or power spectral analyses. Such newer techniques have led to important insights regarding clinical sleep disorders.

At the present time, clinical determination of REM sleep without atonia during overnight polysomnography still requires manual review and visual inspection by a registered polysomnographic scoring technician and/or board-certified sleep clinician. This manual approach is highly burdensome, time consuming, and prone to subjective errors. As such, scoring technicians do not routinely score each phasic EMG event, and ultimately, clinicians make a judgment call as to the binary presence or absence of REM sleep without atonia. Developing an automated method of identifying and quantifying REM sleep without atonia would accelerate the period of interpretation and reduce potential scoring bias. Additionally, quantification of REM sleep without atonia would provide a continuous, rather than binary, output variable by which to explore predictions of phenoconversion from RBD to parkinsonism. Thus, our goal was to develop and validate a more rigorous method to quantify REM sleep without atonia. Herein, we report an automated method of EMG quantification based upon ratios between high and low frequency (HF:LF) EMG spectral power and compare this to the gold standard visual scoring of REM sleep without atonia, manually labeled by a blinded scorer.

2. Materials and Methods

2.1. Participants and Polysomnography Recording

The overnight polysomnography data for this analysis was collected from an approved protocol performed according to the Declaration of Helsinki with approval of the VA Portland Health Care System Institutional Review Board (#3641). All participants provided verbal and written informed consent prior to participation.

Participants in this study were US Veterans and enrolled prospectively in a cross-sectional manner through the VA Portland Health Care System Sleep Clinic. A total of 595 Veterans enrolled with participants excluded based on (1) having <4 h of recorded sleep ($n = 76$), (2) having <10 epochs of recorded REM sleep ($n = 45$), and (3) were otherwise incomplete ($n = 13$). The remaining $n = 461$ participants were included in subsequent analyses (Figure 1). As previously described [31–34], reasons for referral to in-lab polysomnography included suspected obstructive sleep apnea, excessive daytime sleepiness, hypersomnia, insomnia, restless leg syndrome, and abnormal movements during sleep, with suspected obstructive sleep apnea and excessive daytime sleepiness being the most frequently cited.

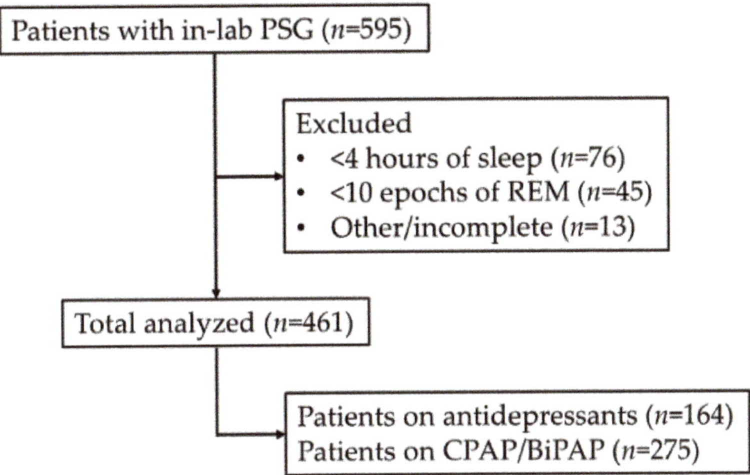

Figure 1. Schematic (CONSORT) overview of our patient population, exclusion criteria, and subsequent sub-analysis groups. Of the total $n = 595$ participants evaluated with in-lab polysomnography (PSG), we excluded $n = 134$ records (due to having <4 h of total sleep time, $n = 76$; <10 epochs of REM sleep, $n = 45$, or were otherwise incomplete, $n = 13$). Of these $n = 461$ participants, $n = 164$ and $n = 275$ were noted to be currently using antidepressant medications or were on CPAP/BiPAP during their PSG, respectively. Since antidepressant use and presence of untreated obstructive sleep apnea have both been associated with increased EMG tone during REM sleep, we wanted to examine these groups as separate subsets, given that these conditions might affect the accuracy of our algorithm.

All subjects completed in-laboratory, American Academy of Sleep Medicine-accredited technician-attended overnight video-polysomnography recordings using Polysmith (NihonKohden, Japan). Standard American Academy of Sleep Medicine parameters were collected, including electroencephalography (6 scalp electrodes), mentalis muscle EMG, bilateral tibialis anterior EMG, bilateral electrooculography, electrocardiography, peripheral blood-oxygen saturation, respiratory movement/effort (thorax and abdominal), airflow (nasal and oral), auditory (snoring), and body positioning (right side, left side, supine, prone) [1]. All EMG analyses described below were derived from the mentalis EMG channel. American Academy of Sleep Medicine-accredited polysomnographic technicians manually performed standard sleep staging for each 30 s epoch according to standard clinical criteria. Each 30 s epoch was scored as Wake, REM, or NREM stages N1, N2, and N3. All sleep staging was validated by a board-certified sleep physician.

All polysomnography records then underwent manual, visual-based scoring of phasic EMG events during REM sleep by a blinded, independent scorer. This training dataset was then used as the gold standard for direct comparison with our automated algorithm using HF:LF ratios.

2.2. HF:LF Analysis

Mentalis EMG, sampled at 200 Hz across the entire polysomnography recording, was analyzed using the spectral decomposition function in MATLAB's signal processing toolbox. Estimates of power spectral density were computed via Welch's method. Thus, we computed both absolute and relative powers of EMG in 2 s sliding windows which were overlapped by one-second, and which produced a time-varying power spectra with a 0.5 Hz resolution (1/2-s). Relative spectra were computed by dividing the absolute power spectra of each 2 s segment by the total spectral power of that segment.

From the second-by-second spectral density functions, we computed the ratio of the integral (sum) of EMG powers in the high and low frequency ranges using the formula:

$$HF:LF\,(t) = \sum_{f=20}^{55\,\text{Hz}} P_{EMG}(t,f) / \sum_{f=2}^{20\,\text{Hz}} P_{EMG}(t,f) \tag{1}$$

where $P_{EMG}\,(t,f)$ is the spectral power density at time t and frequency f (f is between 0 and 55 Hz in 0.5 Hz steps).

Following the computation of HF:LF for every second of recorded EMG, we identified all the 30 s epochs of REM sleep and investigated the level of these indices with regards to the presence of phasic REM sleep without atonia. The following two methods were applied:

2.3. Epoch-Based HF:LF Analysis: Grouping Based on the Number of REM Sleep without Atonia Episodes per REM Epoch

In this method, the HF:LF for every REM epoch was summed and then considered individually. Every epoch of REM sleep was pooled together and separated according to the number of phasic REM sleep without atonia events that were identified during manual/visual review. Using this approach, we identified 3 groups: (E1) average HF:LF in all epochs with REM sleep without atonia; (E2) average HF:LF in every epoch of REM sleep with exactly 1 REM sleep without atonia event; and (E3) average HF:LF in every epoch of REM sleep with $\geq$2 REM sleep without atonia events.

2.4. Window-Based HF:LF Analysis: Compare All EMG-Index Values for Every Second

In this method, the HF:LF for every second of REM sleep was considered individually. Here we examined both every second of REM sleep without REM sleep without atonia and then specifically every REM sleep without atonia event using a 3 s window (average HF:LF in the 1 s before, during, and after). This produced two separate groups:

(W1), the HF:LF in every second of REM sleep with no REM sleep without atonia; and (W2), the average HF:LF in the 3 s window around every REM sleep without atonia event.

3. Results

3.1. Datasets

All 461 participants' polysomnography analyzed met a priori inclusion criteria related to having a threshold level of total sleep time and REM sleep duration (i.e., >4 h and $\geq$10 REM epochs). The overall participant group (Table 1) was predominantly male, middle-aged, obese, and of white racial and non-Hispanic/Latino ethnicity. The majority of participants reported some college education (or greater) and were married living with their spouse/partner. Roughly 1/3 of the participant group reported exercising >90 min/week, had a self-reported history of traumatic brain injury or provisional post-traumatic stress disorder diagnosis based on the PCL-5 [35]. Finally, nearly half of the participant group endorsed the RBD1Q single-question related to dream enactment described by Postuma et al. [36] and a self-reported history of restless legs syndrome.

3.2. EMG Relative Spectral Power

Relative spectral power within the EMG signal was computed on a second-by-second basis across all REM sleep epochs. The median value within all REM sleep epochs was then separated based on whether or not there was a recorded REM sleep without atonia event. These relative EMG spectral power data within periods of no REM sleep without atonia and periods of REM sleep without atonia (i.e., consisting of a 3 s window around the REM sleep without atonia event; 1 s before and 1 s after) are presented in Figure 2. The crossover point occurred between 20.5 and 21 Hz. This crossover provided the subsequent rationale to analyze the ratio between high and low frequency power (HF:LF).

Table 1. Demographic and anthropometric variables.

	Whole Group $n = 461$
Age, years	53.9 ± 15.9
Sex, male	92.0%
Height, cm	177.2 ± 7.8
Weight, kg	102.3 ± 21.1
BMI, kg/m^2	32.6 ± 6.6
Race, white	83.9%
Ethnicity, not Hispanic or Latino	90.2%
Education, at least some college	79.0%
Marital status, married/partnered	62.5%
Living situation, spouse/partner	63.8%
Exercise, >90 min/week	28.0%
TBI, yes	20.8%
PTSD, yes	30.6%
RBD1Q dream enactment, yes	43.4%
RLS, yes	44.3%
Snore, yes	88.3%
CPAP/BiPAP, yes	59.7%
Antidepressant use, yes	35.6%

Data are presented as mean ± standard deviation or as a % of the total sample size. BMI, body mass index; TBI, traumatic brain injury; PTSD, post-traumatic stress disorder; RBD1Q, previously published single question related to dream enactment; RSL, restless legs syndrome; CPAP/BiPAP, continuous positive airway pressure/bidirectional positive airway pressure.

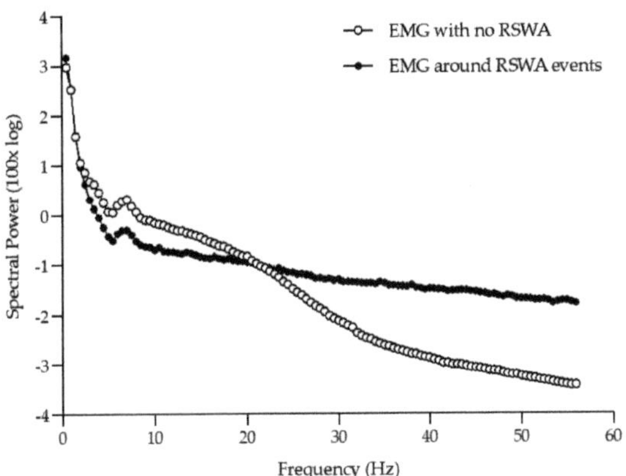

Figure 2. Overall median second-by-second EMG relative spectral power around periods of no REM sleep without atonia events and periods with an REM sleep without atonia event.

3.3. Epoch-Based HF:LF Analysis

In this approach, the HF:LF per second was averaged across every epoch of REM sleep and considered individually (Figure 3). This produced a total of 53,680 epochs of REM sleep, of which there were 47,483 epochs with no REM sleep without atonia events (Group E1), 4246 epochs with exactly 1 REM sleep without atonia event (Group E2), and 1951 epochs with ≥2 REM sleep without atonia events (Group E3). Average HF:LF for groups E1, E2, and E3 were 8.38 ± 9.42, 16.68 ± 17.36, and 23.52 ± 18.35, respectively. Median (25th–75th percentile) HF:LF for groups E1, E2, and E3 were 5.69 (3.66–9.40), 10.80 (6.30–19.47), 17.73 (10.04–30.84), respectively. Statistical power considering each REM epoch individually is likely exaggerated (omnibus ANOVA: $p < 0.0001$, F (2, 53,677)), with Tukey's posthoc analysis illustrating significant differences between each of the comparisons (all $p < 0.0001$).

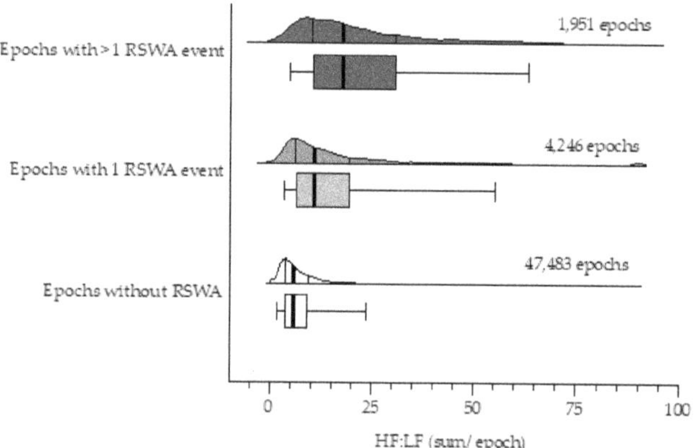

Figure 3. Epoch-based HF:LF analysis: horizontal violin plots (non-truncated/halved) with corresponding box-whisker plots illustrating the spread across Group E1 (open plot; average HF:LF in all epochs of REM sleep without REM sleep without atonia), Group E2 (light shaded plot; average HF:LF in every epoch of REM sleep with exactly 1 REM sleep without atonia event), and Group E3 (heavy shaded plot; average HF:LF in every epoch of REM sleep with ≥ 2 REM sleep without atonia events). The heavy solid line corresponds to the median value with the 25th and 75th percentile indicated by the bracketed lines/box outline. Whiskers indicate the 5th and 95th percentiles.

The same comparison was made within two separate sub-analyses, considering participants who were on/off CPAP/BiPAP during their polysomnography, and considering participants who were/were not currently taking antidepressants at the time of their polysomnography. There was $n = 275$ (corresponding to 28,229 REM sleep epochs) and $n = 186$ (corresponding to 21,106 REM sleep epochs) participants on and off CPAP/BiPAP, respectively. Similarly, there was $n = 164$ (corresponding to 18,705 REM sleep epochs) and $n = 286$ (corresponding to 33,794 REM sleep epochs) participants on and off antidepressant medications, respectively. The subanalyses did not produce different distributions or relative numbers of epochs per grouping (i.e., groups E1, E2, E3) compared to the whole group analysis (data not shown).

3.4. Window-Based HF:LF Analysis

In this approach, the HF:LF for every second of REM sleep was computed and considered individually using a 1 s window (Figure 4). This produced a total of 1,618,842 s of REM sleep, of which there were 1,587,659 s with no REM sleep without atonia events (Group W1), and 31,183 s of REM sleep around REM sleep without atonia events (Group W2). Average HF:LF for Groups W1 and W2 were 0.26 ± 0.30, and 0.77 ± 0.72, respectively. Median (5th-75th percentile) HF:LF for Group W1 and Group W2 were 0.17 (0.10–0.29), and 0.49 (0.21–1.13), respectively. Statistical power considering each second of REM sleep individually is likely exaggerated yet significant via a two-tailed unpaired t-test ($p < 0.0001$; $t = 125.3$, df $= 31,390$) and F-test comparing variances ($p < 0.0001$; F $= 5.89$, DFn $= 31,182$, Dfd $= 1,587,658$).

As in the first method, epoch-based HF:LF analysis, the same comparison was made considering participants who were on/off CPAP/BiPAP during their polysomnography, and considering participants who were/were not currently taking antidepressants at the time of their polysomnography. There was $n = 275$ (corresponding to 848,141 s of REM sleep) and $n = 186$ (corresponding to 639,543 s of REM sleep) participants on and off CPAP/BiPAP, respectively. Similarly, there was $n = 164$ (corresponding to 560,532 s of REM sleep) and $n = 286$ (corresponding to 1,022,332 s of REM sleep) participants on and off antidepressant medications, respectively. The subanalyses did not produce different distributions or relative numbers of seconds per grouping (i.e., Groups W1 or W2) compared to the whole group analysis (data not shown).

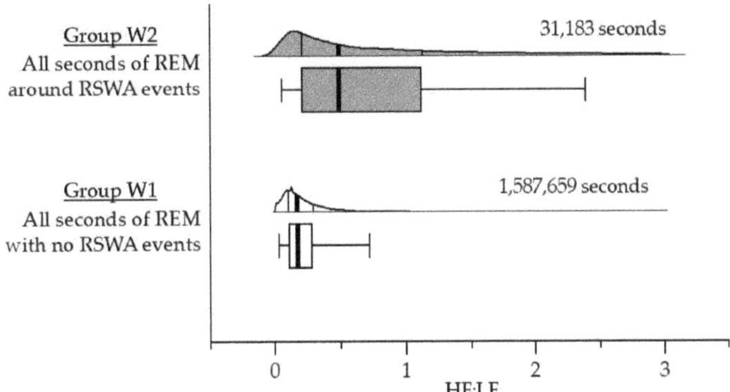

Figure 4. Window-based HF:LF analysis: horizontal violin plots (non-truncated/halved) with corresponding box-whisker plots illustrating the HF:L:F spread across Group W1 (open plot; the HF:LF in every second of REM sleep with no REM sleep without atonia), and Group W2 (light shaded plot; the average HF:LF in the 3 s window around every REM sleep without atonia event). The heavy solid line corresponds to the median value with the 25th and 75th percentile indicated by the bracketed lines/box outline. Whiskers indicate the 5th and 95th percentiles.

4. Discussion

We sought to validate a novel automated algorithm to identify and quantify phasic REM sleep without atonia events in human polysomnography records. Power spectral analyses of manually/visually scored events showed exceptionally strong correspondence, suggesting that the HF:LF approach developed here represented an extremely robust algorithm. We tested this with two separate analyses: (1) epoch-based HF:LF analysis that calculated the HF:LF on a conventional epoch-by-epoch basis (e.g., 30 s epochs are standard for sleep staging in human polysomnography recordings), and (2) window-based HF:LF analysis that calculated HF:LF on a second-by-second basis. Both analyses revealed high comparability to the manual, visually based gold standard scoring rubric, and analyses indicated that our method was highly accurate in discriminating phasic REM sleep without atonia from baseline REM sleep EMG tone.

These two methods that we developed and validated were both highly accurate, and importantly, are complementary to each other. Our rationale to presenting both approaches derives from the fact that, in the study of human sleep, there has been little consensus on the optimal time base with which to analyze the EMG signal. Clinically, scoring set forth by the American Academy of Sleep Medicine guidelines determine human sleep staging at no shorter than 30 s at a time (i.e., 30 s epochs). The epoch-based HF:LF analysis remains faithful to this and although using a sophisticated parametric approach to signal processing (i.e., the HF:LF ratio computed on an individual second-to-second basis), this approach still resolves into being non-parametric, classifying 30 s REM sleep epochs into those with 0 events, 1 event, or 2 or more events. This non-parametric paradigm was retained to maintain translational compatibility with how human sleep recordings are clinically analyzed visually.

In contrast to the epoch-based HF:LF analysis, the window-based HF:LF analysis establishes the relevant time base for application of the HF:LF ratios on a 1 s basis. This approach allows for a more fully parametric use of the measured ratios, because it makes no assumptions about the number of single seconds with activity that are meaningful. Furthermore, the increased temporal resolution of the window-based HF:LF analysis might be useful during sleep stage transitions between epochs and/or during sleep studies using animal models, which are not constrained by the 30 s epoch rule. In fact, the newly published statement from the International REM Behavior Disorder working group [37] has suggested that just such micro-epoch scoring is likely to have the highest yield clinically and produce the more consistent results across laboratories. Although full analyses of clinical material are beyond the scope of the current paper, a potential and immediate benefit of the higher temporal resolution and more parametric approach is seen in Figure 3. These data highlight the benefit of more fine-grained resolution in the analysis of the EMG signals, in this case for determination of cut-point frequency defining the presence or absence of activity.

The epoch- and window-based HF:LF analyses each reveal different aspects of the measurement of the EMG in human sleep. The epoch-based HF:LF analysis allows immediate translation of this particular signal processing strategy to an enormous body of literature that has approached the analyses of EMG signals in sleep using visual techniques. The window-based HF:LF analysis will potentially pave the way for future work in this area. As such, the two analytic approaches are complementary to each other and reveal a more complete story than either could tell alone.

Although other approaches to digitization of surface EMG signals recorded during sleep have been developed, validation of such approaches typically have relied upon evaluating their utility in the service of diagnostic relevance [21,38–43]. Ultimately, such approaches confound complex issues of medical diagnoses, such as incidence, positive predictive value, and sensitivity/specificity of a disease, with the particular signal processing approach being tested. By contrast, our analyses may be considered more elemental and have focused solely on validating the HF:LF approach with the judgments of expert visual scorers about those signals. Such an approach makes fewer epidemiologic assumptions about how well a particular digital feature operates, and maintains a strict focus only on technologic, rather than clinical, utility. Nonetheless, we evaluated the impact of several additional factors impacting EMG activity in sleep.

In clinical practice, EMG tone during REM sleep can be affected by several factors. One of the most common conditions that can cause increased EMG tone during REM sleep is the use of anti-depressants, especially within the selective serotonin reuptake inhibitor (SSRI) class [44,45]. To this end, we assessed whether our HF:LF algorithm could distinguish REM sleep without atonia within those subjects on anti-depressants, versus those not on anti-depressants, and found that the algorithm was still highly accurate and comparable to manual visual scoring. Another common condition that can cause increased EMG tone during REM sleep is obstructive sleep apnea, in which increased apneas during REM sleep can cause muscle artifact through snoring and other respiratory-related movements of the head. In order to address sleep apnea as a potential confound, we assessed whether our HF:LF algorithm could distinguish REM sleep without atonia within those subjects with mild, moderate, and severe sleep apnea, versus those without sleep apnea, and found that the algorithm was still highly accurate in identifying REM sleep without atonia regardless of sleep apnea status.

Strengths of this study include a large sample size comprised of over 450 polysomnographic records yielding over 50,000 epochs of REM sleep for analysis. Additionally, our dataset with manual visual scoring of REM sleep without atonia phasic events in these polysomnography records, in addition to American Academy of Sleep Medicine-standard scoring, represents a valuable gold standard with which to compare other automated methods. Finally, the ratio approach is novel, with strong rationale based upon naturally occurring cutoffs in the power spectra from gold standard REM sleep without atonia versus non-REM sleep without atonia events and may serve to accentuate existing differences between REM sleep without atonia and non REM sleep without atonia EMG signal.

Limitations of our analysis include the following considerations. While the dataset is large, it was obtained from a single site. As this site is an American Academy of Sleep Medicine-accredited sleep laboratory and studies were obtained for clinical indications and not research, no additional EMG leads were acquired beyond the standard chin and leg leads. The addition of arm leads and other muscles may increase the sensitivity for detecting REM sleep without atonia and diagnosis of RBD [46]. Finally, our HF:LF algorithm was validated against gold standard scoring of phasic, and not tonic, events only. As the REM sleep without atonia events in this dataset contained nearly four times the number of phasic events compared to tonic events, this may be a minor limitation. However, some have postulated that sustained tonic EMG activity essentially represents a cumulative summation of shorter duration phasic events [47], and basic science studies have suggested that phasic activity may represent the most fundamental neurobiological substrate of the neurochemical control of REM [48–50]. Nonetheless, there remains a caveat about generalization of our algorithm to tonic events and other cohorts/datasets that may have more tonic activity. Lastly, we note that the proposed signal analytic approaches (i.e., HF:LF ratios) are not inherently novel from a pure signal processing perspective. As described, this mathematical approach has a considerable history of usage. However, we note that the novelty of the proposed analysis lies in the application and potential clinical utility. This approach, while simple, mirrors the human visual experience and sets the stage for future advanced proposals that leverage machine learning and other methodologies beyond the scope of the current report. Additional analytical progressions that combine these signal processing approaches, and clinical application, will include how to define performance of the models. This may be based comparisons to human scoring on a second-by-second basis, agreement on a 30 s epoch basis, agreement on binary diagnosis, or even examining the extent/quantification of RSWA as a continuous variable.

In summary, these data indicate that our automated HF:LF ratio approach to EMG quantification during human sleep recording is practical, feasible, and may provide a much-needed clinical tool for screening of phasic REM sleep without atonia events as relevant to diagnosis of REM sleep behavior disorder and eventual phenoconversion to parkinsonism [51].

5. Conclusions

We report a highly accurate approach to detect and quantify surface EMG activity during sleep using a large dataset of overnight polysomnography containing over 50,000 epochs of REM sleep from over 450 individuals. This digitized, signal processing method utilizes the ratio of high frequency to low frequency (HF:LF) spectral power and validated this method against expert human scorer interpretation of EMG signal. Data demonstrate a significant association between visual interpretation and the spectrally processed signals, indicating a highly accurate approach to detecting and quantifying abnormally high levels of EMG activity during REM sleep. These data indicate that our automated approach to EMG quantification during human sleep recording is practical, feasible, and may provide a much-needed clinical tool for screening of disorders with elevated EMG tone during REM sleep, such as REM sleep behavior disorder and parkinsonism.

Author Contributions: Conceptualization, M.H.M., J.E.E., D.L.B. and M.M.L.; data curation, M.H.M. and J.E.E.; formal analysis, M.H.M. and J.E.E.; funding acquisition, M.H.M., J.E.E., D.L.B. and M.M.L.; investigation, J.E.E., K.B.W., D.P. and M.M.L.; methodology, M.H.M., J.E.E., K.B.W., D.P., D.L.B. and M.M.L.; project administration, M.M.L.; resources, M.M.L.; supervision, J.E.E. and M.M.L.; validation, M.H.M., J.E.E., D.L.B. and M.M.L.; visualization, J.E.E.; writing—original draft, J.E.E. and M.M.L.; writing—review and editing, J.E.E., D.L.B. and M.M.L. All authors have read and agreed to the published version of the manuscript.

Funding: The data in this work was supported with resources and the use of facilities at the VA Portland Health Care System, VA RRD Merit Award #I01 RX002846 to M.H.M. and M.M.L.; VA CSRD Merit Award #I01 CX002022 to M.M.L.; VA RRD Career Development Award #1K2 RX002947 to J.E.E.; NIH NIA R34 AG056639 to J.E.E., D.L.B., and M.M.L. The interpretations and conclusions expressed in this article are those of the authors and do not necessarily reflect the position or policy of the Department of Veterans Affairs, the National Institute of Health, or the United States government.

Institutional Review Board Statement: The project that supported collection of the polysomnography recordings used to examine the novel algorithm described herein was conducted according to the guidelines of the Declaration of Helsinki and approved by the VA Portland Health Care System Institutional Review Board (#3641).

Informed Consent Statement: All participants gave verbal and written informed consent prior to participation.

Data Availability Statement: A de-identified, anonymized dataset will be created and shared, and released once institutional approvals are in place. Where practicable, sharing will take place under a written agreement prohibiting the recipient from identifying or re-identifying (or taking steps to identify or re-identify) any individual whose data are included in the dataset. However, it is permissible for final datasets in machine-readable format to be submitted to and accessed from PubMed Central (and similar sites) provided that care is taken to ensure that the individuals cannot be re-identified using other publicly available information. De-identified data sets will be maintained locally on VA computer drives that are regularly backed up and password protected until central data repositories become available for long-term storage and access. Until that time, the data sets will be available by request to any interested investigators together with a data dictionary to enable interpretation of the data set.

Acknowledgments: The authors would like to express their sincere appreciation and gratitude for the participation of research subjects involved in this project.

Conflicts of Interest: The authors declare no conflict of interest.

References

1. Berry, R.B.; Brooks, R.; Gamaldo, C.E.; Harding, S.M.; Marcus, C.L.; Vaughn, B.V.; Tangredi, M.M. *The AASM Manual for the Scoring of Sleep and Associated Events: Rules, Terminology and Technical Specifications*; Version 2.0.; American Academy of Sleep Medicine: Darien, IL, USA, 2012.

2. Dement, W. The occurence of low voltage, fast, electroencephalogram patterns during behavioral sleep in the cat. *Electroencephalogr. Clin. Neurophysiol.* **1958**, *10*, 291–296. [CrossRef]

3. Jouvet, M.; Michel, M.; Courjon, J. Sur la mise en jeu de deux mechanismes a l'expression electro-encephalographique differente au cours du sommeil physiologique chez le Chat. *Comptes Rendus Seances L'Academie Sci.* **1959**, *248*, 3043–3045.

4. Bliwise, D.; Coleman, R.; Bergmann, B.; Wincor, M.Z.; Pivii, R.T.; Rechtschaffen, A. Facial Muscle Tonus During REM and NREM Sleep. *Psychophysiology* **1974**, *11*, 497–508. [CrossRef]

5. Schenck, C.H.; Bundlie, S.R.; Ettinger, M.G.; Mahowald, M.W. Chronic Behavioral Disorders of Human REM Sleep: A New Category of Parasomnia. *Sleep* **1986**, *9*, 293–308. [CrossRef]

6. Boeve, B.; Silber, M.; Ferman, T.; Lin, S.; Benarroch, E.; Schmeichel, A.; Ahlskog, J.; Caselli, R.; Jacobson, S.; Sabbagh, M.; et al. Clinicopathologic correlations in 172 cases of rapid eye movement sleep behavior disorder with or without a coexisting neurologic disorder. *Sleep Med.* **2013**, *14*, 754–762. [CrossRef]

7. Plazzi, G.; Corsini, R.; Provini, F.; Pierangeli, G.; Martinelli, P.; Montagna, P.; Lugaresi, E.; Cortelli, P. REM sleep behavior disorders in multiple system atrophy. *Neurology* **1997**, *48*, 1094–1096. [CrossRef] [PubMed]

8. Marion, M.-H.; Qurashi, M.; Marshall, G.; Foster, O. Is REM sleep Behaviour Disorder (RBD) a risk factor of dementia in idiopathic Parkinson's disease? *J. Neurol.* **2008**, *255*, 192–196. [CrossRef] [PubMed]

9. Gagnon, J.-F.; Postuma, R.B.; Mazza, S.; Doyon, J.; Montplaisir, J. Rapid-eye-movement sleep behaviour disorder and neurodegenerative diseases. *Lancet Neurol.* **2006**, *5*, 424–432. [CrossRef]

10. Galbiati, A.; Verga, L.; Giora, E.; Zucconi, M.; Ferini-Strambi, L. The risk of neurodegeneration in REM sleep behavior disorder: A systematic review and meta-analysis of longitudinal studies. *Sleep Med. Rev.* **2019**, *43*, 37–46. [CrossRef] [PubMed]

11. Barone, D.A.; Henchcliffe, C. Rapid eye movement sleep behavior disorder and the link to alpha-synucleinopathies. *Clin. Neurophysiol.* **2018**, *129*, 1551–1564. [CrossRef]

12. Iranzo, A.; Stockner, H.; Serradell, M.; Seppi, K.; Valldeoriola, F.; Frauscher, B.; Molinuevo, J.L.; Vilaseca, I.; Mitterling, T.; Gaig, C.; et al. Five-year follow-up of substantia nigra echogenicity in idiopathic REM sleep behavior disorder. *Mov. Disord.* **2014**, *29*, 1774–1780. [CrossRef] [PubMed]

13. Schenck, C.H.; Boeve, B.F.; Mahowald, M.W. Delayed emergence of a parkinsonian disorder or dementia in 81% of older men initially diagnosed with idiopathic rapid eye movement sleep behavior disorder: A 16-year update on a previously reported series. *Sleep Med.* **2013**, *14*, 744–748. [CrossRef]

14. Berg, D.; Postuma, R.B.; Adler, C.H.; Bloem, B.R.; Chan, P.; Dubois, B.; Gasser, T.; Goetz, C.G.; Halliday, G.; Joseph, L.; et al. MDS research criteria for prodromal Parkinson's disease. *Mov. Disord.* **2015**, *30*, 1600–1611. [CrossRef] [PubMed]

15. Lapierre, O.; Montplaisir, J. Polysomnographic features of REM sleep behavior disorder: Development of a scoring method. *Neurology* **1992**, *42*, 1371. [CrossRef]

16. Bliwise, D.L.; He, L.; Pour Ansari, F.; Rye, D.B. Quantification of Electromyographic Activity During Sleep: A Phasic Electromyographic Metric. *J. Clin. Neurophysiol.* **2006**, *23*, 59–67. [CrossRef] [PubMed]

17. Fairley, J.A.; Georgoulas, G.; Mehta, N.A.; Gray, A.G.; Bliwise, D.L. Computer detection approaches for the identification of phasic electromyographic (EMG) activity during human sleep. *Biomed. Signal Process. Control.* **2012**, *7*, 606–615. [CrossRef] [PubMed]

18. Lai, D.; Bin Heyat, B.; Khan, F.I.; Zhang, Y. Prognosis of Sleep Bruxism Using Power Spectral Density Approach Applied on EEG Signal of Both EMG1-EMG2 and ECG1-ECG2 Channels. *IEEE Access* **2019**, *7*, 82553–82562. [CrossRef]

19. Bin Heyat, B.; Akhtar, F.; Khan, M.H.; Ullah, N.; Gul, I.; Khan, H.; Lai, D. Detection, Treatment Planning, and Genetic Predisposition of Bruxism: A Systematic Mapping Process and Network Visualization Technique. *CNS Neurol. Disord. Drug Targets* **2021**, *20*, 755–775. [CrossRef]

20. Ferri, R.; Fulda, S.; Cosentino, F.I.; Pizza, F.; Plazzi, G. A preliminary quantitative analysis of REM sleep chin EMG in Parkinson's disease with or without REM sleep behavior disorder. *Sleep Med.* **2012**, *13*, 707–713. [CrossRef]

21. Figorilli, M.; Ferri, R.; Zibetti, M.; Beudin, P.; Puligheddu, M.; Lopiano, L.; Cicolin, A.; Durif, F.; Marques, A.; Fantini, M.L. Comparison Between Automatic and Visual Scorings of REM Sleep Without Atonia for the Diagnosis of REM Sleep Behavior Disorder in Parkinson Disease. *Sleep* **2017**, *40*, 13–16. [CrossRef] [PubMed]

22. Yildiz, S.; Opel, R.A.; Elliott, J.E.; Kaye, J.; Cao, H.; Lim, M.M. Categorizing Sleep in Older Adults with Wireless Activity Monitors Using LSTM Neural Networks. In Proceedings of the EMBC 2019, Berlin, Germany, 23–27 July 2019; pp. 3368–3372. [CrossRef]

23. Bin Heyat, B.; Lai, D.; Khan, F.I.; Zhang, Y. Sleep Bruxism Detection Using Decision Tree Method by the Combination of C4-P4 and C4-A1 Channels of Scalp EEG. *IEEE Access* **2019**, *7*, 102542–102553. [CrossRef]

24. Ali, L.; He, Z.; Cao, W.; Rauf, H.T.; Imrana, Y.; Bin Heyat, B. MMDD-Ensemble: A Multimodal Data–Driven Ensemble Approach for Parkinson's Disease Detection. *Front. Neurosci.* **2021**, *15*. [CrossRef] [PubMed]

25. Bin Heyat, B.; Akhtar, F.; Ansari, M.; Khan, A.; Alkahtani, F.; Khan, H.; Lai, D. Progress in Detection of Insomnia Sleep Disorder: A Comprehensive Review. *Curr. Drug Targets* **2021**, *22*, 672–684. [CrossRef] [PubMed]

26. Heyat, B.; Akhtar, F.; Khan, A.; Noor, A.; Benjdira, B.; Qamar, Y.; Abbas, S.; Lai, D. A Novel Hybrid Machine Learning Classification for the Detection of Bruxism Patients Using Physiological Signals. *Appl. Sci.* **2020**, *10*, 7410. [CrossRef]

27. Fernández-Arcos, A.; Serradell, M.; Guaita, M.; Santamaria, J.; Iranzo, A.; Gaig, C.; Salamero, M. Diagnostic Value of Isolated Mentalis Versus Mentalis Plus Upper Limb Electromyography in Idiopathic REM Sleep Behavior Disorder Patients Eventually Developing a Neurodegenerative Syndrome. *Sleep* **2017**, *40*, zsx025. [CrossRef] [PubMed]

28. Cesari, M.; Christensen, J.A.; Sixel-Döring, F.; Trenkwalder, C.; Mayer, G.; Oertel, W.H.; Jennum, P.; Sorensen, H. Validation of a new data-driven automated algorithm for muscular activity detection in REM sleep behavior disorder. *J. Neurosci. Methods* **2019**, *312*, 53–64. [CrossRef] [PubMed]

29. Seven, Y.B.; Mantilla, C.; Zhan, W.-Z.; Sieck, G.C. Non-stationarity and power spectral shifts in EMG activity reflect motor unit recruitment in rat diaphragm muscle. *Respir. Physiol. Neurobiol.* **2013**, *185*, 400–409. [CrossRef] [PubMed]

30. Shokrollahi, M.; Krishnan, S. A Review of Sleep Disorder Diagnosis by Electromyogram Signal Analysis. *Crit. Rev. Biomed. Eng.* **2015**, *43*, 1–20. [CrossRef]

31. Elliott, J.E.; Opel, R.A.; Pleshakov, D.; Rachakonda, T.; Chau, A.Q.; Weymann, K.B.; Lim, M.M. Posttraumatic stress disorder increases the odds of REM sleep behavior disorder and other parasomnias in Veterans with and without comorbid traumatic brain injury. *Sleep* **2020**, *43*, zsz237. [CrossRef] [PubMed]

32. Elliott, J.E.; Opel, R.A.; Weymann, K.B.; Chau, A.Q.; Papesh, M.A.; Callahan, M.L.; Storzbach, D.; Lim, M.M. Sleep Disturbances in Traumatic Brain Injury: Associations with Sensory Sensitivity. *J. Clin. Sleep Med.* **2018**, *14*, 1177–1186. [CrossRef]

33. Balba, N.M.; Elliott, J.E.; Weymann, K.B.; Opel, R.A.; Duke, J.W.; Oken, B.S.; Morasco, B.J.; Heinricher, M.M.; Lim, M.M. Increased sleep disturbances and pain in Veterans with co-morbid TBI and PTSD. *J. Clin. Sleep Med.* **2018**, *14*, 1865–1878. [CrossRef]

34. Sandsmark, D.K.; Elliott, J.; Lim, M.M. Sleep-Wake Disturbances after Traumatic Brain Injury: Synthesis of Human and Animal Studies. *Sleep* **2017**, *40*, 44. [CrossRef]

35. Blevins, C.A.; Weathers, F.W.; Davis, M.T.; Witte, T.K.; Domino, J.L. The Posttraumatic Stress Disorder Checklist for DSM-5 (PCL-5): Development and Initial Psychometric Evaluation. *J. Trauma. Stress* **2015**, *28*, 489–498. [CrossRef] [PubMed]

36. Postuma, R.B.; Arnulf, I.; Hogl, B.; Iranzo, A.; Miyamoto, T.; Jennum, P.; Pelletier, A.; Wolfson, C.; Leu-Semenescu, S.; Frauscher, B.; et al. A Single-Question Screen for REM Sleep Behavior Disorder: A Multicenter Validation Study. *Mov. Disord.* **2012**, *27*, 913–916. [CrossRef]

37. Cesari, M.; Heidbreder, A.; Louis, E.K.S.; Sixel-Döring, F.; Bliwise, D.L.; Baldelli, L.; Bes, F.; Fantini, M.L.; Iranzo, A.; Knudsen-Heier, S.; et al. Video-polysomnography procedures for diagnosis of rapid eye movement sleep behavior disorder (RBD) and the identification of its prodromal stages: Guidelines from the International RBD Study Group. *Sleep* **2021**. [CrossRef]

38. Ferri, R.; Rundo, F.; Manconi, M.; Plazzi, G.; Bruni, O.; Oldani, A.; Strambi, L.F.; Zucconi, M. Improved computation of the atonia index in normal controls and patients with REM sleep behavior disorder. *Sleep Med.* **2010**, *11*, 947–949. [CrossRef]

39. Frauscher, B.; Gabelia, D.; Biermayr, M.; Stefani, A.; Hackner, H.; Mitterling, T.; Poewe, W.; Högl, B. Validation of an Integrated Software for the Detection of Rapid Eye Movement Sleep Behavior Disorder. *Sleep* **2014**, *37*, 1663–1671. [CrossRef]

40. Guttowski, D.; Mayer, G.; Oertel, W.; Kesper, K.; Rosenberg, T. Validation of semiautomatic scoring of REM sleep without atonia in patients with RBD. *Sleep Med.* **2018**, *46*, 107–113. [CrossRef]

41. Mayer, G.; Kesper, K.; Ploch, T.; Canisius, S.; Penzel, T.; Oertel, W.; Stiasny-Kolster, K. Quantification of Tonic and Phasic Muscle Activity in REM Sleep Behavior Disorder. *J. Clin. Neurophysiol.* **2008**, *25*, 48–55. [CrossRef]

42. Cooray, N.; Andreotti, F.; Lo, C.; Symmonds, M.; Hu, M.; De Vos, M. Detection of REM sleep behaviour disorder by automated polysomnography analysis. *Clin. Neurophysiol.* **2019**, *130*, 505–514. [CrossRef]
43. Frandsen, R.; Nikolic, M.; Zoetmulder, M.; Kempfner, L.; Jennum, P. Analysis of automated quantification of motor activity in REM sleep behaviour disorder. *J. Sleep Res.* **2015**, *24*, 583–590. [CrossRef]
44. Winkelman, J.W.; James, L. Serotonergic Antidepressants are Associated with REM Sleep without Atonia. *Sleep* **2004**, *27*, 317–321. [CrossRef]
45. Frauscher, B.; Jennum, P.; Ju, Y.-E.; Postuma, R.B.; Arnulf, I.; De Cock, V.C.; Dauvilliers, Y.; Fantini, M.L.; Strambi, L.F.; Gabelia, D.; et al. Comorbidity and medication in REM sleep behavior disorder: A multicenter case-control study. *Neurology* **2014**, *82*, 1076–1079. [CrossRef]
46. Iranzo, A.; Frauscher, B.; Santos, H.; Gschliesser, V.; Ratti, L.; Falkenstetter, T.; Stürner, C.; Salamero, M.; Tolosa, E.; Poewe, W. Usefulness of the SINBAR electromyographic montage to detect the motor and vocal manifestations occurring in REM sleep behavior disorder. *Sleep Med.* **2011**, *12*, 284–288. [CrossRef]
47. Bliwise, P.D.L.; Fairley, P.J.; Hoff, M.S.; Rosenberg, P.R.S.; Rye, M.D.B.; Schulman, D.; Trotti, M.L.M. Inter-rater agreement for visual discrimination of phasic and tonic electromyographic activity in sleep. *Sleep* **2018**, *41*, 41. [CrossRef] [PubMed]
48. Brooks, P.L.; Peever, J. A Temporally Controlled Inhibitory Drive Coordinates Twitch Movements during REM Sleep. *Curr. Biol.* **2016**, *26*, 1177–1182. [CrossRef]
49. Brooks, P.L.; Peever, J.H. Glycinergic and GABAA-Mediated Inhibition of Somatic Motoneurons Does Not Mediate Rapid Eye Movement Sleep Motor Atonia. *J. Neurosci.* **2008**, *28*, 3535–3545. [CrossRef]
50. Iranzo, A. The REM sleep circuit and how its impairment leads to REM sleep behavior disorder. *Cell Tissue Res.* **2018**, *373*, 245–266. [CrossRef]
51. McCarter, S.J.; Sandness, D.J.; McCarter, A.R.; Feemster, J.C.; Teigen, L.N.; Timm, P.C.; Boeve, B.F.; Silber, M.H.; Louis, E.K.S. REM sleep muscle activity in idiopathic REM sleep behavior disorder predicts phenoconversion. *Neurology* **2019**, *93*, e1171–e1179. [CrossRef] [PubMed]

 *sensors*

MDPI

Article

A Coupled Piezoelectric Sensor for MMG-Based Human-Machine Interfaces

Mateusz Szumilas *, Michał Władziński and Krzysztof Wildner

Warsaw University of Technology, Faculty of Mechatronics, Institute of Metrology and Biomedical Engineering, A. Boboli 8 St., 02-525 Warsaw, Poland; michal.wladzinski@pw.edu.pl (M.W.); krzysztof.wildner@pw.edu.pl (K.W.)
* Correspondence: mateusz.szumilas@pw.edu.pl

Abstract: Mechanomyography (MMG) is a technique of recording muscles activity that may be considered a suitable choice for human–machine interfaces (HMI). The design of sensors used for MMG and their spatial distribution are among the deciding factors behind their successful implementation to HMI. We present a new design of a MMG sensor, which consists of two coupled piezoelectric discs in a single housing. The sensor's functionality was verified in two experimental setups related to typical MMG applications: an estimation of the force/MMG relationship under static conditions and a neural network-based gesture classification. The results showed exponential relationships between acquired MMG and exerted force (for up to 60% of the maximal voluntary contraction) alongside good classification accuracy (94.3%) of eight hand motions based on MMG from a single-site acquisition at the forearm. The simplification of the MMG-based HMI interface in terms of spatial arrangement is rendered possible with the designed sensor.

Keywords: mechanomyography; piezoelectric sensor; vibration sensor; human-machine interface; prosthetic control; hand gesture recognition; convolutional neural network

Citation: Szumilas, M.; Władziński, M.; Wildner, K. A Coupled Piezoelectric Sensor for MMG-Based Human-Machine Interfaces. *Sensors* **2021**, *21*, 8380. https://doi.org/10.3390/s21248380

Academic Editor: Francesco Di Nardo, Valentina Agostini and Silvia Conforto

Received: 31 October 2021
Accepted: 13 December 2021
Published: 15 December 2021

Publisher's Note: MDPI stays neutral with regard to jurisdictional claims in published maps and institutional affiliations.

1. Introduction

Mechanomyography (MMG) is a measurement technique used to record muscles activity based on vibrations arising as an effect of muscle fibers mechanical contractions [1–3]. This technique is still less popular, especially in clinical applications, compared to electromyography (EMG). Nevertheless, the area of application of mechanomyography is relatively wide [1,2,4]: from human-machine interfaces (HMI), especially prosthetic devices control [5,6] and gesture recognition [7–9], to investigations of physiological principles of neuro-muscular system functions [10–12]. MMG does not require an electrical connection to the skin, therefore it may be applied without prior skin preparation in unconditioned environments, and the provided response exhibits low variability over time regarding the skin condition, as long as the sensor position is not adjusted [13,14]. Differences between the EMG and MMG signals may be observed, e.g., in their fatigue- and force-related responses, with some examples of MMG showing greater sensitivity than EMG when changes in muscle activation strategies are examined [15,16]. The determinants of proper MMG implementation are actively investigated, among others, in order to eliminate crosstalk from neighboring muscles, improve repeatability and signal-to-noise ratio of acquired signals [17–19].

Mechanical vibrations originating from muscle activity can be converted to electrical signal via various types of transducers: accelerometers, microphones, or laser distance sensors [2,14,20]. The main differences are in the methods of ensuring a stable coupling between the sensor and the muscle, which directly affect the achieved frequency response of the MMG setup. Accelerometers are often used for MMG recording due to their low mass, small dimensions, and availability of sensors with integrated signal conditioning and digital output. Among microphones, either contact or non-contact ones might be used.

For low-cost design, piezoelectric discs in the role of contact microphones sometimes are the choice because of their wide availability, good sensitivity, and very low price. They might be placed directly on the surface of the skin, above the muscle of interest [21,22]. However, with minimal effort, the way they are used can be modified to obtain signals several times higher in amplitude. Piezoelectric elements may be used in a way to benefit from their strong bending, such as in the case of diaphragm piezoelectric microphones or flexural mode accelerometers. Commercial contact microphones based on such an approach have also been used to record mechanomyographic signals previously [2,23,24].

When the myographic signals are to be used for HMI, one of the essential areas of optimization is the number and placement of the sensors used. The interface must be reliable and intuitive but cost-effective at the same time [25]. This enforces endeavours to improve sensing quality, looking for multimodal approaches and new sensors configurations as well. Coupling sensors together is one possible approach, e.g., Silva and Chau [26] coupled microphone and accelerometer, within one sensor, for effective noise reduction in MMG recording; Gregori et al. developed a combined EMG/MMG sensor, which provided improvement, especially in artifact rejection [27]; similarly, Fukuhara and Oka [28] and Wolczowski et al. [5] used a hybrid EMG/MMG sensor. Zhang et al. [29] showed that an introduction of MMG signals might significantly improve the performance of an EMG-pattern recognition-based prosthetic control.

Here, we present a new MMG sensor design, which consists of two coupled piezoelectric discs in a single housing so that two complementary signals related to the muscle activity may be measured simultaneously in a single site. Such an approach brings simplification of the MMG interface in terms of transducers' spatial distribution.

The sensor functionality has been verified in two typical MMG applications: (a) for estimating a force/MMG relation under isometric conditions and (b) for a neural network-based motion recognition, as in [30,31]. To provide a reference to surface EMG (sEMG), a measurement technique widely accepted in clinical practice, the relationship between MMG and sEMG signals recorded simultaneously during isometric measurements is shown.

2. Materials and Methods

All the data were collected from a single subject (male, 34 years old), who was a member of the research team, free of neuromuscular diseases and prior musculoskeletal injuries.

The proposed MMG sensor consists of two piezoelectric discs in a single housing (Figure 1). The discs are excited by the MMG signal from the skin's surface by the direct and indirect transmission of mechanical waves. The direct transmission is achieved by pressing the external disc directly to the area of interest on the skin. The indirect transmission is achieved with a coupler. The coupler consists of a base ring, which supports the external disc at its edge, and a hollow pin that contacts the central area of the internal piezoelectric disc, which is supported at four evenly distributed points. Therefore, the internal disc is working in a bending mode. The coupler is supported on flexible hinges, which allow for the sensor's self-alignment while being attached to the subject. All the mechanical elements of the proposed sensor are 3D-printed with a PLA material (Easy PLA, Fiberlab, Brzezie , Poland) in a fused deposition modeling process. The weight of the sensor is 5.5 g without wires. The photographs of the manufactured sensor are provided in Figure 2. The dimensions of the sensor are determined mainly by the diameters of the piezoelectric discs. Due to the different mechanical interfaces between the skin and both discs, these are intended to exhibit different excitations. The resulting signal from each of them is a superposition of MMG from the acquisition site and a muscle activity coupled from the mounting strap.

The sensor was tested in two experimental setups. In the first one, the force/MMG relation under static conditions was examined with simultaneous acquisition of the sEMG signal. The second test comprised the verification of the sensor's suitability for the classification of hand motion. Both setups employed 20 mm diameter piezoelectric discs (from unspecified manufacturer).

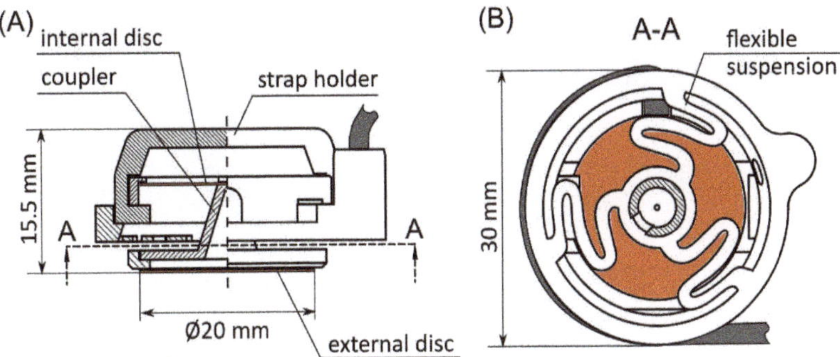

Figure 1. The designed mechanomyograpic (MMG) sensor: (**A**) side view and (**B**) section (A-A) through the pin of the coupler. The piezoelectric discs are marked with an orange color.

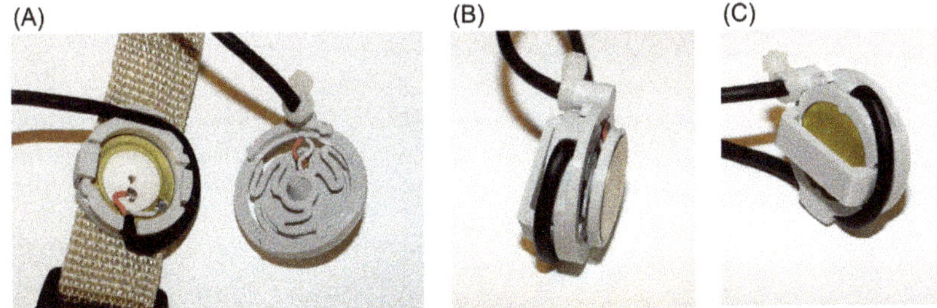

Figure 2. (**A**) The sensor before its final assembly. Two-core shielded microphone cables are used for signal transmission. (**B**) Side view of the assembled sensor with the external piezoelectric disc visible. (**C**) Side view of the assembled sensor with the internal disc visible.

2.1. Evaluation of the Sensor Performance during Step and Ramp Isometric Contractions

All the measurements were performed under static conditions during a single session. The right arm of the subject was fixed in the measurement setup in 90° abduction, 0° flexion with 90° elbow joint flexion. Forearm was fixed in 0° pronation. The measurement setup was based on TAS606 (HT Sensors Technology, Xi'an, China) load cell. For force signal conditioning TBM4 Transbridge amplifier (WPI, Sarasota, FL, USA) was used.

The MMG sensor was placed above the middle part of the biceps brachii belly. A surface EMG signal was acquired with a pair of wet electrodes (Kendall H92SG, Medtronic, Minneapolis, MN, USA) placed along the muscle and symmetrically with respect to the MMG sensor, 70 mm apart. Since sensor–skin interfaces of low quality are expected in real-world HMI applications, no skin preparation was performed. The reference electrode was placed in equal distance from both active electrodes (around 40 mm). Additionally, metal parts of the setup were grounded for interference noise suppression.

For signals amplification and data acquisition a custom made 8-channel MMG/EMG amplifier was used with active headstage × 100 for EMG probes and high impedance channels for the MMG transducers. For both sensors types, filtering was performed in the signal chain, specifically:

- EMG signals were high-pass (HP) filtered at the pre-amplification stage followed by an AC coupling, resulting in a 2nd order filter with 2.44 Hz cut-off frequency; the low-pass (LP) filtering was introduced with a 2nd order Sallen-Key topology with 245 Hz cut-off frequency;
- MMG signals were HP filtered with an AC coupling, which provided 1st order filter with 1.54 Hz cut-off frequency; the LP filter design was the same as for EMG signals, i.e., a 2nd order filter with 245 Hz cut-off frequency.

Signals were digitized using a 12-bit analog-to-digital converter of an STM32L476RG microcontroller (STMicroelectronics, Geneva, Switzerland), sampled at 500 sps from the range $0 \div 3.3$ V. The digitized signals were transmitted using a serial wired connection to the computer and stored with appropriate labeling using a custom-made LabVIEW (NI, Austin, TX, USA) application.

During measurements the subject used a biofeedback provided by means of a scope (MSO2012B, Tektronix, Beaverton, OR, USA) where the force signal along with the line indicating the target force level (for step measurements) were displayed.

The target force levels were determined in relation to the maximal voluntary contraction (MVC). MVC measurements took place approximately 10 min prior to the step and ramp measurements and were performed with the same measurement setup, with the same limb position as during the following procedures. Two short ($\sim$3 s) maximal contractions, with around 3 min of rest to limit fatigue, were recorded on the scope. For each trial, a mean value of the plateau was estimated. Finally, the maximum from two measurements was considered as MVC.

The step measurements were performed for 0%, 20%, 40%, 60%, and 70% of MVC. For every step, the contraction was sustained for 30 s with around 90 s of rest between subsequent steps. In addition to the visual feedback, the subject was provided with voice information on the remaining trial time: every 5 s during the first 20 s after the required force level was reached and then every second during the last 10 s. The ramp measurements were performed by a linear increase in exerted force from 0% to 70% of MVC, which took 50 s to complete. During the ramp measurements, the voice information was initially provided every 10 s and then every second during the last 10 s. An experienced investigator visually inspected compliance of the subject's force profiles with the respective force templates before the trials were assumed successful.

The data analysis for step and ramp isometric measurements was based on the instantaneous root mean square (RMS) values of the signals and performed using the R environment (ver. 3.4.4) [32]. For each of the acquired time series, the corresponding series of RMS values were calculated independently within subsequent 1-second long time windows (without overlapping), similarly for MMG and EMG signals. For step measurements, linear regression models were fit to the natural log-transformed values of RMS to quantify the relationships between the selected signals. The log-transformation of MMG RMS was introduced due to the reported non-linearities in the force-related MMG responses [15].

In the case of the step isometric measurements, data obtained during the experiment were used to evaluate (a) relationships between the response of each of the piezoelectric discs (internal and external) and force exerted by the muscle; (b) the relationship between responses from both of the piezoelectric discs; (c) the gain obtained at the internal disc with respect to the external disc; (d) the relationship between MMG and EMG data for both internal and external discs. Data recorded during the ramp isometric measurement were used to (a) visualize the relationship between EMG and force; (b) visualize the MMG/EMG relationship.

2.2. Classification Task

The piezoelectric discs were connected to charge amplifiers (custom-made amplifiers, with the feedback capacitor value set to correspond closely to the piezoelectric disc capacitance, which was measured to be 13 nF). Signals at the outputs of the amplifiers were digitized using the same setup as in the evaluation of the sensor performance during step and ramp isometric contractions.

For the classification of hand positions and gestures, the signals were recorded from the proximal part of an unsupported forearm of a seated subject. The elbow remained bent at 90 degrees during measurements. The evaluated MMG sensor was placed above the extensor carpi radialis longus (ECRL) muscle, providing two signals: from internal (ECRL_int) and external (ECRL_ext) discs. A second MMG sensor, having a structure similar to the proposed sensor but without the external disc, was placed over the flexor carpi ulnaris muscle, thus providing a single signal from the internal disc (FCU). Both sensors shared a non-elastic, woven polyester strap that held them in place after initial tightening. The location of the sensors on the forearm is marked in Figure 3. The hand motions used in this task are listed in Table 1 and illustrated in Figure 4.

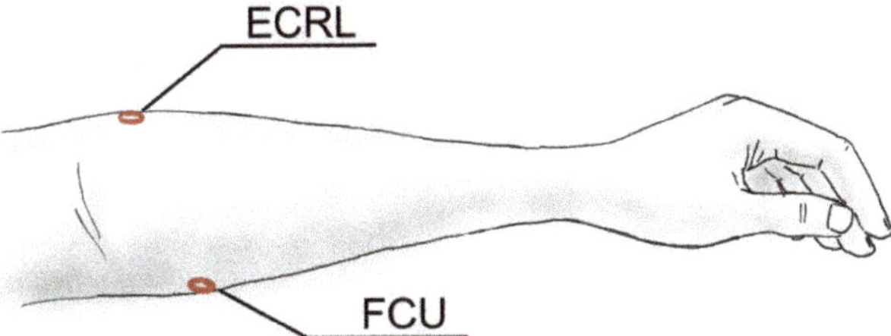

Figure 3. The locations of the sensors on the forearm during acquisition of the signals for the classification tasks. Labels: ECRL—externsor carpi radialis longus , FCU—flexor carpi ulnaris.

Table 1. The hand motions (positions and gestures) used in the classification tasks and their respective labels.

Motion	Label
Positions:	
Normal wrist and hand position	WrNorm
Extended wrist	WrExtHeld
Flexed wrist	WrFlexHeld
Clenched fist	WrFist
Gestures:	
From normal position to extended wrist	WrExt
From extended wrist to normal position	WrDeExt
From normal position to a flexed wrist	WrFlex
From flexed wrist to normal position	WrDeFlex

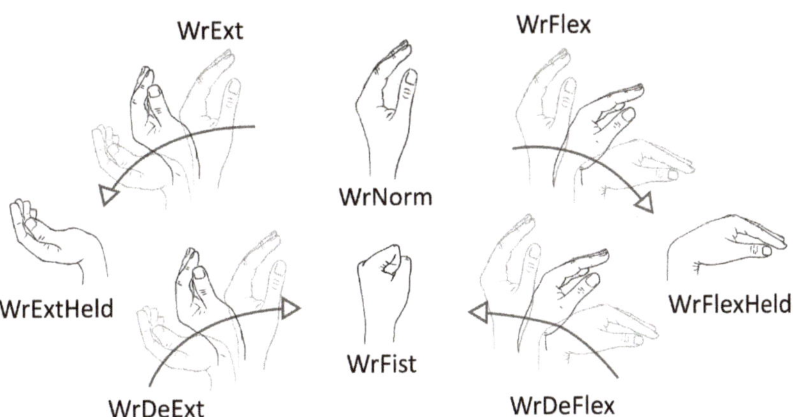

Figure 4. Illustration of the hand motions (positions and gestures) used in the classification tasks and their respective labels.

The subject was free to decide on the order of the motions executed during the measurement session. The proposed gestures are transitions between the classified positions and were included because when the online data processing is anticipated, these transitions cannot be avoided and will probably become misclassified if omitted in the training set. The time of each acquisition was fixed to 3 s, and a counting down timer was displayed to the subject to allow synchronization of the motion start with the acquisition window. The received signals were visually verified before saving to allow for discarding data if they were corrupted.

The Pearson's correlations r between the signal pairs were calculated to evaluate their mutual relationships for each recording. The analysis was extended with the derived signals that had their low-frequency components additionally suppressed by filtering the raw signals in the digital domain with a 5th order Butterworth filter (5 Hz cut-off frequency). Subsequently, the distributions of Pearson's r were compared visually.

The classification was performed using a 1D convolutional neural network (CNN) in the R environment (ver. 3.4.4) [32] with a TensorFlow library [33] and R interface to Keras [34]. A structure of the proposed CNN is shown in Figure 5. It consisted of four convolutional layers, each followed by a max-pooling layer. The rest of the network comprises two dense layers preceded by dropout layers (with the dropout rate set to 0.5). ReLU activation function was used throughout the internal network layers. The final layer employed a softmax activation due to the multinomial classification. The network was trained and tested using all possible combinations of signals, i.e., complete set (ECRL_int, ECRL_ext, and FCU), signal pairs (ECRL_int–ECRL_ext, ECRL_ext–FCU, and ECRL_int–FCU), and individual signals. In total, 70% of the dataset was used for training and validation. A common random seed was set for each of the combinations to ensure the repeatability of the training process. The training data were augmented using a moving window for generating three crops of 1.8 s from each original 3-s recording. A 5-fold, 5-times cross-validation was performed during training to support the choice of the number of training epochs. The final model was trained in the number of epochs which yielded the lowest value of mean loss + 1 standard error in the cross-validation phase. The overall accuracies achieved in the test set with the described CNN were calculated for each combination of signals. The confusion matrices were used for the assessment of per-class accuracies.

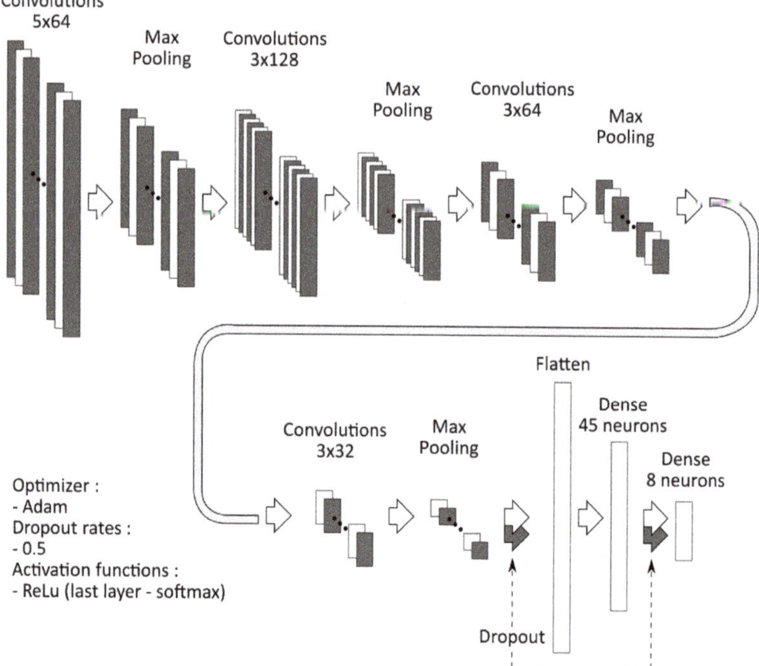

Figure 5. A structure of the convolutional neural network from the classification task.

3. Results

3.1. Evaluation of the Sensor Performance during Step and Ramp Isometric Contractions

The data were collected during a single session in a single trial for each of the intended contractions. Relationships between RMS and force values were modeled using the linear regression with log-transformed values of RMS (Figure 6A). The achieved R^2 coefficients were 0.953 and 0.971 for external and internal disc, respectively.

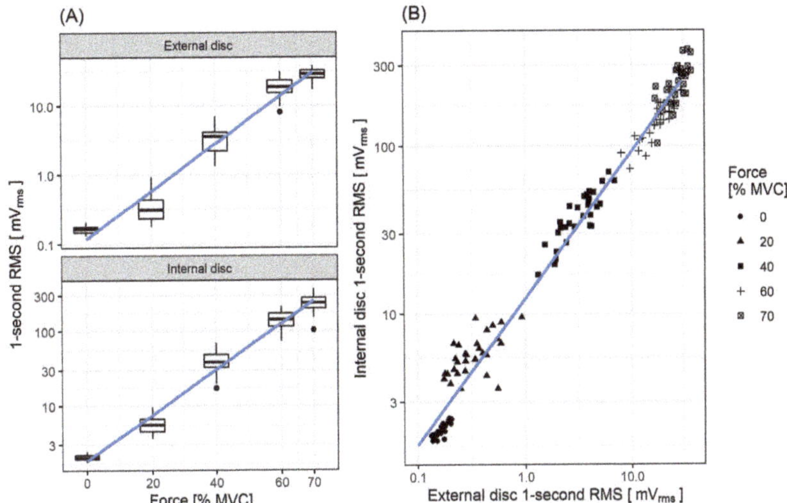

Figure 6. The root mean square (RMS) values of the subsequent 1-second parts (no overlapping) of MMG signals recorded during step isometric measurements. (**A**) RMS values for both of the sensor discs are plotted against the percentage of maximum voluntary contraction (MVC). The regression lines are fitted with log-transformed values of RMS. (**B**) RMS values of both signals are compared and a linear regression model in the log–log space is fitted.

The fitted relationships, after back-transformation of RMS variable to the linear scale, are as follows:

$$RMS_{ext} = 0.118 \, e^{0.0796 \cdot F} \tag{1}$$

$$RMS_{int} = 1.82 \, e^{0.0708 \cdot F} \tag{2}$$

where: F is the force given in [% MVC], RMS_{int} and RMS_{ext} are the RMS values of signals recorded from the internal and external piezoelectric discs, respectively, both given in [mV].

To estimate dependency between RMS values obtained for internal and external discs, a linear regression model was fitted ($R^2 = 0.983$) to the log–log transformed data (Figure 6B). Back-transformation of the model yields the relationship:

$$RMS_{int} = 12.4 \, RMS_{ext}^{0.873} \tag{3}$$

Based on Equations (1) and (2) the signal gain for $F \in [0, 70]$ and $RMS_{ext} \in [0.1, 30]$ was estimated as RMS_{int}/RMS_{ext}. The calculated gain/force relation is presented in Figure 7A. Similarly, based on Equation (3) the gain as a function of RMS_{ext} was calculated and presented in the Figure 7B, along with recalculated data-points.

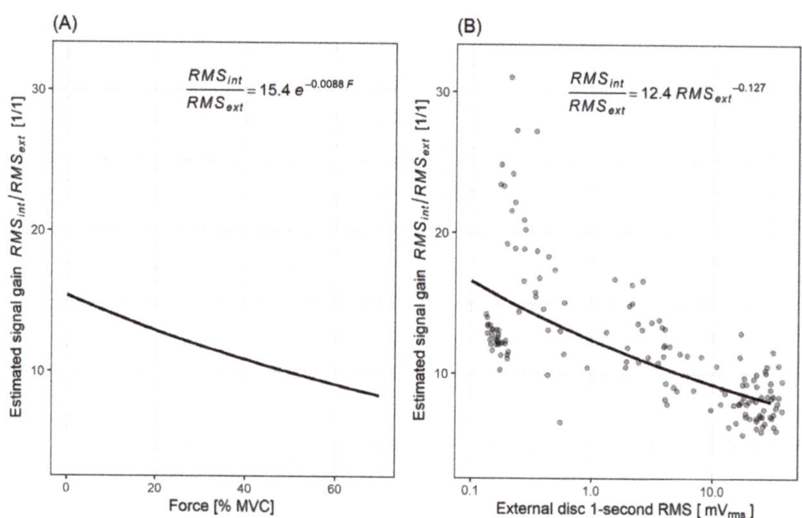

Figure 7. Estimated signal gain (RMS_{int}/RMS_{ext}) as a function of: (**A**) force (F), represented as % MVC, (**B**) RMS_{ext} (data-points recalculated based on experimental data).

In Figure 8 the relationship between EMG and MMG recorded from both external and internal piezoelectric discs during step isometric contractions is shown.

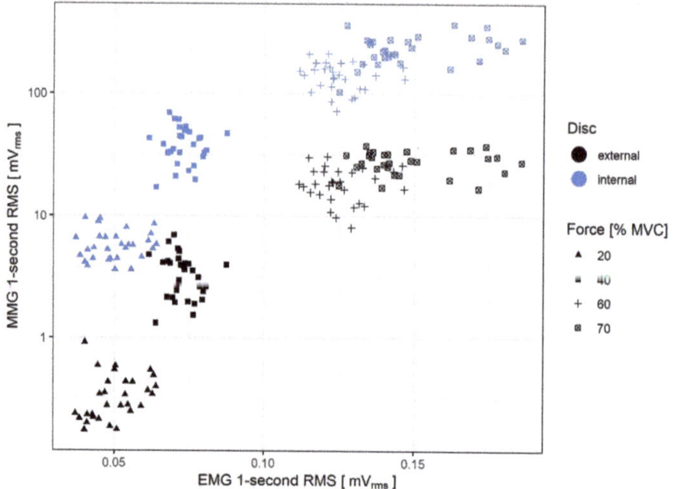

Figure 8. Comparison of MMG and EMG RMS values obtained during step isometric contraction measurement. The 0% MVC level is omitted, as it does not reflect muscle activity. The comparison is made for both internal (blue) and external (black) piezoelectric discs.

In Figure 9 the relationship between EMG and MMG recorded from both external and internal piezoelectric discs during ramp isometric contraction is shown.

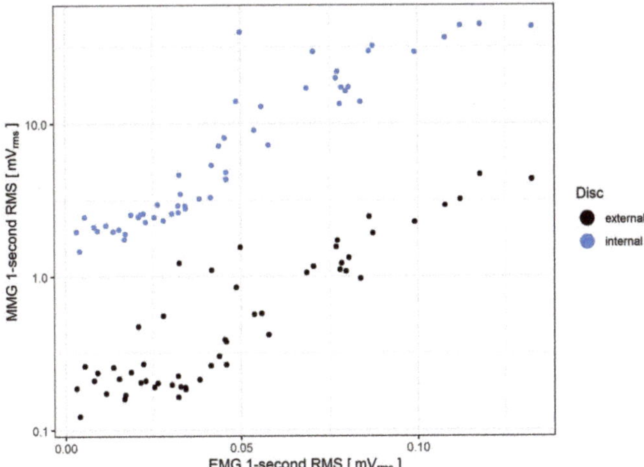

Figure 9. Comparison of MMG and EMG RMS values obtained during the ramp isometric contraction measurement. The comparison is made for both internal (blue) and external (black) piezoelectric discs.

Figure 10 illustrates the relationship between MMG and force signals obtained for the ramp isometric measurement.

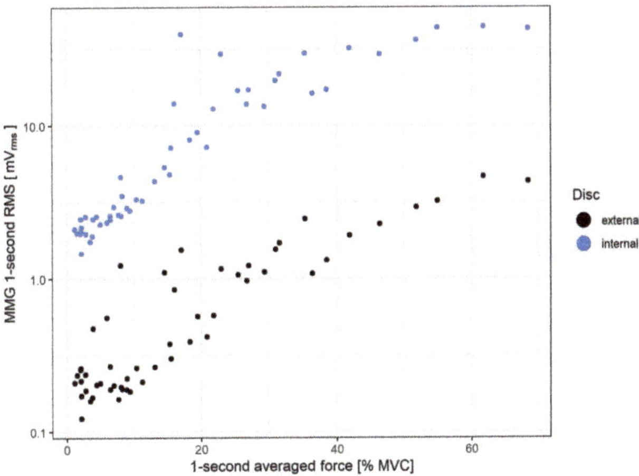

Figure 10. MMG as function of force exerted by the muscle during the ramp (from 0% to 70% MVC in 50 s) isometric contraction. Data are plotted both for internal (blue) and external (black) piezoelectric discs.

3.2. Classification Task

The data were collected during two sessions. For each classified motion, 35 repetitions were acquired, yielding in total 280 records. In each session, the measurements were taken from the left forearm. An example of signals acquired during the wrist de-flexion (WrDeFlex) is provided in Figure 11.

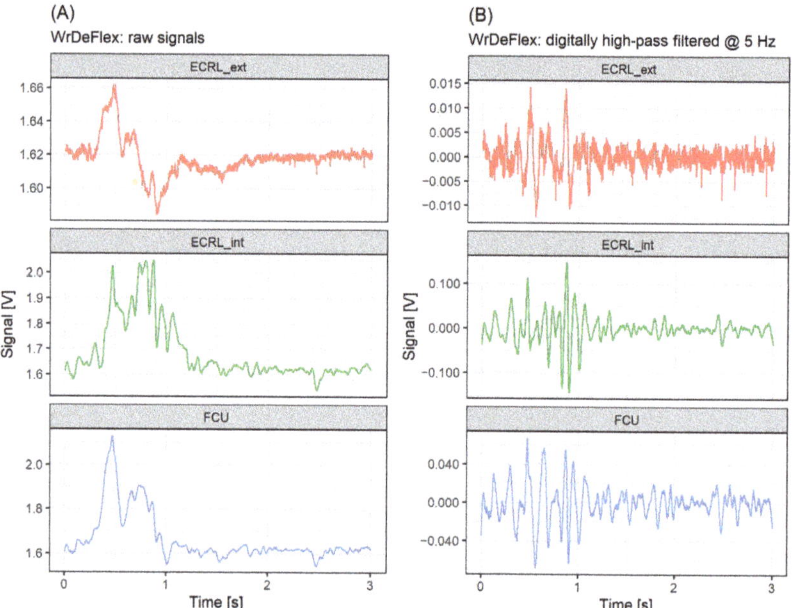

Figure 11. Example of signals acquired during the wrist de-flexion (WrDeFlex): (**A**) raw and (**B**) digitally high-pass filtered at 5 Hz. The signal sites abbreviations: ECRL_ext\ECRL_int–extensor carpi radialis longus, external\internal piezoelectric disc of the sensor, respectively; FCU—flexor carpi ulnaris, internal piezoelectric disc of the sensor.

The Pearson's correlations r between the signal pairs were calculated for each recording to evaluate their mutual relationships. The results for both raw and digitally filtered signals are shown in Figure 12.

For all motions, the correlations between the raw signals from the internal disc of the sensor placed above the ECRL and the internal disc of the sensor placed above the FCU (i.e., ECRL_int–FCU pair) are visibly high, and for most cases, r is between 0.8 and 1.0. The remaining two pairs of signals show relatively lower absolute correlations. Only in the case of gestures WrDeExt and WrExt their absolute r values approach those of the pair ECRL_int–FCU. However, for these gestures, correlations between the ECRL_int–ECRL_ext, and ECRL_ext–FCU pairs are negative.

For the digitally high-pass filtered signals, when compared to the raw signals, the main observed differences involve correlations between the ECRL_int–ECRL_ext, and ECRL_ext–FCU pairs for:

- WrDeExt and WrExt gestures, where absolute r values are diminished significantly;
- WrFlex gesture, where the distributions of r values are narrowed and shifted to the positive values.

For all motions, the ECRL_int–FCU pair correlations are slightly diminished but remain positive and relatively high for the filtered signals.

The accuracies achieved with the described CNN in the test set are presented in Table 2. Confusion matrices for all trained models are provided in Figure 13.

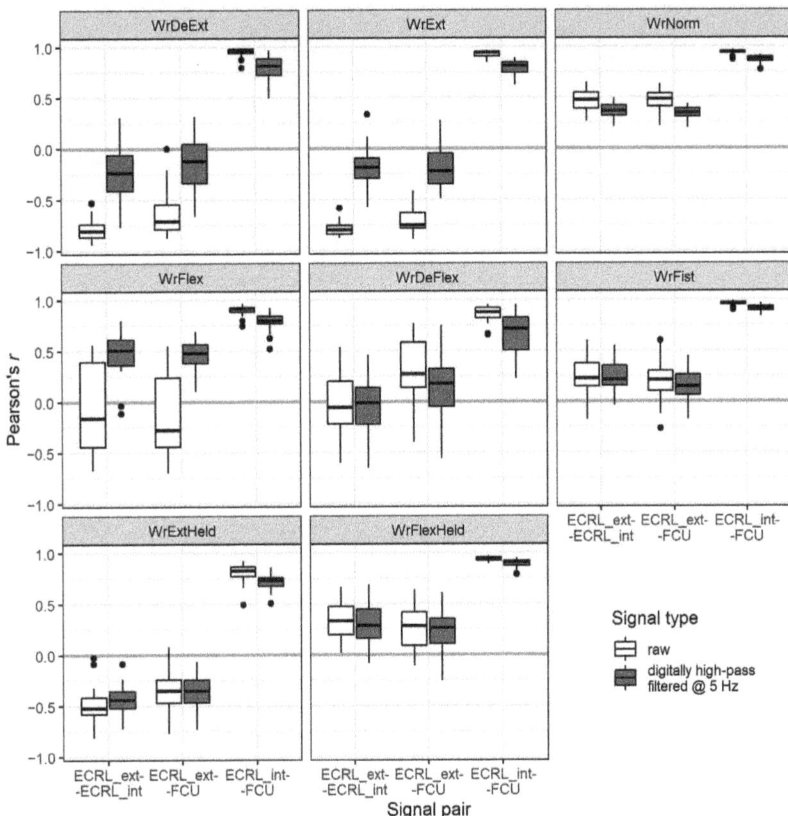

Figure 12. The Pearson's correlations between the signals, grouped by the type of executed motion (see Table 1 for the motion labels description and Figure 11 for signal sites abbreviations). The correlations are given for raw and digitally high-pass filtered signals with a 5 Hz cut-off frequency.

Table 2. Classification task results—test accuracies (see Figure 11 for the description of the signal abbreviations).

Signal Set	Cross-Validated Training Epochs	Overall Test Accuracy
ECRL_ext, ECRL_int, FCU	35	97.7%
ECRL_ext, ECRL_int	35	94.3%
ECRL_int, FCU	38	97.7%
ECRL_ext, FCU	35	94.3%
ECRL_ext	24	75.0%
ECRL_int	33	84.1%
FCU	39	89.8%

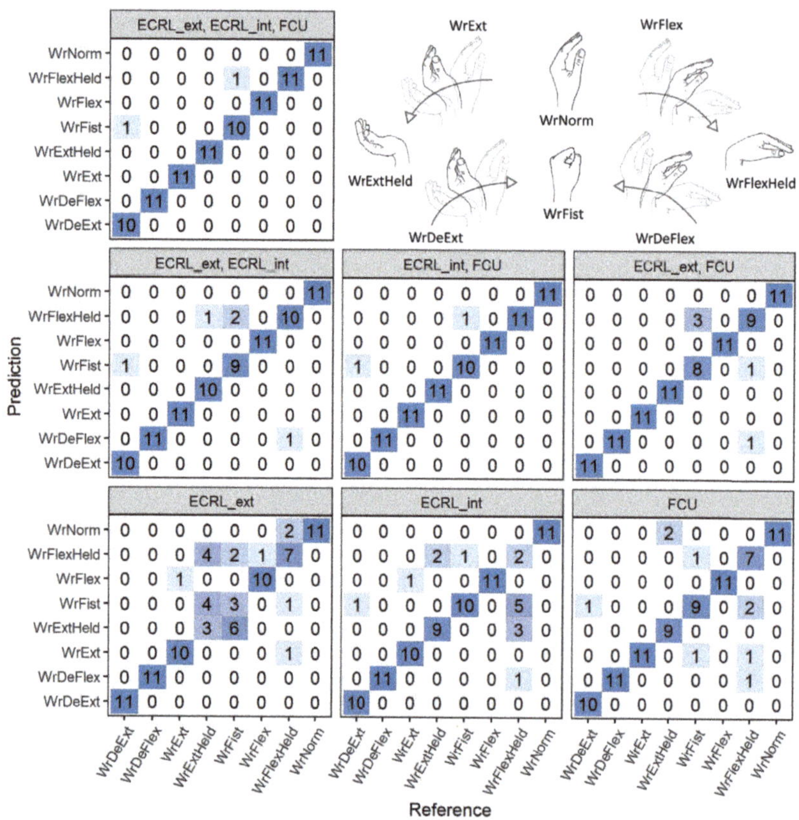

Figure 13. Confusion matrices for all trained models, headers of matrices indicate the signal set used (see Table 1 for the description of the motion labels and Figure 11 for the description of the signal labels). Illustrations of classified hand motions are recalled for convenience.

4. Discussion

4.1. Evaluation of the Sensor Performance during Step and Ramp Isometric Contractions

The MMG RMS (log-transformed) for both piezoelectric discs of the sensor exhibits a marked correlation with the force exerted by the muscle during step contractions, as observed for up to 70% MVC: R^2 coefficient equals 0.953 and 0.971 for the external and internal disc, respectively. However, as the measured force exceeds 60% MVC, the rate of the MMG RMS increase begins to decline for both the step and ramp measurements. Such a change is in line with findings from previous studies of the biceps brachii muscle isometric force generation, where plateaus or decreases of MMG amplitude were observed from approximately 60–80% MVC [10,35]. An interesting feature of the presented sensor is that for force levels below 20% MVC, the rate of MMG RMS increase is lower for the external disc than the internal one. This occurs likewise in step (Figure 6A) and ramp (Figure 10) contractions. Moreover, it has been previously documented that the MMG response to the force varies with the muscle fiber type composition and, for the ramp contractions, also depends on the type of muscle action (increase vs. decrease) [36]. Therefore, the sensor's usage in HMI applications exploiting the force/MMG relationship must account for differences in muscle composition, type of contraction (stable vs. ramp) and may be limited to the force values not exceeding 60% MVC, with additional preference for the internal disc signal if the force remains below 20% MVC. The observed MMG/sEMG relationship shows similar MMG RMS saturation from approximately 60% MVC, contrary to the sEMG RMS, which continues to rise with the exerted force (Figures 8 and 9).

Looking into the sensor-related details, MMG signal obtained from the internal disc has a considerably higher amplitude than the one acquired from the external disc. This is a feature that is beneficial for the design of signal chains, where it enables, for example, a replacement of the electronic amplification of a signal by a mechanical one, or, when combining measurements from two sensors having different gains and placed at a single site (as in the coupled sensor presented here), it extends the covered dynamic range. However, the achieved gain changes with the exerted force level from around 15 V/V for minimal force to around 8 V/V for 70% MVC. Nevertheless, one could observe a high correlation ($R^2 = 0.983$) between log–log transformed RMS of MMG signals from both piezoelectric discs. This should be expected, as both transducers are mechanically coupled. Thus, the advantage of having both discs at a single site is not obvious (beyond the mentioned possibility of extending the dynamic range), at least for operation under static conditions, when the signal's RMS is used. However, they should not be considered redundant for all static applications, as the back-transformation of the regression equation indicates a non-linear relationship (Equation (3)) between both signals. Depending on the signal processing method chosen, the mutual non-linearity at the system input may provide benefits when signals are superimposed.

4.2. Classification Task

The best-performing models in the classification task yielded accuracies in a range from 94.3% to 97.7%, similar to those reported in referenced works: up to 96% [31] and from 85.6% to 95.1% [30]. However, the direct comparison of results was not carried out due to differences in the sets of considered positions and gestures. In the presented setup, the gestures had the highest classification accuracies for all signal combinations, while hand positions were more often misclassified. The performance of CNN models was visibly lower for the 1-signal inputs with the rise in the misclassification rate for the following positions: flexed and extended wrist, clenched fist. The mentioned hand positions are inevitably more challenging to classify than gestures, as the latter yield significantly higher signal levels that result from the large change of muscle shape during the onset of the motion.

Analysis of the correlations of filtered and raw signals showed that when gestures were performed, a significant, low-frequency component below 5 Hz was present, which differentiated ECRL_ext from the rest of acquired raw signals, as shown in Figure 12. Thus, unlike for measurements made for isometric contractions, where combining signals from a single site (ECRL_ext and ECRL_int) should be possible and potentially beneficial, here it would result in a loss of information, and the usability of such a derived signal may vary depending on the characteristics of the considered gestures. However, signals from the ECRL_int and FCU sites were highly correlated when acquired during all chosen hand motions, which indicates their possible interchangeability. Results of the classification task also supported this conclusion, as the models that were based on the combinations: ECRL_ext–ECRL_int (single-site, 2-signal configuration) and ECRL_ext–FCU (double-site, 2-signal configuration), provided similar overall test accuracies. At the same time, the models based on double-site configurations (i.e., yielding 2- and 3-signal sets) provided accuracies better or equal to the single-site, 2-signal one. The single-site, 1-signal configurations resulted in the lowest classification accuracies.

We have shown that attaching the sensor to the examined limb with a strap allows for transmitting usable signals from the encompassed area to the sensor's internal piezoelectric disc. In other words, a source of potential crosstalk has been turned into a valid input to the classification algorithm. Therefore, the proposed coupled single-site sensor may be considered a replacement for the two-site arrangement. As a result, it may simplify the myographic interface in terms of sensors' spatial distribution. In our proof of concept, the substitution of the two-site sensor pair with a single-site sensor resulted only in a slight reduction in the classification accuracy (i.e., it decreased from 97.7% to 94.3%).

4.3. Observations on the Sensor Application

Based on the review of the acquired data, a few points regarding the sensor operation should be highlighted. These are general remarks that may be helpful when it comes to the implementation of the sensor in the MMG interface. The presented sensor design with a coupling pin that is free to move relative to the internal disc sporadically led to artifact generation due to the short-range repositioning of the pin contact point during large muscle motion. Such a movement resembles a stick-slip phenomenon and is observed as a narrow pulse, simultaneously in signals from both discs. However, these pulses occurred infrequently and likely did not pose noticeable issues during classification and did not introduce significant RMS analysis biases. The cardiac interference was the second source of a non-myographic signal, most apparent in the data recorded for stable hand positions. It may be treated as an auxiliary input for some applications and employed, e.g., for heart rate estimation. The next observed property stems from the difference in disc gains. Due to the higher gain achieved for the internal disc, its signal chain was prone to saturation when gestures were executed rapidly. Eventually, these had to be performed with a moderate speed, and acquisitions with saturation were discarded. However, the signal from the external disc had rarely saturated during faster motions. Even in such events, the sensor as a whole continued to provide a partially valid output. This feature may be beneficial for applications that involve both slow and fast movements, as the signal saturation may be easily detected and eventually managed in an algorithm or research protocol.

Another valuable feature of the designed sensor is that it may be manufactured from easily accessible and low-cost materials, and its assembly is not a demanding process. The 3D-printed parts required approximately 3000 mm^2 of a PLA material (1.22 m of 1.75 mm filament), with the cost of manufacturing (including electric energy and material) estimated as EUR 0.2. The price of a single piezo disc was approximately EUR 0.2. Without the wire and an optional connector, the cost of components of the proposed sensor is approximately EUR 0.6.

4.4. Limitations of the Study

As the data were collected from a single subject, no evaluation of an inter-subject variability of the sensor's performance was possible. Moreover, collecting more measurements from the same subject over an extended period would be beneficial to establish repeatability characteristics of the sensor. In this study, the subject was additionally a research team member. Therefore, he may be considered a skilled user of the developed sensor, which may introduce a bias resulting in increased performance if compared with a person having no or limited prior experience with similar instrumentation. Even though omitted in the proof of concept presented here, such analyses would be of significant importance for every specific application of the coupled MMG sensor, where more participants having different skill levels in terms of using the designed interface should be recruited.

5. Conclusions

We have confirmed that the simplification of the MMG interface in terms of transducers' spatial arrangement is possible with the designed sensor. Results of its evaluation under static and dynamic conditions support its potential for application in HMI.

Further investigation is desirable to identify the source of the differences between the vibration transmission to the internal and external disc. These differences seem to arise not only from the sensor's mechanical structure but to a large extent from the sensor-skin interface. A partially related issue of adipose tissue influence on MMG was discussed, e.g., by Santos et al. [18]. In subsequent studies aimed at refining the requirements of the application of the proposed sensor, one should consider measuring the skinfold thickness or the subcutaneous adipose tissue to characterize the application site better. In general, it would be beneficial to obtain a model that would make it possible to optimize the sensor characteristics depending on the application requirements. Furthermore, the sensor's performance should be tested with more muscle groups (which exhibit different recruitment of motor units and firing rates) and muscle actions (e.g., isokinetic contractions). Lastly, introducing a second myographic modality while preserving the proposed mechanical structure is worth considering, like in a work by Ke et al. [37], where a hybrid EMG-FMG sensor for prosthetic control was presented. For example, an sEMG–MMG integration may significantly reduce the EMG interelectrode distance, which had to be relatively large in our study due to the MMG sensor size.

Author Contributions: Conceptualization, M.S. and K.W.; methodology, M.S. and K.W.; software, M.S.; validation, M.S., M.W., and K.W.; formal analysis, M.S.; investigation, M.S. and K.W.; resources, M.S., M.W., and K.W.; data curation, M.S.; writing—original draft preparation, M.S. and K.W.; writing—review and editing, M.S., M.W., and K.W.; visualization, M.S.; project administration, K.W.; funding acquisition, K.W. All authors have read and agreed to the published version of the manuscript.

Funding: This research and APC were funded by the Dean of the Faculty of Mechatronics of the Warsaw University of Technology (with a grant for employees of the Faculty of Mechatronics of the Warsaw University of Technology for supporting scientific activity, grant number 504/04577/1142/44.000000).

Institutional Review Board Statement: The study was conducted according to the guidelines of the Declaration of Helsinki. Ethical review and approval were waived for this study, as the sensors used were non-invasive, non-significant risk devices, applied to a healthy adult person during an intervention brief in duration and harmless.

Informed Consent Statement: Informed consent was obtained from all subjects involved in the study.

Data Availability Statement: The data presented in this study are available on request from the corresponding author.

Conflicts of Interest: The authors declare no conflicts of interest. The funders had no role in the design of the study; in the collection, analyses, or interpretation of data; in the writing of the manuscript, or in the decision to publish the results.

References

1. Islam, M.A.; Sundaraj, K.; Ahmad, R.; Ahamed, N.; Ali, M.A. Mechanomyography Sensor Development, Related Signal Processing, and Applications: A Systematic Review. *IEEE Sens. J.* **2013**, *13*, 2499–2516. [CrossRef]
2. Beck, T.; Housh, T.; Cramer, J.; Weir, J.; Johnson, G.; Coburn, J.; Moh, H.; Mielke, M. Mechnomyographic amplitude and frequency responses during dynamic muscle actions: A comprehensive review. *Biomed. Eng. Online* **2005**, *4*, 67. [CrossRef]
3. Orizio, C. Muscle sound: Bases for the introduction of a mechanomyographic signal in muscle studies. *Crit. Rev. Biomed. Eng.* **1993**, *21*, 201–243. [PubMed]
4. Krueger, E.; Scheeren, E.M.; Nogueira-Neto, G.N.; da Silveira Nantes Button, V.L.; Nohama, P. Advances and perspectives of mechanomyography. *Braz. J. Biom. Eng.* **2014**, *8*, 384–401. [CrossRef]
5. Wolczowski, A.; Błędowski, M.; Witkowski, J. The System for EMG and MMG Singals Recording for the Bioprosthetic Hand Control. *J. Autom. Mob. Robot. Intell. Syst.* **2017**, *11*, 22–29._3-2017/25. [CrossRef]
6. Geng, Y.; Chen, L.; Tian, L.; Li, P. Comparison of electromyography and mechanomyogram in control of prosthetic system in multiple limb positions. In Proceedings of the IEEE-EMBS International Conference on Biomedical and Health Informatics (BHI 2012), Hong Kong, China, 5–7 January 2012, doi:10.1109/BHI.2012.6211702. [CrossRef]
7. Rajamani, Y.; Lam, C.K.; Sundaraj, K.; Zulkefli, N.F.B. Analysis and Classification of Multiple Hand Gestures using MMG Signals. *J. Telecommun. Electron. Comput. Eng.* **2018**, *10*, 67–71.
8. Donnarumma, M.; Caramiaux, B.; Tanaka, A. Muscular Interactions Combining EMG and MMG sensing for musical practice. In Proceedings of the International Conference on New Interfaces for Musical Expression, Daejeon, Korea, 27–30 May 2013.
9. Prociow, P.; Wolczowski, A.; Amaral, T.; Dias, O.; Filipe, J. Identification of Hand Movements based on MMG and EMG Signals. In Proceedings of the 1st International Conference on Bio-inspired Systems and Signal Processing, Funchal, Madeira, Portugal, 28–31 January 2008; Volume 2, pp. 534–539.
10. Talib, I.; Sundaraj, K.; Lam, C.K.; Sundaraj, S. A systematic review of muscle activity assessment of the biceps brachii muscle using mechanomyography. *J. Musculoskelet. Neuronal Interact.* **2018**, *18*, 446–462. [PubMed]
11. Ibitoye, M.; Hamzaid, N.; Zuniga, J.; Hasnan, N.; Abdul Wahab, A. Mechanomyographic Parameter Extraction Methods: An Appraisal for Clinical Applications. *Sensors* **2014**, *14*, 22940–22970. [CrossRef] [PubMed]

12. Guo, J.Y.; Zheng, Y.P.; Xie, H.B.; Chen, X. Continuous monitoring of electromyography (EMG), mechanomyography (MMG), sonomyography (SMG) and torque output during ramp and step isometric contractions. *Med. Eng. Phys.* **2010**, *32*, 1032–1042. 10.1016/j.medengphy.2010.07.004. [CrossRef]
13. Meagher, C.; Franco, E.; Turk, R.; Wilson, S.; Steadman, N.; McNicholas, L.; Vaidyanathan, R.; Burridge, J.; Stokes, M. New advances in mechanomyography sensor technology and signal processing: Validity and intrarater reliability of recordings from muscle. *J. Rehabil. Assist. Technol. Eng.* **2020**, *7*, 2055668320916116. [CrossRef]
14. Islam, M.A.; Sundaraj, K.; Ahmad, R.; Ahamed, N. Mechanomyogram for Muscle Function Assessment: A Review. *PLoS ONE* **2013**, *8*, e58902. [CrossRef] [PubMed]
15. Herda, T.J.; Housh, T.J.; Fry, A.C.; Weir, J.P.; Schilling, B.K.; Ryan, E.D.; Cramer, J.T. A noninvasive, log-transform method for fiber type discrimination using mechanomyography. *J. Electromyogr. Kinesiol.* **2010**, *20*, 787–794. [CrossRef] [PubMed]
16. Beck, T.W.; Housh, T.J.; Fry, A.C.; Cramer, J.T.; Weir, J.P.; Schilling, B.K.; Falvo, M.J.; Moore, C.A. The influence of muscle fiber type composition on the patterns of responses for electromyographic and mechanomyographic amplitude and mean power frequency during a fatiguing submaximal isometric muscle action. *Electromyogr. Clin. Neurophysiol.* **2007**, *47*, 221–232.
17. Castillo, C.S.M.; Wilson, S.; Vaidyanathan, R.; Atashzar, S.F. Wearable MMG-Plus-One Armband: Evaluation of Normal Force on Mechanomyography (MMG) to Enhance Human-Machine Interfacing. *IEEE Trans. Neural Syst. Rehabil. Eng.* **2021**, *29*, 196–205. [CrossRef]
18. Santos, E.; de Fatima Fernandes Vara, M.; Ranciaro, M.; Strasse, W.; Nunes Nogueira Neto, G.; Nohama, P. Influence of sensor mass and adipose tissue on the mechanomyography signal of elbow flexor muscles. *J. Biomech.* **2021**, *122*, 110456. [CrossRef] [PubMed]
19. Szumilas, M.; Lewenstein, K.; Ślubowska, E. Verification of the functionality of device for monitoring human tremor. *Biocybern. Biomed. Eng.* **2015**, *35*, 240–246. [CrossRef]
20. Islam, M.A.; Ahamed, N. Mechanomyography Sensors for Muscle Assessment: A Brief Review. *J. Phys. Ther. Sci.* **2012**, *24*, 1359–1365. [CrossRef]
21. Rodriguez-Labra, J.I.; Narakathu, B.B.; Atashbar, M.Z. Development of a Wireless Robotic Arm Control System Using Piezoelectric Sensors and Neural Networks. In Proceedings of the 2019 IEEE SENSORS, Montreal, QC, Canada, 27–30 October 2019; pp. 1–4. [CrossRef]
22. Booth, R.; Goldsmith, P. A Wrist-Worn Piezoelectric Sensor Array for Gesture Input. *J. Med Biol. Eng.* **2017**, *38*, 284–295. [CrossRef]
23. Watakabe, M.; Itoh, Y.; Mita, K.; Akataki, K. Technical aspects of mechnomyography recording with piezoelectric contact sensor. *Med Biol. Eng. Comput.* **1998**, *36*, 557. [CrossRef] [PubMed]
24. Bolton, C.F.; Parkes, A.; Thompson, T.R.; Clark, M.R.; Sterne, C.J. Recording sound from human skeletal muscle: Technical and physiological aspects. *Muscle Nerve* **1989**, *12*, 126–134. [CrossRef]
25. Ahmadizadeh, C.; Pousett, B.; Menon, C. Investigation of Channel Selection for Gesture Classification for Prosthesis Control Using Force Myography: A Case Study. *Front. Bioeng. Biotechnol.* **2019**, *7*, 331. [CrossRef] [PubMed]
26. Silva, J.; Chau, T. Coupled microphone-accelerometer sensor pair for dynamic noise reduction in MMG signal recording. *Electron. Lett.* **2003**, *39*, 1496 – 1498.:20031003. [CrossRef]
27. Gregori, B.; Galié, E.; Accornero, N. Surface electromyography and mechanomyography recording: A new differential composite probe. *Med. Biol. Eng. Comput.* **2003**, *41*, 665–669. [CrossRef] [PubMed]
28. Fukuhara, S.; Oka, H. A Simplified Analysis of Real-time Monitoring of Muscle Contraction during Dynamic Exercise Using an MMG/EMG Hybrid Transducer System. *Adv. Biomed. Eng.* **2019**, *8*, 185–192. [CrossRef]
29. Zhang, X.; Li, X.; Samuel, O.W.; Huang, Z.; Fang, P.; Li, G. Improving the Robustness of Electromyogram-Pattern Recognition for Prosthetic Control by a Postprocessing Strategy. *Front. Neurorobotics* **2017**, *11*, 51. [CrossRef] [PubMed]
30. Asheghabadi, A.S.; Moqadam, S.B.; Xu, J. Multichannel Finger Pattern Recognition Using Single-Site Mechanomyography. *IEEE Sens. J.* **2021**, *21*, 8184–8193. [CrossRef]
31. Esposito, D.; Andreozzi, E.; Gargiulo, G.D.; Fratini, A.; D'Addio, G.; Naik, G.R.; Bifulco, P. A Piezoresistive Array Armband With Reduced Number of Sensors for Hand Gesture Recognition. *Front. Neurorobotics* **2020**, *13*, 114. [CrossRef]
32. R Core Team. *R: A Language and Environment for Statistical Computing*; R Foundation for Statistical Computing: Vienna, Austria, 2018.
33. Abadi, M.; Barham, P.; Chen, J.; Chen, Z.; Davis, A.; Dean, J.; Devin, M.; Ghemawat, S.; Irving, G.; Isard, M.; et al. TensorFlow: A system for large-scale machine learning. *arXiv* **2016**, arXiv:1605.08695.
34. Allaire, J.; Chollet, F. *Keras: R Interface to 'Keras'*; R Package Version 2.4.0. 2021. Available online: https://keras.rstudio.com/ (accessed on 15 December 2021).
35. Beck, T.W.; Housh, T.J.; Fry, A.C.; Cramer, J.T.; Weir, J.P.; Schilling, B.K.; Falvo, M.J.; Moore, C.A. The influence of myosin heavy chain isoform composition and training status on the patterns of responses for mechanomyographic amplitude versus isometric torque. *J. Strength Cond. Res.* **2008**, *22*, 818–825.

[CrossRef] [PubMed]

36. Trevino, M.A.; Herda, T.J.; Fry, A.C.; Gallagher, P.M.; Vardiman, J.P.; Mosier, E.M.; Miller, J.D. The influence of myosin heavy chain isoform content on mechanical behavior of the vastus lateralis in vivo. *J. Electromyogr. Kinesiol.* **2016**, *28*, 143–151. [CrossRef] [PubMed]

37. Ke, A.; Huang, J.; Chen, L.; Gao, Z.; He, J. An Ultra-Sensitive Modular Hybrid EMG–FMG Sensor with Floating Electrodes. *Sensors* **2020**, *20*, 4775. [CrossRef] [PubMed]

Article

Does the Score on the MRC Strength Scale Reflect Instrumented Measures of Maximal Torque and Muscle Activity in Post-Stroke Survivors?

Pawel Kiper [1,*], Daniele Rimini [2], Deborah Falla [3], Alfonc Baba [4], Sebastian Rutkowski [5], Lorenza Maistrello [6] and Andrea Turolla [6,*]

1 Physical Medicine and Rehabilitation Unit, Azienda ULSS 3 Serenissima, 30126 Venice, Italy
2 Medical Physics Department-Clinical Engineering, Salford Care Organisation, Salford M6 8HD, UK; daniele.rimini@srft.nhs.uk
3 Centre of Precision Rehabilitation for Spinal Pain (CPR Spine), School of Sport, Exercise and Rehabilitation Sciences, University of Birmingham, Birmingham B15 2TT, UK; d.falla@bham.ac.uk
4 Rehabilitation Unit, Azienda Ospedale Università Padova, 35128 Padua, Italy; alfonc.baba@aopd.veneto.it
5 Faculty of Physical Education and Physiotherapy, Opole University of Technology, 45-758 Opole, Poland; s.rutkowski@po.opole.pl
6 Laboratory of Neurorehabilitation Technologies, San Camillo IRCCS, 30126 Venice, Italy; lorenza.maistrello@hsancamillo.it
* Correspondence: pawel.kiper@aulss3.veneto.it (P.K.); andrea.turolla@hsancamillo.it (A.T.)

Citation: Kiper, P.; Rimini, D.; Falla, D.; Baba, A.; Rutkowski, S.; Maistrello, L.; Turolla, A. Does the Score on the MRC Strength Scale Reflect Instrumented Measures of Maximal Torque and Muscle Activity in Post-Stroke Survivors? *Sensors* 2021, 21, 8175. https://doi.org/10.3390/s21248175

Academic Editors: Francesco Di Nardo, Valentina Agostini and Silvia Conforto

Received: 5 November 2021
Accepted: 5 December 2021
Published: 7 December 2021

Publisher's Note: MDPI stays neutral with regard to jurisdictional claims in published maps and institutional affiliations.

Abstract: It remains unknown whether variation of scores on the Medical Research Council (MRC) scale for muscle strength is associated with operator-independent techniques: dynamometry and surface electromyography (sEMG). This study aimed to evaluate whether the scores of the MRC strength scale are associated with instrumented measures of torque and muscle activity in post-stroke survivors with severe hemiparesis both before and after an intervention. Patients affected by a first ischemic or hemorrhagic stroke within 6 months before enrollment and with complete paresis were included in the study. The pre- and post-treatment assessments included the MRC strength scale, sEMG, and dynamometry assessment of the triceps brachii (TB) and biceps brachii (BB) as measures of maximal elbow extension and flexion torque, respectively. Proprioceptive-based training was used as a treatment model, which consisted of multidirectional exercises with verbal feedback. Each treatment session lasted 1 h/day, 5 days a week for a total 15 sessions. Nineteen individuals with stroke participated in the study. A significant correlation between outcome measures for the BB (MRC and sEMG $p = 0.0177$, $\varrho = 0.601$; MRC and torque $p = 0.0001$, $\varrho = 0.867$) and TB (MRC and sEMG $p = 0.0026$, $\varrho = 0.717$; MRC and torque $p = 0.0001$, $\varrho = 0.873$) were observed post intervention. Regression models revealed a relationship between the MRC score and sEMG and torque measures for both the TB and BB. The results confirmed that variation on the MRC strength scale is associated with variation in sEMG and torque measures, especially post intervention. The regression model showed a causal relationship between MRC scale scores, sEMG, and torque assessments.

Keywords: stroke; neurorehabilitation; EMG; MRC; dynamometer; strength

1. Introduction

The primary aim of the post-stroke rehabilitation process is to restore and maintain the patient's ability to perform actives of daily living. Importantly, this process starts within the first days after stroke and often continues over many years [1]. Because one of the most evident consequences of a cerebrovascular injury is hemiparesis, the rehabilitation process requires accurate assessment of residual muscle activity to define rehabilitative requirements [2,3]. Hemiparesis is associated with muscle weaknesses and inability to produce adequate muscle force for task execution. The ability to generate muscle force is determined by neural, muscular, and biomechanical factors. The contraction that is generated following depolarization of the nervous cells, which release calcium ions, giving start to the process of contraction in the main body of the muscle fiber [4], can be evaluated either in terms of muscular activity or the resultant muscular force.

The status or change of a clinical condition is traditionally assessed by completing assessment scales. The validity and reliability of these assessment scales are essential to monitor a patient's recovery and are critical for determining appropriate therapies. A common and widely accepted assessment scale for muscle strength is the Medical Research Council (MRC) scale [5,6]. This scale is commonly used to evaluate patients with stroke suffering from muscle weakness due to hemiplegia [7]. The MRC scale for muscle strength uses manual muscle testing to grade muscle strength, ranging from 0 to 5, according to the maximum force expected for a certain muscle [6]. The grades are as follows: 0 = No contraction, 1 = Flicker or trace contraction, 2 = Active movement, with gravity eliminated, 3 = Active movement against gravity, 4 = Active movement against gravity and resistance, and 5 = Normal power. A modified version of the scale takes into account the evaluation of range of movement (ROM), and the grades are as follows: 0 = No contraction, 1 = Flicker or trace contraction, 2 = Active movement, with gravity eliminated, 2–3 = Active movement against gravity over less than 50% of the feasible ROM, 3 = Active movement against gravity over more than 50% of the feasible ROM, 3–4 = Active movement against resistance over less than 50% of the feasible ROM, 4 = Active movement against resistance over more than 50% of the feasible ROM, 4–5 = Active movement against strong resistance over the feasible ROM, but distinctly weaker than the contralateral side, 5 = Normal power. The reliability of the MRC scale and its modified versions (mMRC) have been investigated [8]. Substantial inter-rater reliability of the MRC scale and the mMRC scale as well as intra-rater reliability of the MRC and the mMRC was observed for forearm muscle evaluation [8]. Additionally, the validity of the MRC scale has been confirmed, thus supporting its use in clinical practice. Jepsen et al. examined the inter-rater reliability of manual muscle tests of maximal voluntary strength and observed that reduced strength was significantly associated with the presence of symptoms; they suggested that manual muscle testing in upper limb disorders has diagnostic potential [9]. However, some limitations exist relating to the interpretation of the MRC scale, for example, the width of the MRC grades are unequal [10]. For instance, when testing elbow flexion strength, the MRC grades overlap between grades 3 and 4, indicating that the MRC grading may be unreliable in quantifying elbow flexion strength. Furthermore, excluding patients with either grade 0 or grade 5 decreases the reliability of the MRC Scale [8]. This may indicate that the assessment of the different grades of impairment can be more difficult than the assessment of a muscle that has no contraction at all or is evaluated with its maximal strength. An additional consideration is that Dupépé et al. suggested that the inter-observer reliability of the mMRC scale has discrepancy among trained observers. Additionally, the reliability of the MRC scale varies depending on whether lower-extremity or upper-extremity muscle groups are tested [11,12]. Importantly, however, it remains unknown whether variation of MRC scale scores is associated with a similar variation obtained with operator-independent techniques such as strength measures obtained with dynamometry and quantitative measures of muscle activity such as surface electromyography (sEMG).

Muscle force can be directly measured with a dynamometer, an electromechanical device that can be used to measure muscular strength of a maximum isometric contraction for most major joints in the human body [13,14]. Baschung Pfister et al. evaluated the reliability and validity of manual muscle testing and hand-held dynamometry (HHD) by measuring maximum isometric strength in eight muscle groups across three measurement points. The correlation between the total score on manual muscle testing and HHD was not satisfactory and raised doubt as to whether manual muscle testing measures the same construct (i.e., isometric strength) as HHD [15]. The total score from manual muscle testing was considered reliable and a time-efficient assessment to consider for the detection of general muscle weakness but not for single muscle groups. On the contrary, HHD could be recommended to evaluate isometric muscle strength of single muscle groups [15]. A study by Aguiar et al. revealed that dynamometry provided adequate inter- and intra-rater reliability when used in the subacute phase of stroke [16]. Additionally, recent studies evidenced the utility and the reliability of dynamometry to evaluate force of the paretic side of post-stroke patients [17]. The available literature reports that HHD is an efficient, objective, sensitive, and affordable alternative for strength quantitation [18].

The sEMG is a non-invasive technique for recording the electrical signal generated by muscular activity [19]. Decoding and extracting information contained in this signal provides information on neuromuscular function, which is not provided by other assessment techniques in neurorehabilitation [20]. This data can enhance the characterization of neuromuscular impairments, while tracking the changes in muscle activity from baseline when neurorehabilitation interventions are administered. Clinically, sEMG is frequently used to obtain a precise and objective measure of muscle activity during motor performance [21–25]. EMG is useful to assess hyperactivity and inactivity of selected muscles [26] and, given that it can be used to evaluate the integrity of neuromuscular system, it is often adopted as a physiological biofeedback in physical therapy [27]. In recent decades, the limitations of analyzing EMG have emerged, including physiologically confounding factors [28]. For this reason, pattern recognition techniques have been widely adopted to classify hand gesture [29–31], gait analysis [32], and upper limb prosthesis control [33–35]. The importance of integrating kinematics and kinetics has also been highlighted [36]. The generation of muscular force assessed by the MRC scale has been associated with the electrical signal observed via sEMG recordings [37]. Furthermore, some mathematical models of motor unit with a parameterization of the electrical and mechanical components of the model were proposed. These models can highlight a physiologically meaningful EMG–force relation in the simulated signals [38]. However, the relationship between muscular force and sEMG during voluntary contractions in pathological conditions (e.g., central nervous system injury) is still poorly understood [39,40].

Thus, in this study we examined whether the scores of the MRC scale are associated with instrumented measures of muscular force and muscle activity pre- and post- an intervention for severe hemiparesis in post-stroke survivors.

Proprioceptive-based training (PBT) was used as a treatment model; PBT is a neuromodulatory treatment modality that has been proposed for the treatment of the upper limb to recover voluntary muscle contraction and strength in stroke survivors [41].

2. Materials and Methods

2.1. Setting and Participants

This study was conducted in the neurorehabilitation hospital and research institute of San Camillo IRCCS (Venice, Italy). Inpatients affected by first ischemic or hemorrhagic stroke within 6 months before enrollment in the study and with an MRC score at baseline between 0 and 1 point for their biceps brachii and triceps brachii were included in this study. The presence of hypertonia, apraxia, global sensory aphasia, neglect, cognitive impairments, severe sensitivity disorders, stroke lesion located in the cerebellum, or refusal to participate resulted in exclusion from the study.

The local Ethics Committee of the IRCCS San Camillo Hospital approved this study (Protocollo 2012.07 BAT v.1.2), which was registered on ClinicalTrials.gov (NCT03155399). Informed, written consent was obtained prior to participation in the study.

2.2. Outcome Measures

The MRC scale for muscle strength, dynamometry measures of maximal elbow flexion and extension torque, and sEMG measures of biceps brachii and triceps brachii activity were implemented pre- and post-intervention. The positions of the upper extremity for dynamometry measurements and for the MRC scale assessment were the same. An elbow splint was used to standardize the position of the patient's arm during sEMG signal acquisition; the elbow joint was fixed to 40° for assessment of the biceps brachii and 90° for the triceps brachii.

2.2.1. MRC Scale for Muscle Strength

Testing was performed by a physiotherapist after assessment of elbow range of motion. The physiotherapist ensured that the wrist flexors were not contracted when assessing biceps brachii and provided stabilization support with a hand placed above the patient's elbow when assessing triceps brachii. All patients firstly underwent an assessment of the biceps brachii followed by assessment of the triceps brachii. The assessment of biceps brachii was performed in the supine (or in sitting in the case of grade 2 or more) position with the forearm supinated and elbow flexed to approximately 45 degrees as the patient was asked to "bend your elbow" (Figure 1). The assessment of triceps brachii was performed in the sitting (or in prone in the case of grade 3 or more) position with the arm supported at 90 degrees of shoulder and elbow flexion [5,42] as the patient was asked to "straighten your arm". For both assessments, the patients performed three attempts and the best result was considered the outcome.

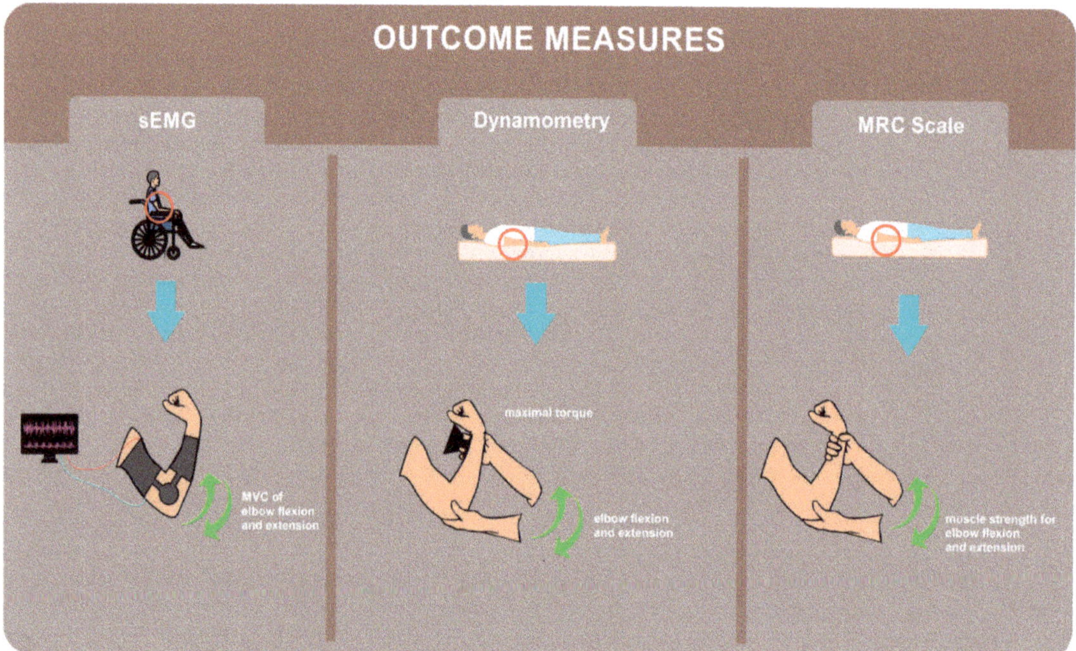

Figure 1. Visualization of the outcome measures applied. sEMG = surface electromyography. MVC = maximal voluntary contraction. MRC Scale = Medical Research Council Scale.

2.2.2. Dynamometry

An electrical dynamometer (CITEC Hand-Held Dynamometer) was used for testing. The participant's positions for assessment of maximal elbow flexion and extension torque were adopted from the MRC scale evaluation. The biceps brachii was assessed first in all participants. Patients were asked to perform three attempts with verbal encouragement to exceed the previous score and the mean value was considered for analysis.

2.2.3. Surface Electromyography

The sEMG was acquired with bipolar electrodes from the long head of the biceps brachii and the lateral head of the triceps brachii, according to published guidelines for electrode placement [43] after skin preparation. The sEMG signal was amplified with a gain of 1000, band-pass filtered (fifth-order Butterworth filter, bandwidth 10–500 Hz), and sampled at 2048 Hz using a multichannel EMG amplifier (EMG-USB2+ OT Bioelettronica SRL, Torino, Italy). The reference electrode was placed around the wrist of the tested arm. Each linear envelope of EMG activity was obtained by full-wave rectifying and then low-pass filtering (Fc = 6 Hz) for each sEMG channel.

The sEMG was acquired during maximal voluntary isometric contraction (MVC) of elbow flexion and extension, each repeated three times. The sEMG was recorded with the following procedure: recording of baseline activity at a resting state followed by the task itself (i.e., elbow flexion or extension MVC) recorded for 2 s each. The peak values of the amplitude of the envelopes of the sEMG of baseline and during the MVC were extracted, and the difference was computed. The mean value from the three repetitions was considered for further analysis.

2.3. Intervention

Participants underwent PBT, which consisted of multidirectional exercises executed synchronously with the unaffected limb and verbal feedback. Patients were asked to move both upper limbs synchronously performing bilateral flexion-extension at the level of their elbow joint. The PBT therapeutic session was divided into the following repetitive phases: proprioceptive stimulation for 3 min with a rest of at least 2 min between stimulations and repeated at least three times for each muscle. Additionally, all participants received individual exercises (passive, active-assisted, or active) for postural control in sitting or standing position. The training protocol lasted 1 h a day, 5 days weekly for a total of 15 sessions [41].

2.4. Statistical Analysis

Data distribution for all the variables was tested through the Shapiro–Wilk test. The Spearman's rank correlation test was used to study potential associations between the MRC scale score and measures of elbow flexion and extension strength and sEMG amplitude of the biceps brachii and triceps brachii muscles both pre- and post-intervention and on the change scores (before–after intervention). A regression model was implemented on the post intervention data to verify the relationship between the MRC strength scale scores and dynamometry measurements of elbow flexion and extension strength and sEMG amplitude of the biceps brachii and triceps brachii. We assessed the MRC models fitting as follows: the overall significance of the regression model with the percentage of variance explained (% Variance explained); the variance of the residuals (Residuals vs. Fitted plot); the normality of the residual distribution (Shapiro–Wilk normality test and Normal QQ-Plot); the presence of outliers (Residuals vs. Leverage plot). Bland–Altman graphs were reported to evaluate the agreement between the measurements made with MRC and those made with sEMG and dynamometry. The statistical significance level was set at $p < 0.05$. All calculations were performed using R Statistical Computing software.

3. Results

Data from 19 patients with a mean age of 61.48 ± 12.77 years (10 female and 9 male) were analyzed in this study. Patients' mean time from stroke onset was 3.19 ± 1.80 months. Twelve patients had ischemic stroke and seven had a hemorrhagic stroke (8 right and 11 left lesion side). Descriptive characteristics of the parameters measured before and after intervention are presented in Table 1.

Table 1. Pre- and post-intervention values.

Clinical Parameters	Before Mean ± SD (95% CI)	After Mean ± SD (95% CI)
MRC (No) (biceps brachii)	0.42 ± 0.51 (0.18–0.67)	2.37 ± 0.96 (1.91–2.83)
MRC (No) (triceps brachii)	0.21 ± 0.42 (0.01–0.41)	2.16 ± 0.90 (1.73–2.59)
Dynamometry (N) (biceps brachii)	4.11 ± 6.04 (1.19–7.02)	23.00 ± 15.89 (15.34–30.66)
Dynamometry (N) (triceps brachii)	2.05 ± 5.45 (−0.58–4.68)	23.68 ± 18.93 (14.56–2.81)
sEMG (mV) (biceps brachii)	7.15 ± 8.89 (2.42–11.88)	40.04 ± 41.43 (17.09–62.98)
sEMG (mV) (triceps brachii)	2.04 ± 2.4 (0.71–3.37)	34.5 ± 43.16 (10.59–58.41)

Values are expressed as mean ± standard deviation (SD); sEMG, surface electromyography; MRC, Medical Research Council scale; No, points; N, newtons; mV, millivolts.

A statistically significant relationship between the outcome measurements was observed pre-intervention between the MRC scale score and dynamometry measures (biceps brachii $p = 0.0000$; triceps brachii $p = 0.0002$) (Table 2), whereas, post-intervention, the MRC scale score was significantly associated with measures of sEMG and dynamometry measures for both biceps brachii (i.e., MRC and sEMG $p = 0.0177$; MRC and Dynamometry $p = 0.0001$) and triceps brachii (i.e., MRC and sEMG $p = 0.0026$; MRC and Dynamometry $p = 0.0001$) (Table 2).

Table 2. Correlation between the MRC scale score and sEMG amplitude and Dynamometry measures.

Clinical Parameters	Before		After		Δ	
	ϱ	*p*-Value	ϱ	*p*-Value	ϱ	*p*-Value
sEMG (Biceps Brachii)	0.342 [A]	0.1953	0.601 [A]	0.0177 *	0.453	0.0898
Dynamometry (Biceps Brachii)	0.954 [A]	0.0000 *	0.867 [A]	0.0001 *	0.795	0.0000 *
sEMG (Triceps Brachii)	0.178 [B]	0.5267	0.717 [B]	0.0026 *	0.677	0.0079 *
Dynamometry (Triceps Brachii)	0.749 [B]	0.0002 *	0.873 [B]	0.0001 *	0.795	0.0000 *

ϱ, correlation coefficient; *, *p*-value < 0.05; sEMG, surface electromyography; MRC, medical research council scale; Spearman's rank correlation Test; A, MRC biceps brachii; B, MRC triceps brachii.

A generalized regression model was used to study the relationship between the MRC scale scores, sEMG amplitude, and dynamometry measures of maximal elbow flexion and extension torque. The regression model showed that an increase of muscular strength by one point on the MRC scale was related to an increase of 59 mV (millivolts) of biceps brachii sEMG amplitude (% of explained variance = 0.50, Figure 2) and 83 mV for the triceps brachii sEMG amplitude (% of explained variance = 0.31, Figure 3). Moreover, the results revealed that a one-point increase on the MRC scale evaluation corresponded to an increase of 20 N (newtons) of elbow flexion torque measured with dynamometry (% of explained variance = 0.70, Figure 4) and 24 N of elbow extension torque (% of explained variance = 0.76, Figure 5).

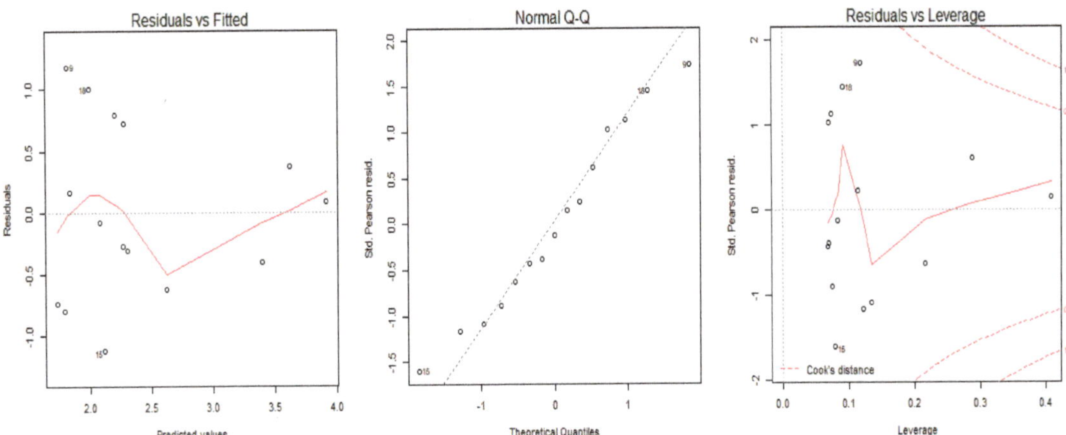

Figure 2. Graphical residual analysis for the MRC model and sEMG for the biceps brachii muscle. The plot on the left shows the residual errors versus their estimated values and the points on the graph should be arranged randomly. The QQ plot in the center shows the distributive normality of the residuals and the points on the plot should follow the diagonal line. The plot on the right identifies any influential data points on the model. In the plot, the Leverage's values of the points and the Cook's distances that measure the influence of each observation on the estimation of the model parameters are present. Cook's distance values greater than 1 are suspect and indicate the presence of a possible outlier or poor model.

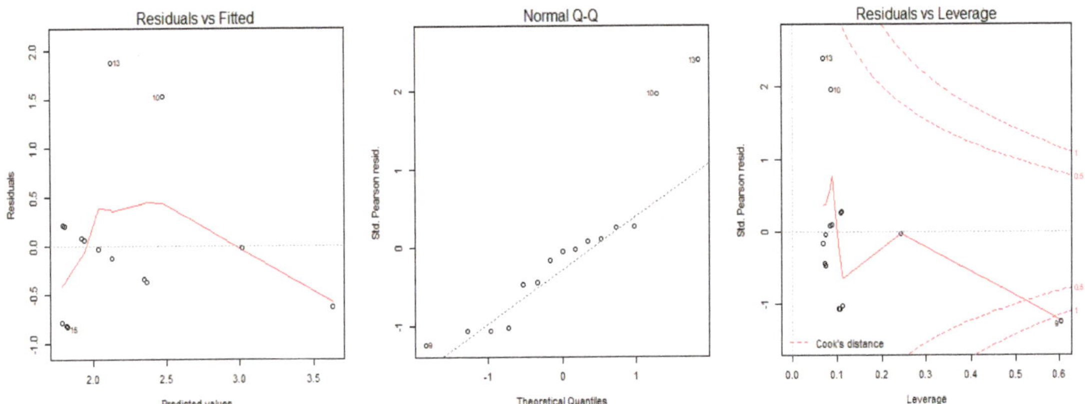

Figure 3. Graphical analysis of residuals for the MRC and sEMG model for Triceps Brachii muscle. The plot on the left shows the residual errors versus their estimated values; the QQ-plot in the center shows the distributive normality of the residuals; the plot on the right identifies any influential data points on the model.

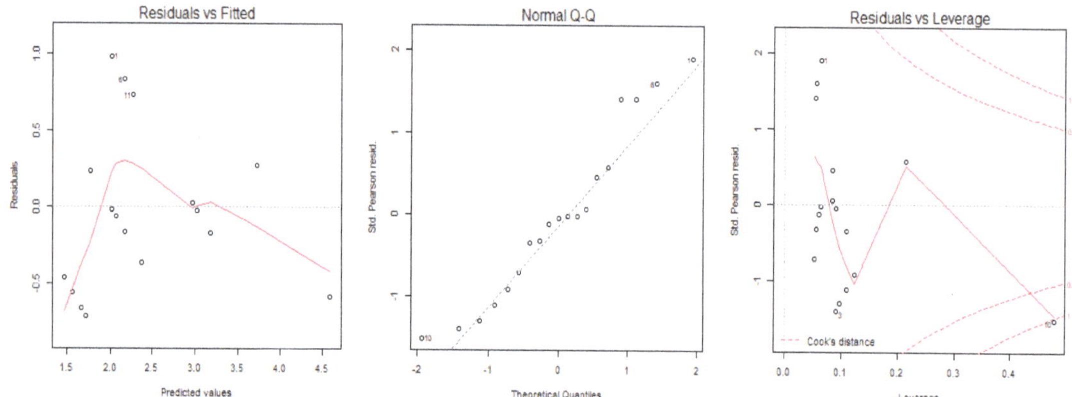

Figure 4. Graphical analysis of residuals for the MRC and Dynamometry model for Biceps Brachii muscle. The plot on the left shows the residual errors versus their estimated values; the QQ-plot in the center shows the distributive normality of the residuals; the plot on the right identifies any influential data points on the model.

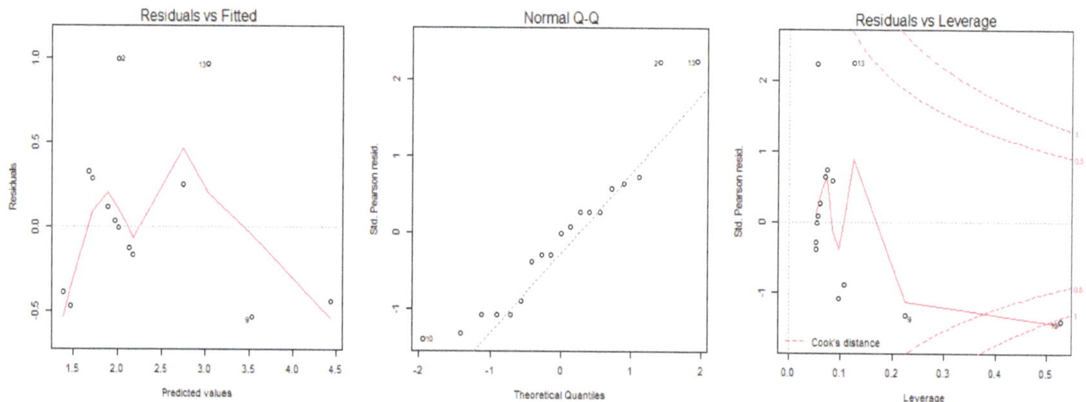

Figure 5. Graphical analysis of residuals for the MRC and Dynamometry model for Triceps Brachii muscle. The plot on the left shows the residual errors versus their estimated values; the QQ-plot in the center shows the distributive normality of the residuals; the plot on the right identifies any influential data points on the model.

The goodness of fit of the first and second models showed normal distribution of residuals, whereas the goodness of fit of the third and fourth models showed non-normal distribution (Table 3). The QQ Plot (Figures 2 and 3) and the Shapiro–Wilk normality test, performed on the two MRC biceps brachii models, confirmed the hypothesis of normality, both for the residuals of the model estimated with sEMG ($W = 0.96$ $p = 0.77$) and for those estimated with dynamometry ($W = 0.93$ $p = 0.17$). On the other hand, the goodness of fit carried out on the models of MRC triceps brachii did not have a normal distribution of residuals for the model estimated with sEMG ($W = 0.83$ $p = 0.009$) or for the model estimated with dynamometry ($W = 0.89$ $p = 0.03$). In all the Residuals versus Fitted graphs (Figures 2–5), the points on the graph were random and did not show any evident pattern, a sign that there was no residual systematic dependence not identified from the estimated model. Some Residuals versus Leverage plots (Figures 3–5) highlighted the presence of observations that could be considered outliers (they exceed the dotted line of Cook's distance) and had an influence on the model estimation as the high Leverage values suggested. Furthermore, the Bland–Altman plots highlight the presence of a linear decreasing dependence, thus excluding the presence of significant bias. These results showed that applied assessment tools (i.e., MRC, Dynamometer, EMG) were comparable; however, differences were also present (Figures 6 and 7).

Table 3. Relationship between the MRC scale scores, sEMG amplitude, and Dynamometry.

Regression Model	% Variance Explained	*p*-Value of Residuals
MRC (biceps brachii) = 0.017 · **sEMG** (biceps brachii)	0.50	$p = 0.766$
MRC (biceps brachii) = 0.050 · **Dynamometry** (biceps brachii)	0.70	$p = 0.165$
MRC (triceps brachii) = 0.012 · **sEMG** (triceps brachii)	0.31	$p = 0.009$ *
MRC (triceps brachii) = 0.041 · **Dynamometry** (triceps brachii)	0.76	$p = 0.033$ *

The outcomes are displayed with equation of the regression models. The model was estimated on post-intervention data. The Normality test was applied on model's residuals and significance was established at $p < 0.05$ *.

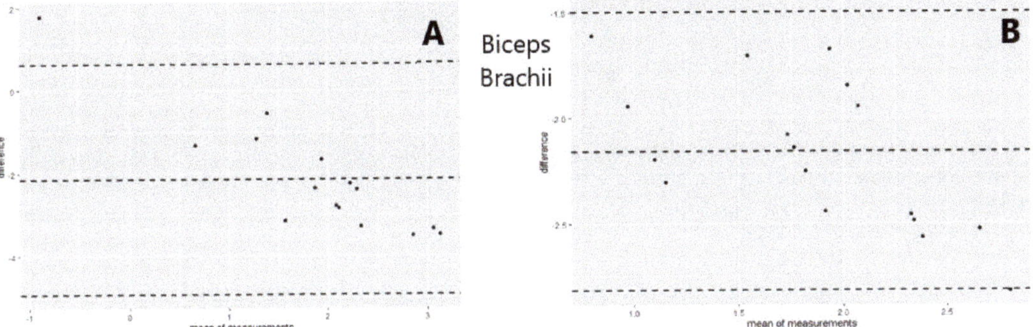

Figure 6. Differences between MRC scale and sEMG (**A**) as well as MRC and Dynamometry (**B**) for Biceps Brachii muscle versus the mean of the two measurements. The central line represents the mean difference (bias A = −2.16; B = −2.16), while the top and bottom lines represent the relative 95% CI (A = −4.93; 0.62) (B = −2.82; −1.50). The agreement between the measures is good when the differences are randomly distributed and fall within the 95% CI. The Bland–Altman plots highlight the differences of measurements performed with the two instruments. When the points (representing the observations) are scattered within the CI, the instruments can be used interchangeably. This mean that there are no significant differences between the measurements obtained from both instruments.

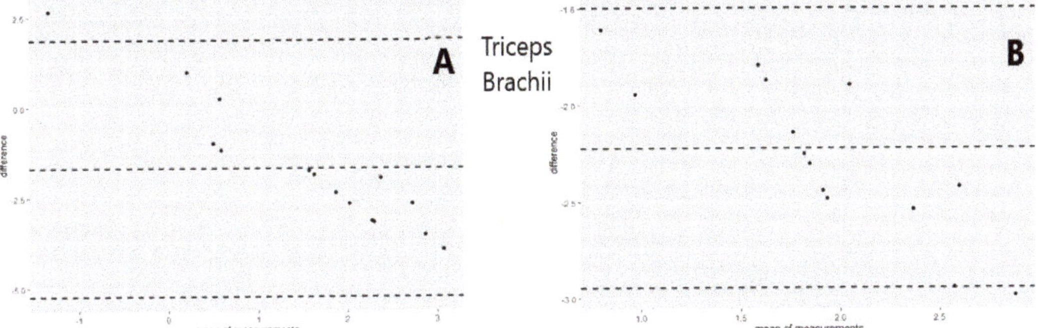

Figure 7. Differences between MRC scale and sEMG (**A**) as well as MRC and Dynamometry (**B**) for Triceps Brachii muscle versus the mean of the two measurements. The central line represents the mean difference (bias A = −1.60; B = −2.22), while the top and bottom lines represent the relative 95% CI (A = −5.23; 1.85) (B = −2.95; −1.49). The Bland–Altman plots highlight the differences of measurements performed with the two instruments. When the points (representing the observations) are scattered within the CI, the instruments can be used interchangeably. This mean that there are no significant differences between the measurements obtained from both instruments.

4. Discussion

The results of this study showed that for the post-intervention data, muscular strength measured by the MRC scale was correlated to both the amplitude of muscle activity measured by sEMG, as well as measures of maximal voluntary torque assessed with dynamometry. Thus, the clinical measure of muscle strength in patients with hemiparesis increased in accordance with the changes observed in sEMG and dynamometry measures. We also observed that the MRC scale score and dynamometry measures were correlated when examining the pre-intervention data, but a similar correlation was not present between the MRC scale score and sEMG measures pre-intervention. This is likely due to the severe weaknesses of the tested muscles pre-intervention, which mostly did not present an active voluntary contraction.

In a study by Deroide et al., which investigated patients with neuropathic conditions, the EMG at baseline and during a MVC were weakly but significantly correlated to the MRC score [44]. Considering that a possible association between EMG and muscular force may help in the assessment of various aspects of muscle physiology, the addition of sEMG measurements for the evaluation of changes in muscle function following an intervention could support the interpretation of the MRC scale scores. This may also help to overcome some of the limitations of the MRC scale; for example, the original MRC scale does not include ROM. Consider an example of a person in the acute phase after stroke who can flex his/herelbow to 30° and after 1month of rehabilitation can flex to 70°. An improvement of 40° of flexion ROM is likely to be functionally significant; however, in both assessments (i.e., baseline and after rehabilitation), a grade 2 in the MRC scale will likely be obtained. Indeed, other authors have suggested that the MRC grading system should not be the sole outcome evaluation for elbow flexion strength, and quantitative measurements, such as using a dynamometer, should be included for outcome comparisons [45]. The results of our study suggest that inter-instrumental variation in muscle strength assessment can partly supplement the MRC scale outcome.

Direct measurement of muscle force using sEMG is not possible, and, although some studies have reported a linear relationship between force and sEMG amplitude [46], several others suggest a non-linear relationship [47]. In the current study, the regression model showed a linear relationship between the MRC scale score and sEMG amplitude as well as between the MRC scale score and dynamometry measures of elbow extension/flexion torque for triceps brachii and biceps brachii.The score on the MRC scale linearly increased with the amplitude observed during the sEMG acquisition and dynamometry assessment. This relationship supports the comparative outcome between the MRC scale and an instrumented assessment of muscle activity/torque during maximum voluntary contractions. Our findings provide new insights into the relationship between the measurements described above applied to plegic muscles resulting from central nervous system injury. This relationship does not explain exactly how much the muscle has recovered, but, from a clinical perspective, it can confirm the appropriateness of the interpretation of the applied MRC test. Despite the wide use of the MRC scale for strength assessment, this tool has been reported as not sufficiently sensitive and with limited accuracy to detect changes. Our results suggest that sEMG can be implemented for accurate assessment of post-stroke individuals when muscular force is evaluated. Thus, this may offer a more precise prediction of functional capabilities in patients with upper limb hemiparesis. The introduction of sEMG assessment can more easily detect and confirm muscle activity and/or residual force. This can be also helpful as a potential predictor of muscle force recovery. Collectively, our findings support the use of the MRC scale to evaluate changes in muscle strength and activity of the biceps and triceps brachii following rehabilitation in patients with severe hemiparesis.

There are some methodological considerations to note when interpreting the findings of this study. Firstly, we enrolled post-stroke patients with severe upper limb hemiparesis and, consequently, the presence of muscular fatigue and hypo-tone introduced non-linear distortions to the force–sEMG relationship, which may have limited this study [48,49]. A further consideration is that the catchment area of the electrode didnot extend sufficiently to detect the signal generated across the entire muscle volume.

Considering that the inter-rater reproducibility of the MRC scale had several limitations [50], future studies should also consider the correlation between MRC scores and instrumental assessments when data are collected from more than one assessor. Moreover, the residuals did not have a normal distribution for the triceps brachii and this may have been due to the small sample size of this study. Therefore, analysis of a larger sample and strength assessment of several muscle groups with both sEMG and dynamometry would provide a better understanding of the relationship between different methods of strength assessment and functional tasks. Another limitation is that we did not consider the potential effects of agonist–antagonist activation, which could have influenced the measures.

5. Conclusions

Variation in scores on the MRC scale was associated with variation in electromyographic activity as well as elbow torque measured with dynamometry. The findings of this study can be used to ensure more precise clinical assessments of patients with stroke.

Author Contributions: Conceptualization, P.K.; methodology, P.K. and A.B.; validation, L.M., D.F. and D.R.; formal analysis, P.K., L.M., A.B., A.T., D.F., D.R. and S.R.; investigation, P.K. and A.B.; resources, A.T.; data curation, P.K., L.M. and A.B.; writing—original draft preparation, P.K., D.R., L.M. and D.F.; writing—review and editing, D.F., A.T., A.B., P.K. and S.R.; visualization, D.R. and S.R.; supervision, A.T. and D.F.; project administration, P.K. All authors have read and agreed to the published version of the manuscript.

Funding: This research received no external funding.

Institutional Review Board Statement: The study was conducted according to the guidelines of the Declaration of Helsinki, approved by Ethical Committee of the IRCCS San Camillo Hospital (Protocollo 2012.07 BAT v.1.2), and registered on ClinicalTrials.gov (NCT03155399).

Informed Consent Statement: Informed consent was obtained from all subjects involved in the study.

Data Availability Statement: The data presented in this study are available on request from the corresponding author.

Acknowledgments: Authors would like to acknowledge Aneta Kiper, Simonetta Rossi, and Michela Agostini, for their support and assistance throughout the study.

Conflicts of Interest: The authors declare no conflict of interest.

References

1. Lui, S.K.; Nguyen, M.H. Elderly Stroke Rehabilitation: Overcoming the Complications and Its Associated Challenges. *Curr. Gerontol. Geriatr. Res.* **2018**, *2018*, 9853837. [CrossRef] [PubMed]
2. Li, Y.; Zhang, X.; Gong, Y.; Cheng, Y.; Gao, X.; Chen, X. Motor Function Evaluation of Hemiplegic Upper-Extremities Using Data Fusion from Wearable Inertial and Surface EMG Sensors. *Sensors* **2017**, *17*, 582. [CrossRef]
3. Kiper, P.; Szczudlik, A.; Mirek, E.; Nowobilski, R.; Opara, J.; Agostini, M.; Tonin, P.; Turolla, A. The application of virtual reality in neuro-rehabilitation: Motor re-learning supported by innovative technologies. *Med. Rehabil.* **2013**, *17*, 29–36. [CrossRef]
4. Kuo, I.Y.; Ehrlich, B.E. Signaling in muscle contraction. *Cold Spring Harb. Perspect. Biol.* **2015**, *7*, a006023. [CrossRef]
5. James, M.A. Use of the Medical Research Council muscle strength grading system in the upper extremity. *J. Hand Surg. Am.* **2007**, *32*, 154–156. [CrossRef]
6. Medical Research Council. *Aids to the Examination of the Peripheral Nervous System*; Pengragon House: London, UK, 1975.
7. Gregson, J.M.; Leathley, M.J.; Moore, A.P.; Smith, T.L.; Sharma, A.K.; Watkins, C.L. Reliability of measurements of muscle tone and muscle power in stroke patients. *Age Ageing* **2000**, *29*, 223–228. [CrossRef] [PubMed]
8. Paternostro-Sluga, T.; Grim-Stieger, M.; Posch, M.; Schuhfried, O.; Vacariu, G.; Mittermaier, C.; Bittner, C.; Fialka-Moser, V. Reliability and validity of the Medical Research Council (MRC) scale and a modified scale for testing muscle strength in patients with radial palsy. *J. Rehabil. Med.* **2008**, *40*, 665–671. [CrossRef] [PubMed]
9. Jepsen, J.; Laursen, L.; Larsen, A.; Hagert, C.G. Manual strength testing in 14 upper limb muscles: A study of inter-rater reliability. *Acta Orthop. Scand.* **2004**, *75*, 442–448. [CrossRef]

10. Dyck, P.J.; Boes, C.J.; Mulder, D.; Millikan, C.; Windebank, A.J.; Dyck, P.J.B.; Espinosa, R. History of standard scoring, notation, and summation of neuromuscular signs. A current survey and recommendation. *J. Peripher. Nerv. Syst.* **2005**, *10*, 158–173. [CrossRef]

11. Hermans, G.; Clerckx, B.; Vanhullebusch, T.; Segers, J.; Vanpee, G.; Robbeets, C.; Casaer, M.P.; Wouters, P.; Gosselink, R.; Van den Berghe, G. Interobserver agreement of medical research council sum-score and handgrip strength in the intensive care unit. *Muscle Nerve* **2012**, *45*, 18–25. [CrossRef] [PubMed]

12. Dupépé, E.B.; Davis, M.; Elsayed, G.A.; Agee, B.; Kirksey, K.; Gordon, A.; Pritchard, P.R. Inter-rater reliability of the modified Medical Research Council scale in patients with chronic incomplete spinal cord injury. *J. Neurosurg. Spine* **2019**, *18*, 515–519. [CrossRef] [PubMed]

13. Samosawala, N.R.; Vaishali, K.; Kalyana, B.C. Measurement of muscle strength with handheld dynamometer in Intensive Care Unit. *Indian J. Crit. Care Med.* **2016**, *20*, 21–26. [CrossRef]

14. Nitschke, J.E. Reliability of isokinetic torque measurements: A review of the literature. *Aust. J. Physiother.* **1992**, *38*, 125–134. [CrossRef]

15. Baschung Pfister, P.; de Bruin, E.D.; Sterkele, I.; Maurer, B.; de Bie, R.A.; Knols, R.H. Manual muscle testing and hand-held dynamometry in people with inflammatory myopathy: An intra- and interrater reliability and validity study. *PLoS ONE* **2018**, *13*, e0194531. [CrossRef] [PubMed]

16. Aguiar, L.T.; Martins, J.C.; Lara, E.M.; Albuquerque, J.A.; Teixeira-Salmela, L.F.; Faria, C.D. Dynamometry for the measurement of grip, pinch, and trunk muscles strength in subjects with subacute stroke: Reliability and different number of trials. *Braz. J. Phys. Ther.* **2016**, *20*, 395–404. [CrossRef] [PubMed]

17. Karthikbabu, S.; Chakrapani, M. Hand-Held Dynamometer is a Reliable Tool to Measure Trunk Muscle Strength in Chronic Stroke. *J. Clin. Diagn. Res.* **2017**, *11*, YC09–YC12. [CrossRef]

18. Kushner, J. *Encyclopedia of the Neurological Sciences*; Choice: Current Reviews for Academic Libraries; Academic Press: Cleveland, OH, USA, 2003; Volume 41, p. 687.

19. Merletti, R.; Parker, P. *Electromyography: Physiology, Engineering, and Noninvasive Applications*; IEEE Press/J Wiley: Hoboken, NJ, USA, 2004.

20. Gilmore, K.L.; Meyers, J.E. Using Surface Electromyography in Physiotherapy Research. *Aust. J. Physiother.* **1983**, *29*, 3–9. [CrossRef]

21. Rong, W.; Tong, K.Y.; Hu, X.L.; Ho, S.K. Effects of electromyography-driven robot-aided hand training with neuromuscular electrical stimulation on hand control performance after chronic stroke. *Disabil. Rehabil. Assist. Technol.* **2015**, *10*, 149–159. [CrossRef]

22. Cesqui, B.; Tropea, P.; Micera, S.; Krebs, H.I. EMG-based pattern recognition approach in post stroke robot-aided rehabilitation: A feasibility study. *J. Neuroeng. Rehabil.* **2013**, *10*, 75. [CrossRef]

23. Li, W.; Luo, Z.; Jin, Y.; Xi, X. Gesture Recognition Based on Multiscale Singular Value Entropy and Deep Belief Network. *Sensors* **2020**, *21*, 119. [CrossRef]

24. Gonzalez, S.; Stegall, P.; Edwards, H.; Stirling, L.; Siu, H.C. Ablation Analysis to Select Wearable Sensors for Classifying Standing, Walking, and Running. *Sensors* **2020**, *21*, 194. [CrossRef] [PubMed]

25. Diez, J.A.; Santamaria, V.; Khan, M.I.; Catalan, J.M.; Garcia-Aracil, N.; Agrawal, S.K. Exploring New Potential Applications for Hand Exoskeletons: Power Grip to Assist Human Standing. *Sensors* **2020**, *21*, 30. [CrossRef] [PubMed]

26. Cappellini, G.; Sylos-Labini, F.; Assenza, C.; Libernini, L.; Morelli, D.; Lacquaniti, F.; Ivanenko, Y. Clinical Relevance of State-of-the-Art Analysis of Surface Electromyography in Cerebral Palsy. *Front. Neurol.* **2020**, *11*, 583296. [CrossRef] [PubMed]

27. Giggins, O.M.; Persson, U.M.; Caulfield, B. Biofeedback in rehabilitation. *J. Neuroeng. Rehabil.* **2013**, *10*, 60. [CrossRef]

28. Merletti, R.; Farina, D. *Surface Electromyography: Physiology, Engineering, and Applications*; John Wiley & Sons, Inc.: Hoboken, NJ, USA, 2016.

29. Zhang, Z.; Yang, K.; Qian, J.; Zhang, L. Real-Time Surface EMG Pattern Recognition for Hand Gestures Based on an Artificial Neural Network. *Sensors* **2019**, *19*, 3170. [CrossRef]

30. Vasanthi, S.M.; Jayasree, T. Performance evaluation of pattern recognition networks using electromyography signal and time-domain features for the classification of hand gestures. *Proc. Inst. Mech. Eng. H* **2020**, *234*, 639–648. [CrossRef]

31. Tsinganos, P.; Cornelis, B.; Cornelis, J.; Jansen, B.; Skodras, A. Data Augmentation of Surface Electromyography for Hand Gesture Recognition. *Sensors* **2020**, *20*, 4892. [CrossRef] [PubMed]

32. Chang, M.; Slater, L.V.; Corbett, R.O.; Hart, J.M.; Hertel, J. Muscle activation patterns of the lumbo-pelvic-hip complex during walking gait before and after exercise. *Gait Posture* **2017**, *52*, 15–21. [CrossRef]

33. Lu, Z.; Tong, K.Y.; Zhang, X.; Li, S.; Zhou, P. Myoelectric Pattern Recognition for Controlling a Robotic Hand: A Feasibility Study in Stroke. *IEEE Trans. Biomed. Eng.* **2019**, *66*, 365–372. [CrossRef]

34. Resnik, L.; Huang, H.H.; Winslow, A.; Crouch, D.L.; Zhang, F.; Wolk, N. Evaluation of EMG pattern recognition for upper limb prosthesis control: A case study in comparison with direct myoelectric control. *J. Neuroeng. Rehabil.* **2018**, *15*, 23. [CrossRef]

35. Parajuli, N.; Sreenivasan, N.; Bifulco, P.; Cesarelli, M.; Savino, S.; Niola, V.; Esposito, D.; Hamilton, T.J.; Naik, G.R.; Gunawardana, U.; et al. Real-Time EMG Based Pattern Recognition Control for Hand Prostheses: A Review on Existing Methods, Challenges and Future Implementation. *Sensors* **2019**, *19*, 4596. [CrossRef] [PubMed]

36. Fang, C.; He, B.; Wang, Y.; Cao, J.; Gao, S. EMG-Centered Multisensory Based Technologies for Pattern Recognition in Rehabilitation: State of the Art and Challenges. *Biosensors* **2020**, *10*, 85. [CrossRef] [PubMed]

37. Kuriki, H.U.; Mícolis de Azevedo, F.; Takahashi, L.S.O.; Mello, E.M.; Filho, R.d.F.N.; Alves, N. The Relationship Between Electromyography and Muscle Force. In *EMG Methods for Evaluating Muscle and Nerve Function*; Schwartz, M., Ed.; InTech: Sao Paulo, Brasil, 2012; p. 532.

38. Petersen, E.; Rostalski, P. A Comprehensive Mathematical Model of Motor Unit Pool Organization, Surface Electromyography, and Force Generation. *Front. Physiol.* **2019**, *10*, 176. [CrossRef]

39. Disselhorst-Klug, C.; Schmitz-Rode, T.; Rau, G. Surface electromyography and muscle force: Limits in sEMG-force relationship and new approaches for applications. *Clin. Biomech.* **2009**, *24*, 225–235. [CrossRef]

40. Zhou, P.; Rymer, W.Z. Factors governing the form of the relation between muscle force and the EMG: A simulation study. *J. Neurophysiol.* **2004**, *92*, 2878–2886. [CrossRef]

41. Kiper, P.; Baba, A.; Agostini, M.; Turolla, A. Proprioceptive based training for stroke recovery. Proposal of new treatment modality for rehabilitation of upper limb in neurological diseases. *Arch. Physiother.* **2015**, *5*, 6. [CrossRef]

42. Hislop, H.J.; Avers, D.; Brown, M.; Daniels, L. *Daniels and Worthingham's Muscle Testing: Techniques of Manual Examination and Performance Testing*, 9th ed.; Elsevier: St. Louis, MO, USA, 2014; 514p.

43. Hermens, H.J.; Freriks, B.; Merletti, R.; Stegeman, D.; Blok, J.; Rau, G.; Disselhorst, C.; Hagg, G. *SENIAM Raccomandazioni Europee per l'Elettromiografia di Superficie*; CLUT: Torino, Italy, 2000; p. 138.

44. Deroide, N.; Bousson, V.; Mambre, L.; Vicaut, E.; Laredo, J.D.; Kubis, N. Muscle MRI STIR signal intensity and atrophy are correlated to focal lower limb neuropathy severity. *Eur. Radiol.* **2015**, *25*, 644–651. [CrossRef] [PubMed]

45. Chia, D.S.Y.; Doi, K.; Hattori, Y.; Sakamoto, S. Elbow flexion strength and contractile activity after partial ulnar nerve or intercostal nerve transfers for brachial plexus injuries. *J. Hand Surg. Eur. Vol.* **2020**, *45*, 818–826. [CrossRef] [PubMed]

46. Bilodeau, M.; Schindler-Ivens, S.; Williams, D.M.; Chandran, R.; Sharma, S.S. EMG frequency content changes with increasing force and during fatigue in the quadriceps femoris muscle of men and women. *J. Electromyogr. Kinesiol.* **2003**, *13*, 83–92. [CrossRef]

47. Madeleine, P.; Bajaj, P.; Sogaard, K.; Arendt-Nielsen, L. Mechanomyography and electromyography force relationships during concentric, isometric and eccentric contractions. *J. Electromyogr. Kinesiol.* **2001**, *11*, 113–121. [CrossRef]

48. Doud, J.R.; Walsh, J.M. Muscle fatigue and muscle length interaction: Effect on the EMG frequency components. *Electromyogr. Clin. Neurophysiol.* **1995**, *35*, 331–339. [PubMed]

49. Woods, J.J.; Bigland-Ritchie, B. Linear and non-linear surface EMG/force relationships in human muscles. An anatomical/functional argument for the existence of both. *Am. J. Phys. Med.* **1983**, *62*, 287–299. [PubMed]

50. Cuthbert, S.C.; Goodheart, G.J., Jr. On the reliability and validity of manual muscle testing: A literature review. *Chiropr. Osteopathy* **2007**, *15*, 4. [CrossRef] [PubMed]

 *sensors*

Article

Muscle Synergies and Clinical Outcome Measures Describe Different Factors of Upper Limb Motor Function in Stroke Survivors Undergoing Rehabilitation in a Virtual Reality Environment

Lorenza Maistrello [1], Daniele Rimini [2,*], Vincent C. K. Cheung [3], Giorgia Pregnolato [1] and Andrea Turolla [1]

1   Laboratory of Rehabilitation Technologies, IRCCS San Camillo Hospital, 30126 Venice, Italy; lorenza.maistrello@hsancamillo.it (L.M.); giorgia.pregnolato@hsancamillo.it (G.P.); andrea.turolla@hsancamillo.it (A.T.)
2   Medical Physics Department—Clinical Engineering, Salford Care Organisation, Salford M6 8HD, UK
3   School of Biomedical Sciences, The Chinese University of Hong Kong, Hong Kong, China; vckc@cuhk.edu.hk
*   Correspondence: daniele.rimini@nca.nhs.uk; Tel.: +44-61620 (ext. 64859)

Abstract: Recent studies have investigated muscle synergies as biomarkers for stroke, but it remains controversial if muscle synergies and clinical observation convey the same information on motor impairment. We aim to identify whether muscle synergies and clinical scales convey the same information or not. Post-stroke patients were administered an upper limb treatment. Before (T0) and after (T1) treatment, we assessed motor performance with clinical scales and motor output with EMG-derived muscle synergies. We implemented an exploratory factor analysis (EFA) and a confirmatory factor analysis (CFA) to identify the underlying relationships among all variables, at T0 and T1, and a general linear regression model to infer any relationships between the similarity between the affected and unaffected synergies (Median-sp) and clinical outcomes at T0. Clinical variables improved with rehabilitation whereas muscle-synergy parameters did not show any significant change. EFA and CFA showed that clinical variables and muscle-synergy parameters (except Median-sp) were grouped into different factors. Regression model showed that Median-sp could be well predicted by clinical scales. The information underlying clinical scales and muscle synergies are therefore different. However, clinical scales well predicted the similarity between the affected and unaffected synergies. Our results may have implications on personalizing rehabilitation protocols.

Keywords: muscle synergies; sEMG; stroke; factor analysis

Citation: Maistrello, L.; Rimini, D.; Cheung, V.C.K.; Pregnolato, G.; Turolla, A. Muscle Synergies and Clinical Outcome Measures Describe Different Factors of Upper Limb Motor Function in Stroke Survivors Undergoing Rehabilitation in a Virtual Reality Environment. *Sensors* **2021**, *21*, 8002. https://doi.org/10.3390/s21238002

Academic Editors: Francesco Di Nardo, Valentina Agostini and Silvia Conforto

Received: 30 October 2021
Accepted: 26 November 2021
Published: 30 November 2021

Publisher's Note: MDPI stays neutral with regard to jurisdictional claims in published maps and institutional affiliations.

1. Introduction

The execution of voluntary movement is based on the functional integration of different areas of the central nervous system (CNS) that send descending neural signals to the spinal interneurons and motoneurons to generate specific motor behaviors. Currently, the mechanisms that allow the CNS to control a large-dimensional system and coordinate many muscles consisting of thousands of motor units are still a matter of debate [1]. In describing voluntary movement, it is common to refer to the term "synergies". However, this term may have several and different meanings, according to the context: indeed, the term *synergy* can refer to a coherent activation of a group of muscles, but it is also used with a negative connotation to describe abnormal motor patterns due to brain lesions [2]. There is also a third way of using the term synergies, commonly used to refer to a motor control model. Indeed, among the many existing models [3,4], it has been proposed that the CNS manages this complexity through a linear combination of fixed spinal modules, each one activating groups of muscles as a single functional unit, called muscle synergies [5–7]. Muscle synergies are obtained by decomposing surface electromyography (sEMG) into two components: vectors of fixed weights, representing the muscle synergies, and time-varying signals, representing the neural command for the synergies [8].

The activation and organization of muscle synergies are altered after stroke, causing a dysfunctional execution of voluntary movement. Early studies demonstrated that, after stroke, muscle synergies remain robust between affected and healthy arms and across subjects [9]. However, a different motor performance is observed, since abnormal motor behaviors are generated through faulty activations of the spinal modules [10,11]. The faulty activations can be generally described in terms of merging [12,13] or fragmentation [14] of the healthy muscle synergies. The degree of merging and fragmentation have been demonstrated to be proportional to the severity of motor impairment and the temporal distance from stroke onset, respectively [15].

Recent studies investigated muscle synergies as a physiological marker to assess the motor performance and recovery after stroke [14,16]. This is required because neural deficits may be masked at the functional and kinematic level by compensatory strategies, and the same motor task can be achieved by many different coordination patterns. However, it remains controversial if the use of muscle synergies can overcome these limitations, or if muscle synergies and clinical observation convey the same information on motor impairment [17]. Early studies provided evidence that muscle synergies were more adequate to capture the complexity of motor behavior than clinical scales [18]. However, some recent studies showed controversial results. In a study where muscle synergies were adopted to stratify stroke patients, synergies distributed coherently according to the Fugl-Meyer scale and Reaching Performance Scale, indicating that synergies convey similar underlying information [19]. Some other studies showed that muscle synergies and clinical scales were weakly correlated, and that stroke does not affect the inner structure of synergies, but rather their temporal recruitment [20,21]. There has also been evidence that synergies can improve in terms of their timing and organization by specific targeted therapies, including robot therapy or virtual reality treatment [15,22,23].

In rehabilitation medicine, the implication of muscle synergies should be considered as a marker of motor recovery, after a specific training for upper limb rehabilitation. Recent studies reveal there is increasing evidence demonstrating the efficacy of VR-based treatment for recovery of upper limb motor functions that facilitate the motor recovery thanks to the reinforced feedback mechanism [24]. Furthermore, it was demonstrated that after a VR treatment, patterns of cortical activation became physiologically more similar to the healthy ones, because the patterns of activations in the lesioned hemisphere were less sparse and more focused on the proper motor areas [25]. These results call into question if an underlying latent information is shared between muscle synergies and clinical scales.

The aim of this study was to identify and describe, in stroke survivors referred to upper limb treatment, whether motor output, as described by muscle synergies, and motor performance, quantified by clinical scales, convey the same information or provide a complementary one. For this purpose, a new set of variables was obtained as a combination of all the original ones by using a factor analysis. Then, we investigated whether synergies and clinical parameters belong to different components or if they convey to shared ones. In the former case, it may indicate that the two groups of variables provide different information, whereas in the latter case, there may be some muscle synergies and clinical-scale parameters that convey the same information.

2. Materials and Methods

2.1. Participants

A cohort of post-stroke patients from San Camillo IRCCS s.r.l. Hospital was recruited from a sample enrolled to participate in a multicenter clinical trial (Clinical Trial identifier: NCT03530358). We considered all patients hospitalized with diagnosis of ischemic or hemorrhagic first stroke in the territory of the middle cerebral artery (MCA).

Specifically, the following criteria were defined to recruit the patients able to perform a virtual reality treatment for upper limb motor recovery. The study included patients with a motor arm sub-score of the Italian version of the National Institute of Health Stroke Scale [26] between 1 and 3, that indicated the maintenance of residual voluntary motor activity. The following conditions were considered as exclusion criteria: (1) moderate cognitive decline defined by a Mini Mental State Examination (MMSE) [27] score lower than 20/30 points; (2) severe verbal comprehension deficits, defined by a number of errors >13 on the Token Test [28]; (3) evidence of apraxia and visuospatial neglect that could interfere with movements of the upper limb in all directions within the visual field, evaluated by neurological examination; (4) history of behavioral disorders (e.g., depression, aggressiveness, apathy) and neurological or vascular comorbidity (e.g., diabetes, myocardial infarction, Parkinson's disease) that could affect the compliance with the rehabilitation programs.

The study was reviewed and approved by the local Ethical Committee of the IRCCS San Camillo Hospital s.r.l. All participants were adequately informed about aims and modalities of the study and provided an informed written consent.

2.2. Study Design

We designed a single-group longitudinal study. At the enrolment time point, a detailed review of the medical history of each patient was collected. Then, we administered a treatment consisting of 20 sessions of upper limb exercises in a virtual reality environment. To define the effect of therapy, residual motor functions were clinically and instrumentally evaluated before and after the treatment: clinical assessment consisted of standardized scales to quantify residual motor capabilities, whereas instrumental assessment consisted of the surface electromyography (sEMG) recording during the execution of motor tasks to compute muscle synergies.

2.2.1. Clinical Assessment

The motor capabilities were clinically assessed with the following three outcome measures: the Modified Ashworth Scale (MAS) [29] to assess the muscle spasticity; the Fugl-Meyer Assessment scale for the upper limb (UE-FMA) [30], to determine the severity of motor impairment in hemiparetic limb, and the Reaching Performance Scale (RPS) [31] to identify and quantify movements patterns during reach-to-grasp tasks.

2.2.2. sEMG Recording and Muscle Synergies

To extract the upper limb muscle synergies, we recorded the sEMG from 16 muscles from both the unaffected and the stroke-affected sides during the execution of a standard section of seven visuo-motor tasks in a virtual environment. Indeed, subjects executed seven standardized motor tasks, each repeated 10 times, by interacting with a Virtual Reality Rehabilitation System (VRRS®, Khymeia Group Ltd., Noventa Padovana, Italy). In VRRS®, the patients interacted with a VR environment by means of a 3D motion-tracking system (Polhemus 3Space FasTrack, Polhemus, Colchester, VT, USA, sampling frequency of 120 Hz) fixed on the back of the hand. At the beginning of each VRRS® exercise, a trigger signal was sent to an sEMG amplifier (EMG-USB2+, OT Bioelettronica, Torino, Italy, sampling frequency of 2000 Hz) instrumented with 72001-K/12 electrodes (AMBU Neuroline, Ballerup, Denmark) to synchronize the kinematics with the sEMG [24,32,33]. The same seven tasks were proposed for both arms, except that the trajectories were mirrored according to the limb side. To facilitate the comprehension of the tasks and to reduce possible subject's frustration, the unaffected arm was recorded first, followed by the affected arm.

Electrodes were placed according to the Surface Electromyography for the Non-Invasive Assessment of Muscles (SENIAM) recommendations for skin preparation, placement, fixation, and testing of the sensor and its connection [34]. sEMG was recorded from the following 16 muscles: triceps brachii (medial head; lateral head); biceps brachii (short head; long head); deltoideus anterior; deltoideus medius; deltoideus posterior; trapezius superior; rhomboid major; brachioradialis; supinator; brachialis; pronator teres; pectoralis major (clavicular head); infraspinatus; teres major. In the case that SENIAM recommendations were not available for a muscle, standard clinical procedures were followed [35].

The sEMG preprocessing and muscle synergies extraction followed the procedure fully described elsewhere [9,14]. Muscle synergies were extracted for the affected and unaffected arms separately. Initially, sEMG of each task were combined into an $m \times t$ matrix, where m indicates the number of muscles and t indicates the time samples. sEMG of each row in the matrix were preprocessed as follows: band-pass filtered (10–500 Hz), normalized to the unit variance, rectified, low-pass filtered to 12 Hz. Muscle synergies were extracted from 1 to 16 iteratively by decomposing the processed sEMG with the nonnegative matrix factorization (NMF) algorithm [36]. The number of synergies was chosen with a cross-validated EMG reconstruction factor R^2 for > 90%.

From the muscle synergies of each subject, we computed the following parameters: (i) the number of synergies of the affected limb (N-aff); (ii) the number of synergies of the unaffected limb (N-ctrl); (iii) the number of synergies of the affected limb and of the unaffected limb analytically similar, with values of scalar product above 0.8 recognized as similar [37] ("Synergies shared", N-sh); (iv) the ratio between N-sh and N-aff (Nsh-naff), (v) the ratio between N-sh and N-ctrl (Nsh-nctrl); (vi) the median scalar product between the affected and unaffected synergies (Median-sp); (vii) the mean number of unaffected synergies merging in every affected synergy (P1) (See [14]; Supporting Information).

2.2.3. Rehabilitation Treatment

The rehabilitation treatment consisted of 20 sessions of one hour each, five sessions per week, 4 weeks total. To avoid discontinuity and comparable treatment intensity, at least three sessions per week were administered. Patients who performed less than 80% of the planned sessions (<16/20 sessions) were excluded from the subsequent analysis.

During each session, patients were asked to perform a defined set of exercises, including shoulder flexion–extension, abduction–adduction, internal–external rotation, circumduction, elbow flexion–extension, forearm pronation–supination, and hand–digit motion. The physical therapist was constantly present during the session, providing instructions according to specific patients' residual abilities and needs.

2.3. Statistical Analysis

2.3.1. Sample Characteristics

Initially, to define the sample size of the trial, we consulted previous proof-of-concept studies: they demonstrated that a sample of 20 patients are appropriate to obtain significant results [9,14]. Thus, with the aim to improve the statistical power of our analysis, we proposed to enroll 50 patients at least.

Firstly, descriptive statistics (i.e., median, interquartile range, mean, standard deviation, and percentage) were used to describe the demographic, clinical, and muscle synergies characteristics of the sample.

To verify whether there was a change in motor performance, we compared the values of the pretreatment (T0) with post-treatment (T1) clinical and instrumental variables by a paired t-test or Wilcoxon test, according to data normality distribution tested by the Shapiro–Wilk test.

Furthermore, we investigated potential associations among clinical outcomes (i.e., MAS, UE-FMA, RPS scores) and synergy parameters (i.e., N-aff, N-ctrl, N-sh, Nsh-nctrl, Nsh-naff, Median-sp, P1) by means of correlation test at T0 and T1 (i.e., Pearson correlation test or Spearman's rank correlation test) with a significant level of correlation defined at $R^2 > 0.3$.

Finally, the factorability of the data was examined by studying data sphericity with the Bartlett's test ($p < 0.05$) [38] and data multicollinearity with the Kaiser–Meyer–Olkin measure of sampling adequacy (MSA, threshold of acceptability MSA > 0.50) [39].

2.3.2. Exploratory Factor Analysis

We implemented an exploratory factor analysis (EFA) to identify the underlying relationships among all variables (EFA-All). Moreover, to investigate if the time of assessment (i.e., T0 and T1) was a parameter influencing the results, two independent models were implemented for variables acquired at T0 (EFA0) and T1 (EFA1).

To obtain each EFA model, we chose the number of latent factors [40] with the following two methods [41]: principal component analysis (PCA) [42] and principal axis factoring (PAF) [43]. We selected the most informative factors by means of the Gorsuch approach, which includes Horn's parallel analysis, Cattell's scree plots, and Kaiser criterion [44]. Once we found the number of factors, a common factor model was extracted with the principal axis (PA) method [45,46]. The model was rotated with oblique rotation methods (e.g., promax) [47] according to the presence of correlation between the factors [48]. Finally, we selected the most significative variables that comprised each factor according to the following criteria [43,49]: 1) factor loadings (FL) (FL > 0.3); 2) communalities, namely, common variance ($h^2 > 0.20$); and 3) factors correlations (correlations r < 0.85).

2.3.3. Confirmatory Factor Analysis

For each EFA model, a confirmatory factor analysis (CFA) was conducted to verify the factor structure of the observed variables (i.e., CFA-All, CFA0, CFA1). For this purpose, structural equation modeling (SEM) with a maximum likelihood estimation model and standardized coefficients (significative factor loadings FL > 0.3 or FL < −0.3) were used. Observations with missing values were excluded [50]. We assessed SEM model fitting by using the following indices [51,52]: the χ^2 test, the comparative fit index (CFI), Tucker-Lewis index (TLI) [53,54], and root mean-squared error of approximation (RMSEA) [55,56]. The CFA model fitted the original data if the indices met the following criteria: a significant χ^2 value indicating a bad model fit; a RMSEA value ≤ 0.05 was considered indicative of "good fit"; the CFI and TLI were considered acceptable for values >0.95 [57–59].

2.3.4. General Linear Regression Model

As a final analysis, we implemented a general linear regression model to infer any potential causal relationships between the synergy parameters (dependent variable) and clinical outcomes (MAS, UE-FMA, RPS as independent variables) falling in the same factor. The statistical power of the CFA analysis was calculated with a post hoc power analysis based on the RMSEA.

The statistical significance level was set to 0.05 for all tests. All statistical analyses were performed using the free software R Studio [60].

3. Results

3.1. Sample characteristics

After the enrolment, 50 subjects completed the entire treatment. Table 1 summarizes demographic characteristics of the entire sample.

Table 1. Demographic and clinical characteristics of the patients.

Patients (N = 50)	
Sex, males/females, n (%)	33 (66%)/17 (34%)
Age, years, mean ± SD	63.62 ± 12.29
Diagnosis, ischemic/hemorrhagic, n (%)	45 (90%)/5 (10%)
Hemisphere, left/right, n (%)	25 (50%)/25 (50%)
Time-stroke, months, mean ± SD	6.99 ± 13.07
0–3 months, n, mean ± SD	15, 2.32 ± 0.42
3–6 months, n, mean ± SD	17, 4.25 ± 0.87
>6 months, n, mean ± SD	18, 20.61 ± 19.83

Values are expressed as mean ± standard deviation (SD) for quantitative measures, and frequency percentages (%) for all discrete variables.

After inferential analysis, two out three clinical outcomes improved significantly: the UE-FMA score improved by 6% (T0, UE-FMA mean = 117.2; T1, UE-FMA mean = 124.26), and the RPS score improved by 4% (T0, RPS mean = 25.4; T1, RPS mean = 26.46). Conversely, synergies parameters revealed no significant change after the treatment. Table 2 reports the clinical outcomes and parameters related to the muscle synergies.

Table 2. Clinical outcomes and parameters related to synergies.

Clinical Parameters	T0		T1		*p* Value
	Median [IQR]	Mean $\pm$ SD	Median [IQR]	Mean $\pm$ SD	
MAS	1 [2.75]	1.92 $\pm$ 2.69	0.5 [2]	1.60 $\pm$ 2.44	0.098
UE-FMA	125.5 [34.75]	117.20 $\pm$ 24.57	131.5 [33.25]	124.26 $\pm$ 25.41	<0.001 *
RPS	30 [6]	24.4 $\pm$ 11.19	17 [6]	26.46 $\pm$ 12.25	<0.001 *
Synergies Parameters	T0		T1		*p* Value
	Median [IQR]	Mean $\pm$ SD	Median [IQR]	Mean $\pm$ SD	
N-aff	8 [1]	8.42 $\pm$ 1.40	8 [2]	8.20 $\pm$ 1.47	0.289
N-ctrl	8 [2]	7.86 $\pm$ 1.31	8 [1.75]	7.84 $\pm$ 1.22	0.855
N-sh	6 [2]	6.24 $\pm$ 1.39	6 [2]	6.12 $\pm$ 1.36	0.456
Nsh-naff	0.75 [0.13]	0.74 $\pm$ 0.12	0.78 [0.22]	0.75 $\pm$ 0.13	0.616
Nsh-nctrl	0.79 [0.16]	0.79 $\pm$ 0.12	0.78 [0.14]	0.78 $\pm$ 0.12	0.432
Median-sp	0.93 [0.04]	0.92 $\pm$ 0.04	0.94 [0.05]	0.93 $\pm$ 0.03	0.056
P1	1.19 [0.58]	1.25 $\pm$ 0.39	1.24 [0.44]	1.24 $\pm$ 0.34	0.913

Values are expressed as median [IQR] and mean $\pm$ SD; IQR = interquartile range; SD = standard deviation; * *p* values < 0.05; Wilcoxon test.

Moreover, correlation analysis showed that some of the clinical scales were significantly correlated with some muscle synergy parameters. Specifically, there was a positive correlation between MAS and N-aff (R^2 = 0.37 at T0; R^2 = 0.34 at T1). Moreover, the clinical scale UE-FMA and RPS correlated positively both with Nsh-naff after treatment (UE-FMA, R^2 = 0.37 and RPS, R^2 = 0.53 at T1) and with Median-sp (UE-FMA, R^2 = 0.47 and RPS, R^2 = 0.49 at T0) (UE-FMA, R^2 = 0.51 and RPS, R^2 = 0.54 at T1). Figure 1 summarizes the correlation coefficients between clinical outcomes and synergies parameters, at T0 and T1, separately.

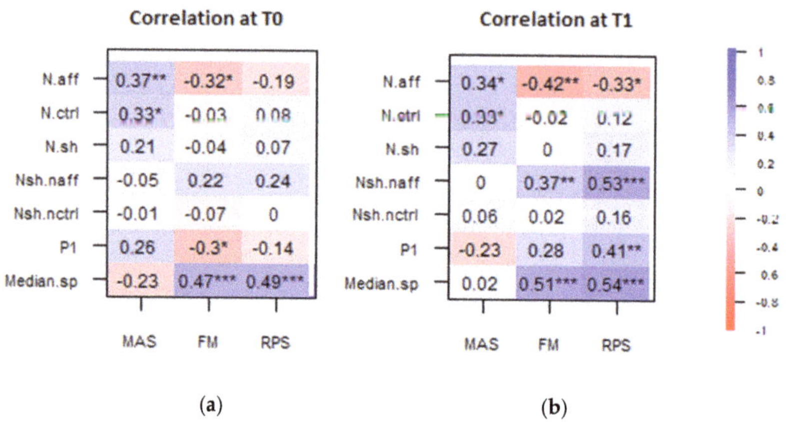

Figure 1. Correlation between clinical outcomes and parameters related to synergies at T0 (**a**) and at T1 (**b**). Significant correlation indices are indicated with * *p* < 0.05, ** *p* < 0.01, and *** *p* < 0.001, respectively.

Both the Bartlett's test of sphericity (χ^2 = 1396.80, df = 190, *p* < 0.001) and the Kaiser–Meyer–Olkin test (MSA = 0.62) indicated that the correlation matrix was factorable. Then, we proceeded with the factor analysis at T0, T1, and with all the variables.

3.2. Exploratory Factor Analysis

3.2.1. Exploratory Factor Analysis with All Variables

At first, we implemented the exploratory factorial analysis on the whole sample (EFA-All) and we obtained structures with three to five factors (Supplementary Material, Figure S1a). All these factor solutions were sequentially examined, with a total explained variance equal to 66%. Specifically, one factor was linked to the clinical variables (both pre- and post-treatment) and two factors, linked to the parameters of the synergies derived from pre- and post-treatment EMGs, respectively.

Table 3 reports the loadings and the corresponding communalities of the EFA-All. It can be observed that Factor 1 was linked to the clinical variables (both pre- and post-treatment) and Median-sp-T0. Factor 2 and Factor 3 were linked to parameters of the synergies only.

Table 3. EFA-All, with promax rotation for all variables.

Outcome	Factor 1	Factor 2	Factor 3	h^2
MAS-T0	−0.579			0.528
UE-FMA-T0	0.914			0.839
RPS-T0	0.948			0.885
MAS-T1	−0.554			0.467
UE-FMA-T1	0.988			0.920
RPS-T1	0.918			0.882
Median-sp-T0	0.616			0.441
N-aff-T0		0.913		0.848
N-ctrl-T0		0.922		0.669
N-sh-T0		0.972		0.849
Nsh-ctrl-T0		0.301		0.218
N-ctrl-T1		0.537		0.415
N-sh-T1			0.780	0.847
Nsh-aff-T1			0.881	0.769
Nsh-ctrl-T1			0.921	0.687
Median-sp-T1			0.589	0.503
% variance of the factor	33.7%	16.5%	15.9%	

Table shows the factor loadings for the 3 factors and the communalities for each variable (h^2).

Factor correlations were r = −0.30 between Factor 1 and Factor 2, r = 0.145 between Factor 1 and Factor 3, and r = 0.37 between Factor 2 and Factor 3.

3.2.2. Exploratory Factor Analysis with T0 Variables

Secondly, we implemented an EFA model with two factors for the variables at T0 time point (EFA0). The factor structure of the sample indicated the presence of more than one unique factor, suggesting that two factors should be retained (Supplementary Material, Figure S1b). The correlation between the two factors was very low (r = −0.0013); therefore, a promax oblique rotation method was used. Table 4 reports the loadings of the factor matrix.

These factors collectively accounted for 70.2% of the variance in the responses. Factor correlation was r = −0.14.

3.2.3. Exploratory Factor Analysis with T1 Variables

Finally, we implemented an EFA model with two factors for the variables at T1 time point (EFA1). The factors were rotated with promax oblique rotation methods as the correlation between the two factors was r = 0.023 (Supplementary Material, Figure S1c). Table 5 reports the loadings of the factor matrix.

Table 4. EFA0, with promax rotation for variables at T0.

Outcome	Factor 1	Factor 2	h^2
MAS	−0.618		0.420
UE-FMA	0.847		0.705
RPS	0.886		0.775
N-aff		0.631	0.759
Nsh-aff		0.811	0.751
N-ctrl		0.674	0.538
Nsh-ctrl		0.811	0.751
N-sh		1.067	1.157
% variance of the factor	39.3%	30.9%	

Table shows the factor loadings for Factor 1 and Factor 2, the communalities for each variable (h^2), and the percentage of variance explained by each factor (%).

Table 5. EFA1, with promax rotation for variables at T1.

Outcome	Factor 1	Factor 2	h^2
MAS		−0.562	0.627
UE-FMA		0.889	0.786
RPS		0.948	0.948
N-ctrl	0.550		0.310
Nsh-ctrl	0.505		0.255
N-sh	1.390		1.933
% variance of the factor	42.8%	33.4%	

Table shows the factor loadings for Factor 1 and Factor 2, the communalities for each variable (h^2), and the percentage of variance explained by each factor (%).

These factors collectively accounted for 76.2% of the variance in the responses and they had a factor correlation of r = −0.068.

3.3. Confirmatory Factor Analysis

After EFA analysis, we proceeded to confirmatory factor analysis, for all sample variables and then for variables at T0 and T1.

3.3.1. Confirmatory Factor Analysis with All Variables

In the CFA analysis carried out on all variables (CFA-All), a latent three-factor model was specified, as suggested by the results obtained in the EFA-All analysis.

Based on the content of their variables, we named the three factors clinical scale, T0 synergies, and T1 synergies (Figure 2a).

The CFA-All model indicated the presence of a correlation between two factors, the T0 synergies and T1 synergies factors (r = 0.37), while no correlations related to the clinical scale factor were detected. Furthermore, all factor loadings were significant. Goodness-of-fit statistics demonstrated that all indices were outside the set cut-offs: RMSEA index between 0.30 and 0.36 and a value of $\chi^2 = 442.84$, with df = 74 and $p = 0.000$. The values of CFI and TLI were 0.55 and 0.45, respectively.

3.3.2. Confirmatory Factor Analysis T0 Variables

The EFA0 suggested a two-factor solution, and we estimated a latent two-factor model (CFA0). According to the content of their items, we named the two factors as clinical scale and synergies parameters (Figure 2b). After estimating the model, goodness-of-fit statistics were obtained. All FL were significant, but the model demonstrated that all indices were outside the set cut-offs, with RMSEA index between 0.45 and 0.57 and $\chi^2 = 255$, with df = 19 and $p = 0.000$. Moreover, the values of CFI and TLI were 0.57 and 0.37, respectively.

3.3.3. Confirmatory Factor Analysis T1 Variables

The EFA1 suggested a two-factor solution, and we estimated a latent two-factor model (CFA1). According to the content of their items, we named the two factors as clinical scale and synergy parameters (Figure 2c). After estimating the model, it did not show a very good fit, with an RMSEA index between 0.11 and 0.29 and $\chi^2 = 23.58$, with df = 8 and $p = 0.003$. Moreover, the values of CFI and TLI were 0.95 and 0.91, respectively.

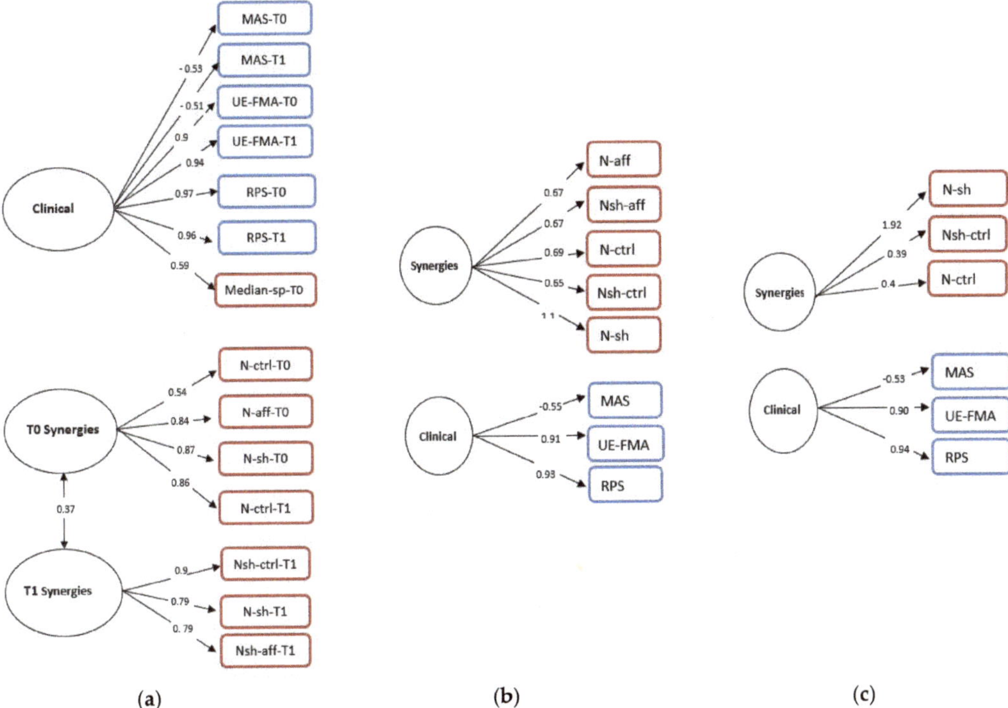

Figure 2. Confirmatory factor analysis for all variables (**a**) at T0 (**b**) and at T1 (**c**). Single-headed arrows indicate direct relationships. The numbers on each represent standardized factor loadings ranging from 1.0 to −1.0. Squares represent measured variables and circles represent latent factors. The figures in blue represent the clinical variables, while those in red represent the synergies parameters. The double-headed arrows represent correlations between the factors.

3.4. General Linear Regression Models

To investigate the reason why the Median-sp at T0 was associated with the clinical factor in both EFA-All and CFA-All, we estimated a general linear regression model with the Median-sp as dependent variable and the clinical variables (i.e., MAS, RPS, UE-FMA) at T0 as independent variables.

In the general linear regression model, the T0 variable Median-sp was significantly associated with RPS at T0 ($\hat{\beta} = 0.002$, $p < 0.001$). The model determination coefficient was $R^2 = 0.97$.

4. Discussion

In the present study, we investigated the statistical relationships among the clinical variables and muscle synergy parameters in a cohort of post-stroke patients enrolled in a specific treatment for upper limb motor recovery provided in a virtual environment. Specifically, the main objective of this study was to identify whether motor output, as described by muscle synergies parameters recorded using sEMG, and motor performance, quantified by clinical scales, convey the same information or provide a complementary one.

In our study, the pre- and post-treatment analysis evidenced a significant improvement in almost all clinical outcomes, whereas no significant differences in muscle synergies parameters were observed. This suggests that they have a different sensitivity to the recovery level after stroke, and that the number of muscles synergies and merging alone are not sensitive enough to describe the effectiveness of treatment. This seems to be in contrast with some previous studies, where changes in the number of synergies and their structure indicated improvement of motor control and movement quality. However, it was limited to patients with low level of residual motor functions [15,61].

There may be several mechanisms that could better describe how muscle synergies change after motor therapy. For instance, rather than counting the number of muscle synergies, it has been shown that modifications of clusters and shifting from one cluster to another can provide insights for assessing the efficacy of the therapy [62]. After stroke, functional and structural recovery processes take place within the brain. Since these processes are mainly related to the reorganization of cortical maps [63], we may hypothesize that changes in the modulation of synergies may be associated with these mechanisms of neural plasticity. This could trigger changes in muscle coactivation within synergies, resulting in synergy merging, for instance [64]. Indeed, it was suggested that the merging phenomena of muscle synergies may depend on neural changes at the cortical level or at the level of the brainstem in the spinal cord [14].

Despite that clinical and instrumental assessments provided different information in terms of motor recovery, some strong relationships emerged between some muscle-synergy parameters and clinical scales. Indeed, the correlation analysis indicated that the number of muscle synergies of the affected limb was positively correlated with clinical outcomes: the higher the N-aff, the higher the level of spasticity to upper limb muscles (i.e., MAS). Conversely, the higher the N-aff, the lower the level of motor ability (i.e., UE-FMA and RPS). In line with our findings, Pan et al. [13] found that muscle synergies were significantly positively correlated with the Brunnstrom stage ($R^2 = 0.52$, $p < 0.01$). This is in good agreement with our study, because the Brunnstrom scale describes the stages of stroke recovery by a progressive decrease in spasticity. Furthermore, there was a strong positive correlation between the Median-sp values and the motor ability of the patients (i.e., UE-FMA and RPS). Since Median-sp describes the similarity between the affected and unaffected synergies, it seems to provide some useful and objective information about the degree of "true recovery" of the paretic arm (i.e., the extent to which the original muscle coordinative structures are restored) after intervention. Moreover, the correlation index increased after the treatment (i.e., T1 assessment), meaning that, after motor treatment, Median-sp is more informative about the motor performance.

The second objective of the study was to group all the variables into one or more clusters and to describe the nature of the underlying relationships among variables as described by the latent factors. More precisely, we used EFA and CFA to explore the information shared between muscle synergies and clinical scales of stroke survivors referred to upper limb treatment.

The EFA model we implemented with all variables (i.e., EFA-All) identified three factors: one linked to the clinical variables (both pre- and post-therapy), and two linked to the pre- and post-treatment parameters of the synergies, respectively. Median-sp was the only muscle-synergy parameter which was associated with the cluster of clinical scales, and thus with the level of a patient's motor performance. This achievement confirms the correlation results, and it represents a potential continuous index of similarity that can provide information also from a clinical point of view. Our finding that clinical variables and muscle-synergy parameters were mostly linked to separable factors argues that muscle synergies and clinical variables provided complementary information, both related to the motor ability of the patient.

The subsequent models, EFA0 and EFA1, highlighted the same structure of EFA-All, with the variables clustered according to their nature: clinical scales and synergy parameters. It should be noted that a synergy parameter (N-ctrl T1), despite referring to post treatment outcome, is attributed to the pretherapy cluster. This is an expected result: since N-ctrl was obtained from the healthy limb, it did not vary due to the motor therapy. On the other hand, in the CFA-All model, the variables were not distributed clearly among the factors according to their nature, as we would expect. Similar to the EFA-All model results, in the clinical factor, there was a parameter of the Median-sp-T0 synergies, while in the two factors related to the synergies (T0 synergies and T1 synergies) there were parameters that did not follow the temporal subdivision of their nature. Moreover, in CFA0 and CFA1, the variables were not all represented by latent factors. Indeed, both in the CFA0 and in the CFA1 models, the clinical factor collects all the clinical variables considered (i.e., MAS, UE-FMA, and RPS).

By considering the differences between CFA0 and CFA1 models, it was highlighted how, after the therapy, the parameters linked to the stroke-affected limb (i.e., N-aff and N-sh-aff) disappeared, probably because after the therapy the differences between healthy limb and affected limb were less marked, and therefore the affected limb no longer provided information.

Since in both EFA-All and CFA-All models, Median-sp-T0 was the only synergy parameter with an underlying structure in common with the clinical scales, a regression model was estimated to determine whether there is a causal relationship between the similarity parameter (Median-sp) (dependent variable) and all clinical variables (independent variables) at the same time point, that is, the pretreatment evaluation (i.e., T0). Our model evidenced a relationship between the similarity of affected and healthy synergies at the beginning of the treatment and the upper limb movement quality during a reach-to-grasp target, assessed by RPS [65]. Indeed, the presence of the RPS scale in the model is consistent with the indication that the reaching movement may be the best predictor of motor recovery. Recently, Pan et al. defined that severe patients had the lowest range of motion and speed during reaching movement. Specifically, they found three muscle synergies that may explain reaching movement. Moreover, severe patients changed one of these muscle synergies; meanwhile, the mild-to-moderate patients were more similar to the control template [66]. Thus, individualized training may be developed to make the patients' features more similar to the ones in control subjects so as to improve similarity values (i.e., Median-sp) [67].

The present study has several limitations that should be addressed in future research. In terms of generalizability, the relatively small sample size used to conduct the factor analysis (i.e., both EFA and CFA methods) should not be considered to obtain valuable results as good-of-fitness index [68]. For this reason, the goodness-of-fit statistics for all CFA models showed that none of them had good overall fit, with RMSEA never dropping below 0.11 and CFI and TLI remaining relatively low, despite a high post hoc statistical power ($1 - \text{beta} > 0.99$). Moreover, to obtain an effect on the synergies, we need to test a longer treatment period or to tailor the upper limb treatment based on the stratification of patients. Indeed, the sample of patients was not homogeneous in terms of timing from lesion and initial level of motor impairment. Actually, more investigation is needed concerning which neurophysiological parameters may help classify patients based on different recovery potential [69,70]. In our study, our results demonstrate that muscle-synergy parameters showed a potential to contributing to discriminate between patients with different recovery potential: the relationship between neurophysiological parameters (i.e., Median-sp) and clinical variables at the beginning of therapy gave some indication about the potential patient-tailored treatment. More trials will be needed to define the real contribution of muscle-synergy parameters to distinguish between fitters and no-fitters of reactive neurobiological recovery [66,71]. Moreover, we may consider using the similarity parameter (i.e., Median-sp) to build patient-specific prediction models to improve clinical decisions, and, ultimately, recovery and outcome after stroke [72].

Finally, the types of movements and the kinematics were not considered in our study, making other biomechanical interpretation of our results possible.

5. Conclusions

In our study, we investigated whether there is a relationship between clinical scales and muscle-synergy parameters in individuals with stroke who underwent a specific upper limb motor training. Specifically, after statistical analysis, we found that there exists a relationship between the similarity of muscle synergy parameters of the affected and unaffected limb and clinical variables, in particular at the beginning of the therapy. The correlation found between Median-sp and clinical variables indicates that there is a related, but complementary, information provided by both different type of parameters. Finally, future analyses may be conducted to investigate the use of similarity parameter as a biomarker of different levels of motor impairment to tailor the upper limb motor treatment to stroke survivors.

Supplementary Materials: The following are available online at https://www.mdpi.com/article/10.3390/s21238002/s1, Supplementary file "Supplementary Materials.docx" contains the Cattell's scree plots based on both PAF and PCA for the EMA models presented in the main text. Figure S1 Cattell's scree plot based on both PAF and PCA for EFA-All (**a**), EFA0 (**b**) and EFA1 (**c**). Cattell's rule states that number of factors and components, corresponding to eigenvalues to the up of the straight line should be retained. Abbreviations: FA = Factor Analysis; PC = Principal Component.

Author Contributions: Conceptualization, L.M., D.R., A.T.; methodology, L.M., V.C.K.C. and A.T.; validation, L.M., D.R., V.C.K.C., G.P. and A.T.; formal analysis, L.M. and D.R.; investigation, D.R., G.P. and A.T.; resources, A.T.; data curation, L.M., D.R., G.P. and A.T.; writing—original draft preparation, L.M., D.R. and G.P.; writing—review and editing, L.M., D.R., G.P., V.C.K.C. and A.T.; supervision, D.R. and A.T.; project administration, A.T.; funding acquisition, A.T. All authors have read and agreed to the published version of the manuscript.

Funding: V.C.K.C. was supported by The CUHK Faculty of Medicine Faculty Innovation Award FIA2016/A/04 and Group Research Scheme NL/JW/rc/grs1819/0426/19hc to V.C.K.C., and The Hong Kong Research Grants Council Grants 24115318 (ECS) and CUHK R4022-18 (RIF) to V.C.K.C. AT was funded by Italian Ministry of Health, Ricerca Finalizzata, Grant No. RF-2019-12371486.

Institutional Review Board Statement: The study was conducted according to the guidelines of the Declaration of Helsinki, and approved by the Institutional Ethics Committee of IRCCS San Camillo Hospital s.r.l. (protocol 2015.14).

Informed Consent Statement: Informed consent was obtained from all subjects involved in the study.

Data Availability Statement: The clinical protocol of this study is reposed and publicly available at the following address: https://www.clinicaltrials.gov/ct2/show/NCT03530358. The data presented in this study are available on request from the corresponding author.

Conflicts of Interest: The authors declare no conflict of interest.

Abbreviations

CNS	Central nervous system
sEMG	Surface electromyography
MCA	Middle cerebral artery
MMSE	Mini Mental State Examination
MAS	Modified Ashworth Scale
UE-FMA	Upper Extremity Fugl-Meyer Assessment Scale
RPS	Reaching Performance Scale
VRRS	Virtual Reality Rehabilitation System
SENIAM	Surface Electromyography for the Non-Invasive Assessment of Muscles
NMF	Nonnegative matrix factorization
N-aff	Number of affected synergies
N-ctrl	Number of unaffected synergies

N-sh	Number of shared synergies
Nsh-naff	Percentage of synergies shared in the affected arm
Nsh-nctrl	Percentage of synergies shared in the unaffected arm
Median-sp	Median of the calar product between the affected and unaffected arm
P1	Merging parameter
T0	Pretherapy variable
T1	Posttherapy variable
R2	Correlation coefficient
P	P-value
MSA	Measure of sampling adequacy
EFA	Exploratory factor analysis
EFA0	Exploratory factor analysis at T0
EFA1	Exploratory factor analysis at T1
EFA-All	Exploratory factor analysis with all variables at T0 and T1
PCA	Principal component analysis
PAF	Principal axis factoring
PA	Principal axis
FL	Factor loadings
h2	Communalities
r	Factors correlation coefficient
CFA	Confirmatory factor analysis
CFA0	Confirmatory factor analysis at T0
CFA1	Confirmatory factor analysis at T1
CFA-All	Confirmatory factor analysis with all variables at T0 and T1
CFI	Comparative fit index
χ^2	Chi-squared
TLI	Tucker–Lewis index
RMSEA	Root mean-squared error of approximation
Df	Degrees of freedom
SD	Standard deviation
IQR	Interquartile range
FA	Factor analysis
PC	Principal component
$\hat{\beta}$	Estimate regression coefficient.

References

1. Cheung, V.C.K.; Seki, K. Approaches to Revealing the Neural Basis of Muscle Synergies: A Review and a Critique. *J. Neurophysiol.* **2021**, *125*, 1580–1597. [CrossRef]
2. Solnik, S.; Furmanek, M.P.; Piscitelli, D. Movement Quality: A Novel Biomarker Based on Principles of Neuroscience. *Neurorehabil Neural Repair* **2020**, *34*, 1067–1077. [CrossRef]
3. Latash, M.L.; Scholz, J.P.; Schöner, G. Motor Control Strategies Revealed in the Structure of Motor Variability. *Exerc. Sport Sci. Rev.* **2002**, *30*, 26–31. [CrossRef]
4. Loeb, G.E. Learning to Use Muscles. *J. Hum. Kinet.* **2021**, *76*, 9–33. [CrossRef]
5. Bizzi, E.; Mussa-ivaldi, F.A.; Giszter, S. Computations Underlying the Execution of Movement: A Biological Perspective. *Science* **1991**, *253*, 287–291. [CrossRef]
6. D'Avella, A.; Saltiel, P.; Bizzi, E. Combinations of Muscle Synergies in the Construction of a Natural Motor Behavior. *Nat. Neurosci.* **2003**, *6*, 300–308. [CrossRef]
7. Saltiel, P.; Wyler-Duda, K.; D'Avella, A.; Tresch, M.C.; Bizzi, E. Muscle Synergies Encoded within the Spinal Cord: Evidence from Focal Intraspinal NMDA Iontophoresis in the Frog. *J. Neurophysiol.* **2001**, *85*, 605–619. [CrossRef] [PubMed]
8. Levine, A.J.; Hinckley, C.A.; Hilde, K.L.; Driscoll, S.P.; Poon, T.H.; Montgomery, J.M.; Pfaff, S.L. Identification of a Cellular Node for Motor Control Pathways. *Nat. Neurosci.* **2014**, *17*, 586–593. [CrossRef] [PubMed]
9. Cheung, V.C.K.; Piron, L.; Agostini, M.; Silvoni, S.; Turolla, A.; Bizzi, E. Stability of Muscle Synergies for Voluntary Actions after Cortical Stroke in Humans. *Proc. Natl. Acad. Sci. USA* **2009**, *106*, 19563–19568. [CrossRef]
10. Singh, R.E.; Iqbal, K.; White, G.; Hutchinson, T.E. A Systematic Review on Muscle Synergies: From Building Blocks of Motor Behavior to a Neurorehabilitation Tool. *Appl. Bionics Biomech.* **2018**, *2018*, 3615368. [CrossRef] [PubMed]
11. Taborri, J.; Agostini, V.; Artemiadis, P.K.; Ghislieri, M.; Jacobs, D.A.; Roh, J.; Rossi, S. Feasibility of Muscle Synergy Outcomes in Clinics, Robotics, and Sports: A Systematic Review. *Appl. Bionics Biomech.* **2018**, *2018*, 3934698. [CrossRef] [PubMed]
12. Clark, D.J.; Ting, L.H.; Zajac, F.E.; Neptune, R.R.; Kautz, S.A. Merging of Healthy Motor Modules Predicts Reduced Locomotor Performance and Muscle Coordination Complexity Post-Stroke. *J. Neurophysiol.* **2010**, *103*, 844–857. [CrossRef] [PubMed]

13. Pan, B.; Sun, Y.; Xie, B.; Huang, Z.; Wu, J.; Hou, J.; Liu, Y.; Huang, Z.; Zhang, Z. Alterations of Muscle Synergies during Voluntary Arm Reaching Movement in Subacute Stroke Survivors at Different Levels of Impairment. *Front. Comput. Neurosci.* **2018**, *12*, 1–11. [CrossRef] [PubMed]

14. Cheung, V.C.K.; Turolla, A.; Agostini, M.; Silvoni, S.; Bennis, C.; Kasi, P.; Paganoni, S.; Bonato, P.; Bizzi, E. Muscle Synergy Patterns as Physiological Markers of Motor Cortical Damage. *Proc. Natl. Acad. Sci. USA* **2012**, *109*, 14652–14656. [CrossRef] [PubMed]

15. Hesam-Shariati, N.; Trinh, T.; Thompson-Butel, A.G.; Shiner, C.T.; McNulty, P.A. A Longitudinal Electromyography Study of Complex Movements in Poststroke Therapy. 2: Changes in Coordinated Muscle Activation. *Front. Neurol.* **2017**, *8*. [CrossRef] [PubMed]

16. Irastorza-Landa, N.; García-Cossio, E.; Sarasola-Sanz, A.; Brötz, D.; Birbaumer, N.; Ramos-Murguialday, A. Functional Synergy Recruitment Index as a Reliable Biomarker of Motor Function and Recovery in Chronic Stroke Patients. *J. Neural Eng.* **2021**, *18*, 46061. [CrossRef]

17. Safavynia, S.; Torres-Oviedo, G.; Ting, L. Muscle Synergies: Implications for Clinical Evaluation and Rehabilitation of Movement. *Top. Spinal Cord Inj. Rehabil.* **2011**, *17*, 16–24. [CrossRef]

18. Bowden, M.G.; Clark, D.J.; Kautz, S.A. Evaluation of Abnormal Synergy Patterns Poststroke: Relationship of the Fugl-Meyer Assessment to Hemiparetic Locomotion. *Neurorehabilit. Neural Repair* **2010**, *24*, 328–337. [CrossRef]

19. Scano, A.; Chiavenna, A.; Malosio, M.; Tosatti, L.M.; Molteni, F. Muscle Synergies-Based Characterization and Clustering of Poststroke Patients in Reaching Movements. *Front. Bioeng. Biotechnol.* **2017**, *5*, 62. [CrossRef]

20. Abdullahi, A.; Truijen, S.; Umar, N.A.; Useh, U.; Egwuonwu, V.A.; Van Criekinge, T.; Saeys, W. Effects of Lower Limb Constraint Induced Movement Therapy in People With Stroke: A Systematic Review and Meta-Analysis. *Front. Neurol.* **2021**, *12*, 638904. [CrossRef]

21. Van Criekinge, T.; Vermeulen, J.; Wagemans, K.; Schröder, J.; Embrechts, E.; Truijen, S.; Hallemans, A.; Saeys, W. Lower Limb Muscle Synergies during Walking after Stroke: A Systematic Review. *Disabil. Rehabil.* **2020**, *42*, 2836–2845. [CrossRef] [PubMed]

22. Pellegrino, L.; Coscia, M.; Muller, M.; Solaro, C.; Casadio, M. Evaluating Upper Limb Impairments in Multiple Sclerosis by Exposure to Different Mechanical Environments. *Sci. Rep.* **2018**, *8*, 2110. [CrossRef]

23. Lencioni, T.; Fornia, L.; Bowman, T.; Marzegan, A.; Caronni, A.; Turolla, A.; Jonsdottir, J.; Carpinella, I.; Ferrarin, M. A Randomized Controlled Trial on the Effects Induced by Robot-Assisted and Usual-Care Rehabilitation on Upper Limb Muscle Synergies in Post-Stroke Subjects. *Sci. Rep.* **2021**, *11*, 5323. [CrossRef] [PubMed]

24. Laver, K.E.; George, S.; Thomas, S.; Deutsch, J.E.; Crotty, M. Virtual Reality for Stroke Rehabilitation. *Cochrane Database Syst. Rev.* **2011**, *9*, CD008349. [CrossRef]

25. You, S.H.; Jang, S.H.; Kim, Y.-H.; Hallett, M.; Ahn, S.H.; Kwon, Y.-H.; Kim, J.H.; Lee, M.Y. Virtual Reality–Induced Cortical Reorganization and Associated Locomotor Recovery in Chronic Stroke: An Experimenter-Blind Randomized Study. *Stroke* **2005**, *36*, 1166–1171. [CrossRef] [PubMed]

26. Pezzella, F.R.; Picconi, O.; De Luca, A.; Lyden, P.D.; Fiorelli, M. Development of the Italian Version of the National Institutes of Health Stroke Scale It-NIHSS. *Stroke* **2009**, *40*, 2557–2559. [CrossRef]

27. Kim, K.S.; Lee, S.J.; Suh, J.C. Numerical Simulation of the Vortical Flow around an Oscillating Circular Cylinder. In Proceedings of the International Offshore and Polar Engineering Conference, Seoul, Korea, 19–24 June 2005; pp. 162–167. [CrossRef]

28. Huber, W.; Poeck, K.; Willmes, K. The Aachen Aphasia Test. *Adv. Neurol.* **1984**, *42*, 291–303.

29. Bohannon, R.W.; Smith, M.B. Interrater Reliability of a Modified Ashworth Scale of Muscle Spasticity. *Phys. Ther.* **1987**, *67*, 206–207. [CrossRef]

30. Fugl Meyer, A.R., Jaasko, L.; Leyman, I. The Post Stroke Hemiplegic Patient. I. A Method for Evaluation of Physical Performance. *Scand. J. Rehabil. Med.* **1975**, *7*, 13–31.

31. Lèvin, M.F.; Desrosiers, J.; Beauchemin, D.; Bergeron, N.; Rochette, A. Development and Validation of a Scale for Rating Motor Compensations Used for Reaching in Patients with Hemiparesis: The Reaching Performance Scale. *Phys. Ther.* **2004**, *84*, 8–22. [CrossRef] [PubMed]

32. Piron, L.; Turolla, A.; Agostini, M.; Zucconi, C.S.; Ventura, L.; Tonin, P.; Dam, M. Motor Learning Principles for Rehabilitation: A Pilot Randomized Controlled Study in Poststroke Patients. *Neurorehabilit. Neural Repair* **2010**, *24*, 501–508. [CrossRef] [PubMed]

33. Todorov, E.; Shadmehr, R.; Bizzi, E. Augmented Feedback Presented in a Virtual Environment Accelerates Learning of a Difficult Motor Task. *J. Mot. Behav.* **1997**, *29*, 147–158. [CrossRef]

34. Merletti, R.; Hermens, H. Introduction to the Special Issue on the SENIAM European Concerted Action. *J. Electromyogr. Kinesiol.* **2000**, *10*, 283–286. [CrossRef]

35. Katirji, B. Anatomical Guide for the Electromyographer: The Limbs and Trunk, 3rd Ed. *Neurology* **1994**, *44*, 2221. [CrossRef]

36. Lee, D.D.; Seung, H.S. Learning the Parts of Objects by Non-Negative Matrix Factorization. *Nature* **1999**, *401*, 788–791. [CrossRef]

37. Saito, A.; Tomita, A.; Ando, R.; Watanabe, K.; Akima, H. Muscle Synergies Are Consistent across Level and Uphill Treadmill Running. *Sci. Rep.* **2018**, *8*, 5979. [CrossRef]

38. Bartlett, M.S. A Further Note on Tests of Significance in Factor Analysis. *Br. J. Stat. Psychol.* **1951**, *4*, 1–2. [CrossRef]

39. Kaiser, H.F.; Rice, J. Little Jiffy, Mark Iv. *Educ. Psychol. Meas.* **1974**, *34*, 111–117. [CrossRef]

40. Thompson, B. *Exploratory and Confirmatory Factor Analysis*; American Psychological Association: Washington, DC, USA, 2004; pp. 245–248. [CrossRef]

1. Carroll, J.B. How Shall We Study Individual Differences in Cognitive Abilities?—Methodological and Theoretical Perspectives. *Intelligence* **1978**, *2*, 87–115. [CrossRef]
2. Fabrigar, L.R.; Wegener, D.T. *Exploratory Factor Analysis*; Oxford University Press: Oxford, UK; New York, NY, USA, 2012; ISBN 0199813515.
3. Wood, P. *Confirmatory Factor Analysis for Applied Research*; The Guilford Press: New York, NY, USA, 2015; ISBN 978-1462515363.
4. Gorsuch, R.L. Exploratory Factor Analysis. In *Handbook of Multivariate Experimental Psychology*; Springer: Boston, MA, USA, 1988; pp. 231–258.
5. Cudeck, R. Exploratory Factor Analysis. In *Handbook of Applied Multivariate Statistics and Mathematical Modeling*; Academic Press: San Diego, CA, USA, 2000; pp. 265–296. ISBN 0-12-691360-9. (Hardcover).
6. Briggs, N.E.; MacCallum, R.C. Recovery of Weak Common Factors by Maximum Likelihood and Ordinary Least Squares Estimation. *Multivar. Behav. Res.* **2003**, *38*, 25–56. [CrossRef]
7. Hendrickson, A.E.; White, P.O. Promax: A Quick Method for Rotation to Oblique Simple Structure. *Br. J. Stat. Psychol.* **1964**, *17*, 65–70. [CrossRef]
8. Emerson, R.W. Exploratory Factor Analysis. *J. Vis. Impair. Blindness* **2017**, *111*, 301–302. [CrossRef]
9. Hair, J.; Black, W.; Babin, B.; Anderson, R. *Multivariate Data Analysis: A Global Perspective*; Pearson Education: Upper Saddle River, NJ, USA; London, UK, 2010; Volume 7, ISBN 0135153093.
50. Beran, T.N.; Violato, C. Structural Equation Modeling in Medical Research: A Primer. *BMC Res. Notes* **2010**, *3*, 267. [CrossRef] [PubMed]
51. Bentler, P.M. Fit Indexes, Lagrange Multipliers, Constraint Changes and Incomplete Data in Structural Models. *Multivar. Behav. Res.* **1990**, *25*, 163–172. [CrossRef] [PubMed]
52. Hooper, D.; Coughlan, J.; Mullen, M. Structural Equation Modeling: Guidelines for Determining Model Fit. *Electron. J. Bus. Res. Methods* **2007**, *6*, 53–60. [CrossRef]
53. Bentler, P.M.; Bonett, D.G. Significance Tests and Goodness of Fit in the Analysis of Covariance Structures. *Psychol. Bull.* **1980**, *88*, 588–606. [CrossRef]
54. Tucker, L.R.; Lewis, C. A Reliability Coefficient for Maximum Likelihood Factor Analysis. *Psychometrika* **1973**, *38*, 1–10. [CrossRef]
55. Steiger, J.H. Structural Model Evaluation and Modification: An Interval Estimation Approach. *Multivar. Behav. Res.* **1990**, *25*, 173–180. [CrossRef]
56. Steiger, J.H.; Lind, J.C. Statistically Based Tests for the Number of Common Factors. In Proceedings of the Annual Meeting of the Psychometric Society, Iowa City, IA, USA, 30 May 1980.
57. Kline, R.B. *Principles and Practice of Structural Equation Modeling*; The Guilford Press: New York, NY, USA, 2015; ISBN 1462523358.
58. Hu, L.T.; Bentler, P.M. Cutoff Criteria for Fit Indexes in Covariance Structure Analysis: Conventional Criteria versus New Alternatives. *Struct. Equ. Model.* **1999**, *6*, 1–55. [CrossRef]
59. Byrne, B.M. *Structural Equation Modeling with AMOS: Basic Concepts, Applications, and Programming*, 2nd ed.; Routledge: New York, NY, USA, 2013. [CrossRef]
60. R Core Team. *R: A Language and Environment for Statistical Computing*; R Foundation for Statistical Computing: Vienna, Austria, 2017; Available online: http://www.R-project.org. (accessed on 1 June 2019).
61. Thompson-Butel, A.G.; Lin, G.G.; Shiner, C.T.; McNulty, P.A. Two Common Tests of Dexterity Can Stratify Upper Limb Motor Function after Stroke. *Neurorehabilit. Neural Repair* **2014**, *28*, 788–796. [CrossRef]
62. Scano, A.; Chiavenna, A.; Caimmi, M.; Malosio, M.; Tosatti, L.M.; Molteni, F. Effect of Human-Robot Interaction on Muscular Synergies on Healthy People and Post-Stroke Chronic Patients. In Proceedings of the IEEE International Conference on Rehabilitation Robotics (ICORR), London, UK, 17–20 July 2017; pp. 527–532. [CrossRef]
63. Carmichael, S.T. Cellular and Molecular Mechanisms of Neural Repair after Stroke: Making Waves. *Ann. Neurol.* **2006**, *59*, 735–742. [CrossRef]
64. Cheung, V.C.K.; Cheung, B.M.F.; Zhang, J.H.; Chan, Z.Y.S.; Ha, S.C.W.; Chen, C.-Y.; Cheung, R.T.H. Plasticity of Muscle Synergies through Fractionation and Merging during Development and Training of Human Runners. *Nat. Commun.* **2020**, *11*, 4356. [CrossRef] [PubMed]
65. Levin, M.F.; Hiengkaew, V.; Nilanont, Y.; Cheung, D.; Dai, D.; Shaw, J.; Bayley, M.; Saposnik, G. Relationship Between Clinical Measures of Upper Limb Movement Quality and Activity Poststroke. *Neurorehabilit. Neural Repair* **2019**, *33*, 432–441. [CrossRef] [PubMed]
66. Pan, B.; Huang, Z.; Jin, T.; Wu, J.; Zhang, Z.; Shen, Y. Motor Function Assessment of Upper Limb in Stroke Patients. *J. Healthc. Eng.* **2021**, *2021*, 6621950. [CrossRef] [PubMed]
67. Cheung, V.C.K.; Niu, C.M.; Li, S.; Xie, Q.; Lan, N. A Novel FES Strategy for Poststroke Rehabilitation Based on the Natural Organization of Neuromuscular Control. *IEEE Rev. Biomed. Eng.* **2019**, *12*, 154–167. [CrossRef]
68. Mundfrom, D.J.; Shaw, D.G.; Ke, T.L. Minimum Sample Size Recommendations for Conducting Factor Analyses. *Int. J. Test.* **2005**, *5*, 159–168. [CrossRef]
69. Levin, M.F.; Liebermann, D.G.; Parmet, Y.; Berman, S. Compensatory Versus Noncompensatory Shoulder Movements Used for Reaching in Stroke. *Neurorehabilit. Neural Repair* **2016**, *30*, 635–646. [CrossRef]

70. Bernhardt, J.; Hayward, K.S.; Kwakkel, G.; Ward, N.S.; Wolf, S.L.; Borschmann, K.; Krakauer, J.W.; Boyd, L.A.; Carmichael, S.T.; Corbett, D.; et al. Agreed Definitions and a Shared Vision for New Standards in Stroke Recovery Research: The Stroke Recovery and Rehabilitation Roundtable Taskforce. *Int. J. Stroke* **2017**, *12*, 444–450. [CrossRef]

71. Winters, C.; Kwakkel, G.; van Wegen, E.E.H.; Nijland, R.H.M.; Veerbeek, J.M.; Meskers, C.G.M. Moving Stroke Rehabilitation Forward: The Need to Change Research. *NeuroRehabilitation* **2018**, *43*, 19–30. [CrossRef]

72. Douiri, A.; Grace, J.; Sarker, S.-J.; Tilling, K.; McKevitt, C.; Wolfe, C.D.A.; Rudd, A.G. Patient-Specific Prediction of Functional Recovery after Stroke. *Int. J. Stroke* **2017**, *12*, 539–548. [CrossRef]

 *sensors*

Article

Evaluation of Muscle Function by Means of a Muscle-Specific and a Global Index

Samanta Rosati, Marco Ghislieri *, Gregorio Dotti, Daniele Fortunato, Valentina Agostini, Marco Knaflitz and Gabriella Balestra

Department of Electronics and Telecommunications and PoliToBIOMed Lab, Politecnico di Torino, 10129 Torino, Italy; samanta.rosati@polito.it (S.R.); gregorio.dotti@polito.it (G.D.); daniele.fortunato@polito.it (D.F.); valentina.agostini@polito.it (V.A.); marco.knaflitz@polito.it (M.K.); gabriella.balestra@polito.it (G.B.)
* Correspondence: marco.ghislieri@polito.it; Tel.: +39-011-090-4136

Abstract: Gait analysis applications in clinics are still uncommon, for three main reasons: (1) the considerable time needed to prepare the subject for the examination; (2) the lack of user-independent tools; (3) the large variability of muscle activation patterns observed in healthy and pathological subjects. Numerical indices quantifying the muscle coordination of a subject could enable clinicians to identify patterns that deviate from those of a reference population and to follow the progress of the subject after surgery or completing a rehabilitation program. In this work, we present two user-independent indices. First, a muscle-specific index (MFI) that quantifies the similarity of the activation pattern of a muscle of a specific subject with that of a reference population. Second, a global index (GFI) that provides a score of the overall activation of a muscle set. These two indices were tested on two groups of healthy and pathological children with encouraging results. Hence, the two indices will allow clinicians to assess the muscle activation, identifying muscles showing an abnormal activation pattern, and associate a functional score to every single muscle as well as to the entire muscle set. These opportunities could contribute to facilitating the diffusion of surface EMG analysis in clinics.

Keywords: gait analysis; EMG; muscle activation patterns; movement analysis

Citation: Rosati, S.; Ghislieri, M.; Dotti, G.; Fortunato, D.; Agostini, V.; Knaflitz, M.; Balestra, G. Evaluation of Muscle Function by Means of a Muscle-Specific and a Global Index. *Sensors* **2021**, *21*, 7186. https://doi.org/10.3390/s21217186

Academic Editor: Georg Fischer

Received: 23 September 2021
Accepted: 27 October 2021
Published: 29 October 2021

Publisher's Note: MDPI stays neutral with regard to jurisdictional claims in published maps and institutional affiliations.

1. Introduction

The assessment of the muscle activation during human locomotion is necessary to perform a comprehensive gait analysis. In previous studies, instrumented gait analysis proved to be a powerful tool to quantitatively assess muscle activation during locomotion [1–3]. In the last decades, the activation of muscles during gait was studied through surface Electromyography (sEMG), which allows for the determination of the timing and extent of muscle activation [4–6] without relevant patient discomfort. A typical dynamic sEMG evaluation session consists of three subsequent phases. Phase I: Patient preparation, preliminary tests on the correct positioning of the probes, and patient instruction. Biomedical engineers, physiotherapists, and kinesiologists/human motion scientists usually carry out this phase [7]. Phase II: Signal acquisition, processing, and quality control. Biomedical engineers and kinesiologists/human motion scientists usually carry out this phase [7]. Phase III: Analysis and interpretation of signals and data obtained as output of the previous phase. A multidisciplinary team comprising kinesiologists/human motion scientists, clinical neurophysiologists, physiatrists, biomedical engineers, and physical therapists evaluate signals and data obtained in the previous phase and jointly prepare the clinical report [7].

Phase I: It depends on the ability of operators in: (i) correctly positioning the sEMG probes; (ii) performing a preliminary check of the signal quality; (iii) correctly instructing the patient on how to perform the movement to be studied. SEMG probe positioning may be standardized following existing protocols (i.e., see "Results of the Seniam European project" http://www.seniam.org/ (accessed on 18 October 2021)). Moreover, when looking at muscle activation intervals, a probe displacement as large as 20 mm along the muscle causes a timing error smaller than 1% of the gait cycle. Hence, probe positioning is not a major cause of variability of the results. Preliminary tests on the patient or poor patient instruction may be controlled by using very simple protocols and trained operators. Hence, this phase is not considered as a major cause of variability of the test results.

Phase II: Signal processing methods adopted are the major cause of poor repeatability of results. It has been shown in the past that pre-processing—usually denoising and band-pass filtering—as well as the choice of parameters to obtain the linear envelope of the signal and, in some cases, the activation intervals, are important causes of poor repeatability of results among different gait analysis laboratories. In fact, these are "user-dependent" choices, which can be replaced by automated algorithms that do not require any inter-action by operators and that are generally referred to as "user-independent". Several user-independent algorithms have been published in the past in specialized journals and thoroughly characterized and validated [8–16]. These algorithms are aimed at standardizing signal processing methods and signal quality control. The aim is to warrant that results obtained in different laboratories are comparable. The two indices we present in this paper are intended to be used in Phase II to quantify the adherence of the activation pattern of a muscle belonging to a specific subject to the activation prototype of that muscle obtained on a specific reference population as well as in Phase III to facilitate the interpretation of results.

Phase III: Analysis and interpretation of signals and data obtained through Phase II is generally carried out by a team of different professionals. This phase is highly subjective and results strongly depend on the team qualification, which may differ in different laboratories. Nonetheless, to facilitate the exchange of the results of gait analysis sessions among different laboratories, a common standard for test reports should be developed and gain a large consensus. At this time, to our best knowledge, a general consensus has not been reached on any of the report prototypes proposed. The two indices presented in this paper may help the team of professionals carrying out the interpretation of the test results and hence in finding consensus on the coordination of muscle activations.

The high variability of sEMG signals collected during gait, even in healthy subjects, makes it difficult to compare the muscle activity of different subjects and to find similarities or differences that could be of clinical interest [17]. Statistical Gait Analysis (SGA) has been proved to lessen this limitation, through the acquisition and processing of a large number of gait cycles [18–21]. This methodology allows for an automatic and user-independent analysis of sEMG, goniometric, and foot-switch signals collected during walking sessions lasting several minutes, and hence containing up to some hundreds of strides. In the literature, there is evidence of a great variability of the muscle activation patterns, both intra- and inter-subjects [8,22]. In previous works, the CIMAP (Clustering for Identification of Muscle Activation Patterns) algorithm was proposed to cope with intra-subject variability [9,10]. This algorithm allows for grouping strides showing similar sEMG activation patterns and, as a spin-off of this procedure, to obtain the subject's Principal Activations (PAs), as the intersection of the cluster prototypes [9,10]. PAs have been defined, from a biomechanical point of view, as those muscle activations that are strictly necessary for accomplishing a specific motor task: they describe the essential contributions of a specific muscle to kinematics, and they are reasonably repeatable among normal subjects [9,10]. This concept is complementary to that of Secondary Activations (SAs), which are activations that have auxiliary functions and are not repeatable during a single walk also within the same subject. The concept of Principal and Secondary Activations applied to the analysis of sEMG signals may significantly simplify the understanding of muscle contribution to the biomechanics of movement, and it has also been applied to the study of muscle synergies [23,24].

The terms *"muscle function"*, *"muscle activation"*, and *"EMG signal"* are crucial for the understanding of basic muscle physiology. The term *"muscle function"* refers to the force production of an active muscle that, in turn, causes its biomechanical action and its contribution to the movement. The term *"muscle activation"* means the physiological active state of a muscle: muscle fibers are activated by the release of acetylcholine underneath the end-plates which, in turn, release the neurotransmitter when they are reached by the depolarization spikes traveling along the second motor neuron that innervates the motor unit the muscle fibers belong to. When a muscle fiber is activated, depolarization arises underneath the end-plate and travels along the muscle fiber, thus causing its contraction. The *"EMG signal"*, that may be detected invasively or by means of surface electrodes (sEMG), is due to the summation of the action potentials of the active muscle fibers that are within the detection volume of the probe. Then, the (s)EMG signal (a physiological variable) is a sign of muscle activation (a physiological state of the muscle) that, in turn, is responsible for the biomechanical action of the muscle (muscle function). There is a large consensus on the fact that sEMG provides information on the neuromuscular function that is not provided by other assessment techniques [7].

In recent years, the objective assessment, based on gait data, of locomotion dysfunctions has become a research field of great interest. In literature, several works showed that it is possible to take advantage of gait parameters to improve the diagnostic process of different conditions [25–33]. Numerous studies proposed indices for an objective gait assessment based on spatio-temporal and/or joint kinematics parameters [11,32–36], but only a few works used the information extracted from sEMG signals to this purpose [12,37–39]. In particular, in the work by Castagneri et al. [12], the asymmetry level of lower limb muscles in healthy, orthopedic, and neurological subjects was assessed by combining the SGA and CIMAP algorithm, suggesting that appropriate indices can be successfully used in clinics for an objective assessment of the muscle activation asymmetry during locomotion. In this context, the definition of a quantitative and reliable index for measuring the similarity of the dynamic muscle activation of a pathological subject with that of a reference population can be extremely useful for the assessment of the disease progression and for the evaluation of treatment outcomes. At this time, to the best of our knowledge, an index with these properties has not yet been presented in the literature.

The aim of this study is twofold. First, to present a Muscle Functional Index (MFI) that quantifies the similarity of the activation pattern of a muscle of a specific subject with that of the corresponding muscle of a reference population. Second, to present a Global Functional Index (GFI) to quantify the overall muscle activation similarity of a muscle set of a specific subject compared to that of a reference population. In this paper, we defined the two indices considering a reference population of healthy children and then we assessed the behavior of the proposed indices using two groups, one consisting of healthy children (not belonging to the reference population and referred to as "controls") and a second one consisting of hemiplegic children.

2. Definition of the Indices

To assess the similarity between the activation pattern of the muscle(s) of a specific subject with respect to a reference population, we introduced two indices. The first index (MFI) quantifies the similarity of the activation intervals of a specific muscle of a subject with respect to the corresponding muscle of a reference healthy population. The second one (GFI) quantifies the overall similarity of the muscle activation patterns of a specific group of muscles with respect to those of the reference healthy population. The definition of the functional indices consists of two phases: (1) the characterization of the muscle activation of the reference healthy population and (2) the computation of the muscle functional indices. Figure 1 describes the various steps of each phase.

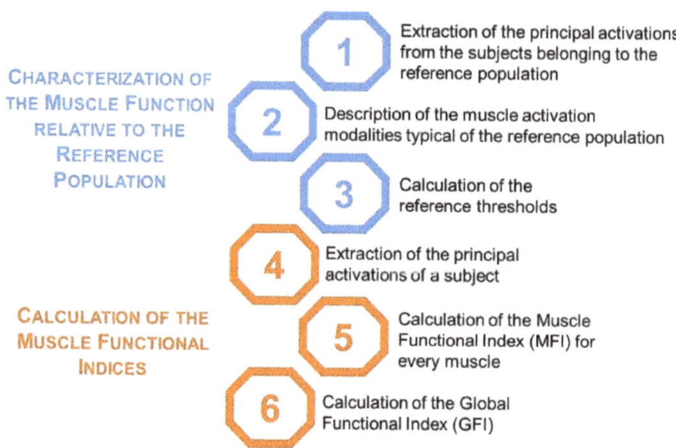

Figure 1. Pipeline for the definition of the muscle functional indices.

Both phases are based on the measure of similarity $Sim_{A,B}$ between the binary vectors A and B, of equal length (n bits), as calculated in Equation (1):

$$Sim_{A,B} = 1 - \frac{\sum_{i=1}^{n}|A_i - B_i|}{n} \tag{1}$$

where A_i and B_i are the values of the i-th bit in A and B, respectively. This measure evaluates the percentage of bits that are similar between vectors A and B and ranges from 0, if A and B are completely different, to 1, if the two vectors are equal.

2.1. Characterization of the Muscle Function Relative to the Reference Population

The characterization of the muscle function relative to the reference population consists of three steps: Section 2.1.1 extraction of the Principal Activations from the myoelectric signals collected on the subjects belonging to the reference population during the task to be studied, Section 2.1.2 description of the muscle activation modalities found in the reference population, and Section 2.1.3 calculation of the reference thresholds.

2.1.1. Extraction of the Principal Activations from the Subjects Belonging to the Reference Population

First, the PAs of each muscle are extracted from all the subjects belonging to the reference population using the optimized version of the CIMAP algorithm [10]. To apply this algorithm, the sEMG signal acquired from a specific muscle is pre-processed as follows:

- The sEMG signal is segmented into separate gait cycles by using foot-switch signals and time-normalized to 1000 samples [13];
- The onset–offset activation intervals are detected by using a two-threshold statistical detector [14];
- The onset–offset activation intervals lasting less than 3% of the gait cycle are removed, while activation intervals separated by less than 3% of the gait cycle are joined together [40];
- Every i-th gait cycle is described through a vector containing m couples of onset–offset activation intervals (ON_i, OFF_i):

$$stride_i = \{ON_{i,1}, OFF_{i,1}, \ldots, ON_{i,m}, OFF_{i,m}\} \tag{2}$$

where m is the number of onset-offset activation intervals within the same gait cycle and generally differs from muscle to muscle.

The CIMAP algorithm [10] is then applied to all the gait cycles of each investigated muscle to obtain clusters showing similar muscle activation patterns. For each cluster, the strides belonging to right and left sides are separated and the prototype of each group is calculated as the median of the time instants (ON_i, OFF_i) (Figure 2a). The prototypes are coded as strings of 1000 bits (0 = no muscle activation; 1 = muscle activation). Then, the intersection of the corresponding cluster's prototypes constitutes the PA (Figure 2b). At the end of this phase, every subject within the reference population is characterized by two PAs (one for each side).

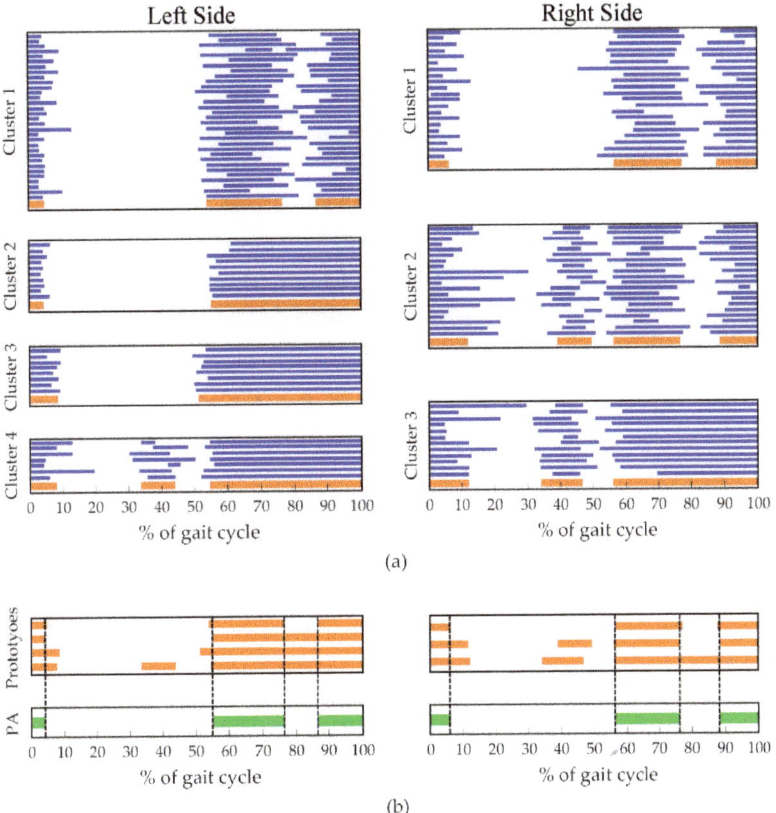

Figure 2. Example of PA extraction (Tibialis Anterior muscle). (**a**) Clusters resulting from the application of CIMAP. Strides belonging to the clusters are represented in blue, clusters' prototypes are represented in orange. (**b**) PAs, obtained as the intersection of the cluster prototypes, are represented in green.

2.1.2. Description of the Muscle Activation Modalities Typical of the Reference Population

This step aggregates the information contained in the PAs extracted from the reference population for each investigated muscle. For a specific muscle, pairwise comparisons among a PA A and all the other PAs B in the reference population are performed using Equation (1). Then, the median of all the obtained similarity values (R_A) is calculated for the PA A as detailed in (3):

$$R_A = median(Dist_{A,B}), \forall B \neq A \tag{3}$$

where B represents every principal activation in the reference population except for A, and $Sim_{A,B}$ is the measure of the similarity as described by Equation (1).

After computing the R values for a given muscle (R_A), the maximum across the reference population R_{max} is used to normalize every R-value:

$$R_{A,norm} = \frac{R_A}{R_{max}} \tag{4}$$

This normalization step allows for the obtaining of comparable values for different muscles, since the R_{max} values generally differ in different muscles. At the end of this phase, a set of $R_{i,norm}$ values, representing the similarity of each i-th PA compared to the other PAs, describes the behavior of the population for each investigated muscle.

2.1.3. Computation of the Reference Thresholds

A reference threshold R_{Th} calculated for each muscle allows for the comparing of the muscle activation of a specific subject with that of the reference population. In particular, R_{Th} was obtained as the 5th percentile of all $R_{A,norm}$ across the reference population. This means that 95% of the PAs in the reference population have a similarity higher than R_{Th} compared to the other PAs. At the end of this phase, a reference threshold R_{Th} is associated with each specific muscle.

2.2. Calculation of the Muscle Functional Indices

Given a subject that does not belong to the reference population, the extraction of the Muscle Functional Index (MFI) and the Global Functional Index (GFI) consists of three steps: Section 2.2.1 extraction of the PAs of the subject, Section 2.2.2 calculation of the MFI for every muscle, and Section 2.2.3 computation of the GFI.

2.2.1. Extraction of the Principal Activations of a Subject

First, the two PAs (one for each side) are extracted for each muscle belonging to the muscle pool of interest using the optimized version of the CIMAP algorithm [10], following a procedure similar to that described above with respect to the reference population (Section 2.1.1 Extraction of the Principal Activations from the subjects belonging to the reference population).

2.2.2. Calculation of the MFI for Every Muscle

For each muscle (left side and right side separately), the MFI is computed as detailed in Equation (5):

$$MFI = \frac{median(Sim_{S,A})}{R_{max}}, \forall A \text{ in the reference population} \tag{5}$$

where $Sim_{S,A}$ is the similarity among the PA of the subject S and all PAs in the reference population as obtained by Equation (1), and R_{max} is the maximum R-value computed within the reference population.

The obtained MFI value can be compared with the corresponding reference threshold R_{Th} to assess the muscle function with respect to a reference population: an MFI value below the reference threshold represents an abnormal muscle function, while an MFI value above the threshold represents a muscle function comparable to that of the reference population.

2.2.3. Calculation of the GFI

The GFI is the average of the MFI values (one for each muscle) for a given muscle pool of a specific subject (6):

$$GFI = \frac{\sum_{i=1}^{M} MFI_i}{M} \tag{6}$$

where M is the total number of observed muscles. The GFI quantifies the overall similarity of the activation patterns of a pool of muscles of a subject compared to the reference population.

3. Demonstration of the Applicability and Proper Behavior of the Indices

3.1. Subjects

In this study, we retrospectively analyzed gait data acquired from 105 school-age children [16,21]: 55 healthy children, without known neurological or orthopedic disorders, were used as reference population; 25 healthy and 25 hemiplegic children were used as test sets to evaluate the behavior of MFI and GFI. Table 1 reports the average anthropometric parameters of the populations.

Table 1. Anthropometric parameters of the populations.

	Number of Subjects	Age (Years) (Median and Range)	Gender [1]	Height (cm) (mean $\pm$ S.D.)	Body Mass (kg) (mean $\pm$ S.D.)
Healthy Children (Ref. population)	55	9 (7–11)	28M/27F	133.1 $\pm$ 9.7	30.3 $\pm$ 6.2
Healthy Children (Test Set)	25	9 (6–11)	12M/13F	133.8 $\pm$ 9.1	31.1 $\pm$ 7.4
Hemiplegic Children (Test Set)	25	8 (4–14)	15M/10F	129.7 $\pm$ 18.8	30.2 $\pm$ 11.7

[1] M = Male; F = Female.

3.2. Acquisition System and Experimental Protocol

To acquire sEMG, goniometric, and foot-switch signals, we used the wearable system STEP32 (Medical Technology, Turin, Italy), CE certified for clinical gait analysis. Participants were equipped bilaterally with:

- Three foot-switches (size: 10 mm $\times$ 10 mm $\times$ 0.5 mm; activation force: 3 N) attached beneath the heel, the first, and the fifth metatarsal heads of each foot;
- Two electrogoniometers (accuracy: 0.5°) attached to the lateral side of the knee joints;
- Five sEMG active probes in single differential configuration (two Ag-disks with a diameter equal to 4 mm per probe; inter-electrode distance: 12 mm; probe size: 27 mm $\times$ 19 mm $\times$ 7.5 mm) attached, after skin preparation, on the belly of each muscle. Specifically, we recorded signals from Tibialis Anterior (TA), Gastrocnemius Lateralis (LGS), Vastus Medialis (VM), Rectus Femoris (RF), and Lateral Hamstring (LH) muscles on both body sides. An expert user visually inspected signals to exclude the presence of crosstalk.

The signal amplifier had an adjustable gain (60–94 dB) and a 3 dB bandwidth ranging from 10 to 400 Hz. Gain was adjusted to fit the signal amplitude to the input dynamic range of the A/D converter as much as possible, but avoiding its saturation. The sampling rate was equal to 2 kHz, and signals were converted by a 12-bit A/D converter and stored on the hard disk of the host computer.

Figure 3 shows the acquisition system composed of the sEMG active probes, the foot-switch sensors, and the electrogoniometers. Figure 4 shows an example of sensor placement on a healthy subject.

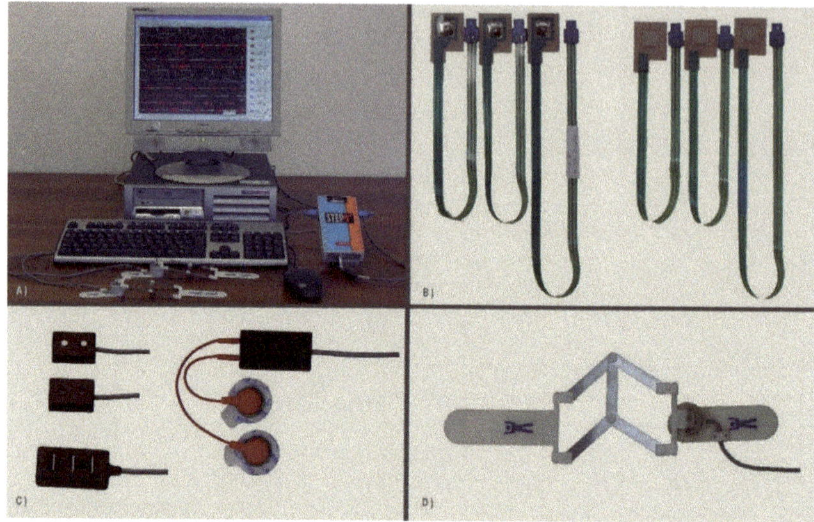

Figure 3. Details of the acquisition system: (**A**) the host computer, the patient unit, and two electrogoniometers; (**B**) two different kinds of foot-switches (on the left, a less sensitive set, for adults; on the right, a more sensitive set, for children); (**C**) different kinds of sEMG probes: two different versions of single differential probes (upper left); a three-bar double differential probe (lower left); a variable geometry probe (right); (**D**) a knee electrogoniometer.

Subjects walked barefoot at self-selected speed over a 10 m walkway, back and forth, for approximately 2.5 min. The experimental protocol conformed to the Helsinki declaration on medical research involving human subjects and was carried out in a clinical environment with continuous medical supervision. Subject assent and signed parental informed consent were obtained for each subject.

3.3. Signal Pre-Processing

Using the SGA routines included in the software of the acquisition system (which is CE certified), we obtained, for each lower limb, the following gait phases: Heel contact (H), Flat foot contact (F), Push off (P), and Swing (S) [13]. The sEMG signals were then segmented in separate gait cycles and time-normalized to the stride duration [13]. For healthy children, we considered only the gait cycles showing the typical sequence of gait phases (i.e., H, F, P, and S phase). For hemiplegic children, since a very small number of HFPS gait cycles was available, we analyzed the strides of the most represented sequence of gait phases of each subject [15,16].

A multivariate statistical filter was used to discard strides related to deceleration, acceleration, and changes of direction [13].

Subjects whose sEMG signals had an SNR value lower than 6 dB for even a single muscle were discarded from the analysis, since we considered the signal quality not suitable to warrant reliable data.

Finally, the onset–offset muscle activation intervals were detected for each muscle and each side through a double-threshold muscle activation detector specifically developed for gait analysis [14].

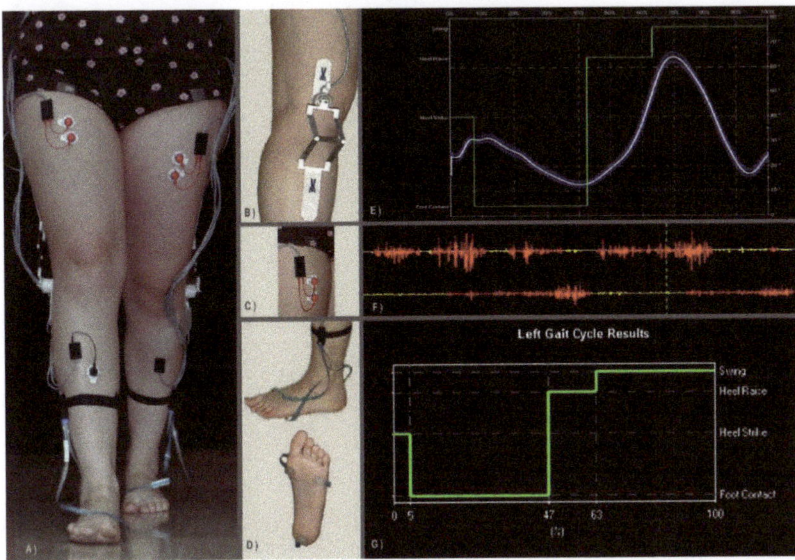

Figure 4. Sensor placement and recorded signals. SEMG active probes are positioned over the main muscles of the lower limb, bilaterally. Electrogoniometers are attached to the lateral aspect of the knee joints. Foot-switches are placed beneath the heel, the first, and the fifth metatarsal heads of each foot. (**A**) Subject performing an evaluation session. (**B**) Detail of the electrogoniometer attached to the lateral aspect of the knee to measure the knee joint angles during gait. (**C**) Detail of a variable geometry sEMG probe attached over the Rectus Femoris muscle of the subject. (**D**) Detail of the foot-switches attached underneath the first and fifth metatarsal heads and the heel (lower picture); how the foot-switches are attached to their connector (upper figure). (**E**) Example of the average variation of the knee joint angle over a given number of strides superimposed to the correspondent four-level coded foot-switch signal. (**F**) Example of two sEMG signals (Tibialis Anterior, upper trace; Gastrocnemius Lateralis, lower trace) collected during gait and processed by the user-independent activation detector: the yellow color means that the muscle is not electrically active and red color means that the muscle is electrically active. (**G**) Example of a four-level coded foot-switch signal: the four levels correspond to Heel strike (H phase), Flat foot contact (F phase), heel raise or Push off (P phase), and Swing (S phase); the sequence of foot-contact phases here represented corresponds to that observed in normal subjects during level walking.

3.4. Characterization of the Muscle Function Relative to the Reference Population

The three steps described in the previous section were applied to data relative to the healthy children included in the reference population to obtain, for each muscle, the $R_{A,norm}$ value and the corresponding reference threshold R_{Th}.

3.5. Calculation of the Muscle and Global Muscle Functional Indices

Onset–offset activation intervals of the groups of healthy and hemiplegic children were used to compute the MFI for each muscle and subject, as well as the corresponding GFI. We believe that using a radar diagram is a simple and effective way for visually inspecting the behavior of a specific subject against the average behavior of the reference population. For each muscle, the MFI values can be compared with the corresponding reference threshold R_{Th} of the healthy population, to visually check their similarity. Figure 5 shows an example of this representation used to inspect the behavior of specific subjects.

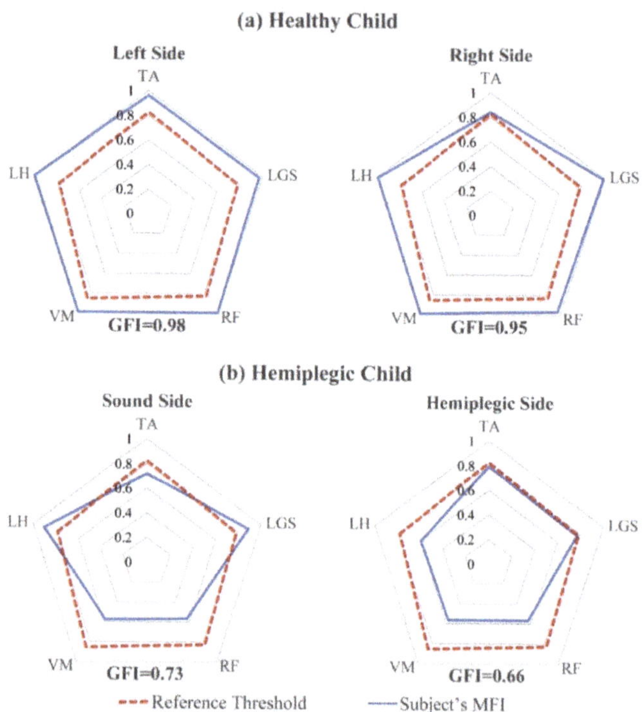

Figure 5. Radar diagram representation of MFI values for (**a**) a healthy child and (**b**) a hemiplegic child, both sides. The corresponding GFIs are reported under each diagram. The dotted red lines join the reference thresholds R_{Th} for each muscle. The blue lines join the MFI values of the subject. Muscles: Tibialis Anterior (TA), Gastrocnemius Lateralis (LGS), Vastus Medialis (VM), Rectus Femoris (RF), and Lateral Hamstring (LH).

3.6. Statistical Analysis

We applied the Lilliefors test to assess the normality of the MFI and GFI distributions of hemiplegic children, both for the hemiplegic and the sound sides, and healthy children, both for the left and right sides. Based on the Lilliefors test result, a two-tailed paired Student *t*-test (α = 0.05) (in case of normal distributions) or a Wilcoxon signed-rank test (α = 0.05) (for non-normal distributions) was used to compare: (a) hemiplegic and sound side of hemiplegic children, (b) left and right side of healthy children. The statistical analysis was carried out using the Statistical and Machine Learning Toolbox of MATLAB® release 2020b (The MathWorks Inc., Natick, MA, USA).

4. Results

The data of 31 children out of 105 were discarded due to the low SNR of the myoelectric signals: 15 children belonging to the reference population, 7 healthy children belonging to the control population, and 9 hemiplegic children.

An average of 168 ± 27 gait cycles were collected for each child of the reference population and an average of 167 ± 25 and 133 ± 35 gait cycles were collected for each child of the two test groups (healthy and hemiplegic children, respectively).

From the reference population of 40 healthy children, we obtained the following threshold values: R_{Th} = 0.86 for VM, R_{Th} = 0.83 for TA and RF, and R_{Th} = 0.78 for LGS and LH. Figure 5 reports the MFI and the GFI values for two representative subjects of the test set (panel a: a typically developing child; panel b: a hemiplegic child). The dotted red lines join the reference threshold R_{Th} for each muscle. The blue lines join the MFI values of specific subjects. The radar diagram allows for the easy highlighting of muscles with an abnormal behavior and significantly simplifies the interpretation of the MFI and GFI values. As an example, Figure 5a shows that, for the representative healthy child chosen from the reference population, all the muscles (both sides) present an MFI value above the reference threshold R_{Th}, and the GFI values are equal to 0.98 and 0.95 for the left and right sides, respectively. Differently, for the hemiplegic child, whose results are reported by Figure 5b, only the Gastrocnemius Lateralis (LGS) and the Lateral Hamstring (LH) muscles of the sound side show MFI values above the reference threshold, meaning that their behavior is similar to that of 95% of the reference healthy population. For the hemiplegic side, the MFI values are below the respective thresholds for all the muscles studied. Tibialis Anterior (TA) and Gastrocnemius Lateralis (LGS) show MFI values close to those of the reference population, thus demonstrating minimal dysfunction of these muscles, while proximal muscles (RF, VM, and VL) show MFI values close to 0.6, thus demonstrating a noticeable dysfunction. Consequently, GFI values of the hemiplegic child are 0.73 and 0.66 for the sound and hemiplegic side, respectively. This demonstrates that, in this specific child, both lower limbs cannot be considered as normally functioning and, as expected, the hemiplegic side shows a more severe condition than the other side. In addition, the non-affected side may not be considered to have a normal function. Figure 5 shows how the two indices may quantify the degree of functionality of the investigated muscles in a specific subject, either normal or pathological. This is the most important use of the two indices in clinics. As an example, considering the healthy child (Figure 5a), the radar plot clearly shows that, on both body sides, all the observed muscles are above the threshold that represents the minimum value of the index obtained on the 95% of subjects belonging to the reference population. Hence, this specific subject may not be distinguished by 95% of subjects belonging to the reference population. Moreover, it is clear that, while on the right side all the five observed muscles show a value of the MFI index close to 1 (the best possible match to controls), on the left side the TA muscle shows a value of MFI that is only slightly higher than that corresponding to the threshold. This could be a suggestion for clinicians to investigate the behavior of the TA muscle more in depth, to decide whether to prescribe a rehabilitation program to the subject or simply to repeat the exam after 6–12 months, to document possible changes. More than one third of typically developing children show mild gait abnormalities when they undergo a gait analysis test; in most cases, these abnormalities have no clinical meaning or disappear when the subject grows up, but in some cases they are worthy of being treated, since they could cause problems in adulthood or in the elderly. Figure 5b is a clear example of how MFI can very simply indicate which muscles of the observed muscle pool show an altered activation. On the hemiplegic side, TA and LGS (dorsi and plantar flexors of the ankle) show an MFI value very close to the threshold, thus demonstrating their almost normal activation timing. On the contrary, RF, VM, and LH show MFI values definitely below the threshold, thus demonstrating that muscle timing is compromised at the level of proximal muscles, that control both knee (LH, VM, and RF) and hip (RF and LH). The left radar plot of Figure 5b shows that, on the sound side, the MFI value of TA is slightly lower than the threshold. More interestingly from a clinical point of view, it is evident that knee extensors (RF and VM) show MFI values close to those of the affected side, while the LH shows a timing compatible with that of 95% of the control population. The considerations above show how the MFI values can be very effective in outlining the inappropriate timing of some of the muscles belonging to the considered muscle pool. The value of the GFI quantifies how the timing of the considered muscle pool is close to that of the normal population.

Figure 6 shows the MFI value for each muscle and for each of the two test groups (Panel a: healthy children belonging to the control group; Panel b: hemiplegic children).

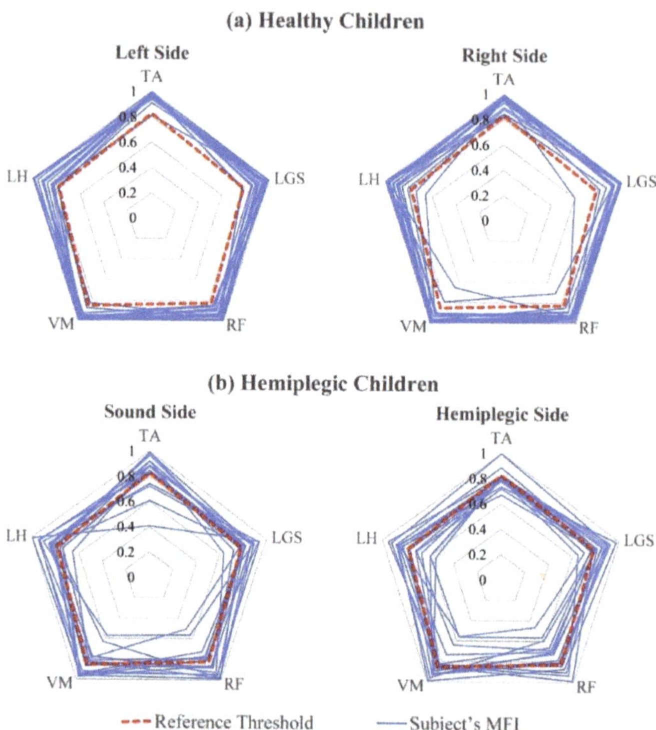

Figure 6. Radar diagram of MFI values for (**a**) the 18 healthy children and (**b**) the 16 hemiplegic children, both sides. The dotted red lines join the reference threshold R_{Th} for each muscle. The blue lines join the MFI values for each subject in the two test groups. Muscles: Tibialis Anterior (TA), Gastrocnemius Lateralis (LGS), Vastus Medialis (VM), Rectus Femoris (RF), and Lateral Hamstring (LH).

The dotted red lines join the reference thresholds R_{Th} computed over the reference population for each muscle. The blue lines join the MFI values for each specific subject in the two test groups. It is evident that the MFI values for healthy children are mostly above the thresholds for all the muscles and both sides. Only in 2 out of 40 cases, on the left side, there are subjects whose MFI values relative to one or two muscles are slightly outside the behavior of 95% of the subjects belonging to the healthy population. For the hemiplegic group, on the contrary, the distribution of the MFI values is wider, showing an abnormal behavior, definitely more evident on the hemiplegic side. We included Figure 6 for two different reasons. First, for demonstrating that the control population (that was not used to obtain the threshold values) shows values of the MFIs that are almost always above the thresholds computed for each specific muscle, while this is not the case—as expected—for hemiplegic children. Hence, the behavior of MFI matches our expectations. Second, when considering hemiplegic children, it is evident that every subject shows a different pattern of MFI values, thus demonstrating the capability of the index to capture differences among different subjects.

Table 2 reports the mean, the first and third quartile of the MFI values for the two test groups and for each muscle and side. The last column of the table contains the values of the reference threshold R_{Th} for the five muscles.

Figure 7 shows the boxplots of the MFI values for the five muscles observed in this study, for the test populations of healthy (in violet) and hemiplegic children (in orange), for the two sides, separately. Since most of the distributions resulted non-normal according to the Lilliefors test, the Wilcoxon signed-rank test was used for comparing the MFI values of each muscle separately. In particular, the values of the MFI are not statistically different between the left and right side of healthy subjects for all muscles ($p > 0.05$), as expected. For the hemiplegic children, MFI values are not statistically different between left and right lower limbs for all the muscles, except for the RF ($p = 0.02$). This is not surprising, because to compensate the deficiency of muscles on the anatomically affected side also muscles belonging to the non-affected side must modify and adapt their activation modality. Comparing the MFI values of each lower limb of the healthy children and the two sides of the hemiplegic population, it emerges that values are statistically different for all comparisons, except for: (i) the right side of the healthy children with respect to the sound side of the hemiplegic children for the RF muscle ($p = 0.08$) and (ii) the left side of the healthy children with respect to the sound side of the hemiplegic children for the LH muscle ($p = 0.11$).

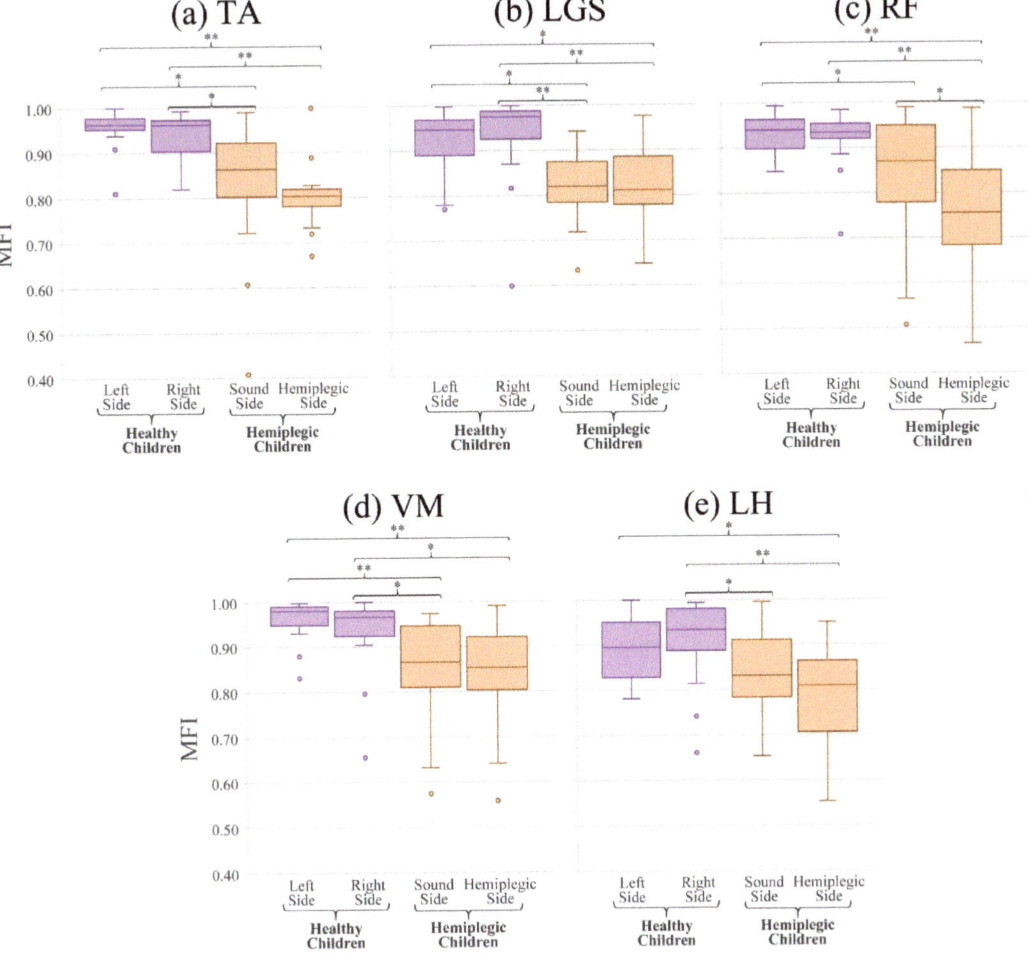

Figure 7. Boxplots of the MFI values of healthy and hemiplegic children of the test set, for the 5 muscles: (**a**) Tibialis Anterior (TA), (**b**) Gastrocnemius Lateralis (LGS), (**c**) Rectus Femoris (RF), (**d**) Vastus Medialis (VM), and (**e**) Lateral Hamstring (LH). Asterisks highlight statistically significant differences between groups or side (*: $p < 0.05$ and **: $p < 0.001$). Outliers are represented by circles.

Table 2. MFI and GFI values for the test groups (mean value [first and third quartile]) and the reference thresholds.

		Healthy Children		Hemiplegic Children		Reference Threshold R_{Th}
		Left Side	**Right Side**	**Sound Side**	**Hemiplegic Side**	
MFI	TA	0.96 [0.95 ÷ 0.98]	0.94 [0.90 ÷ 0.97]	0.83 [0.80 ÷ 0.92]	0.80 [0.78 ÷ 0.82]	0.83
	LGS	0.92 [0.89 ÷ 0.97]	0.93 [0.92 ÷ 0.99]	0.82 [0.78 ÷ 0.87]	0.82 [0.78 ÷ 0.88]	0.78
	RF	0.93 [0.90 ÷ 0.96]	0.92 [0.92 ÷ 0.95]	0.84 [0.78 ÷ 0.95]	0.75 [0.68 ÷ 0.85]	0.83
	VM	0.96 [0.95 ÷ 0.99]	0.93 [0.92 ÷ 0.98]	0.85 [0.81 ÷ 0.95]	0.82 [0.80 ÷ 0.92]	0.86
	LH	0.89 [0.83 ÷ 0.95]	0.91 [0.89 ÷ 0.98]	0.84 [0.78 ÷ 0.91]	0.78 [0.71 ÷ 0.86]	0.78
GFI		0.93 [0.92 ÷ 0.95]	0.93 [0.90 ÷ 0.95]	0.83 [0.82 ÷ 0.86]	0.80 [0.78 ÷ 0.83]	-

Figure 8 reports the boxplots of the GFI values relative to the test populations of healthy and hemiplegic children (for the two sides, separately). The last row of Table 2 reports the mean, and the first and third quartile of the GFI values for the two test groups. Since all distributions resulted normal, the two-tailed Student *t*-test was applied for the comparison of the GFI values. In particular, the values of the GFI are not statistically different between the left and right side of healthy subjects ($p = 0.47$), as expected. Comparing the GFI values relative to each lower limb of the healthy population and the hemiplegic and sound sides of the hemiplegic children it is evident that values are statistically different ($p < 0.001$, all comparisons). Finally, when considering the GFI values of the hemiplegic and sound sides of hemiplegic children they are statistically different ($p = 0.02$), as expected.

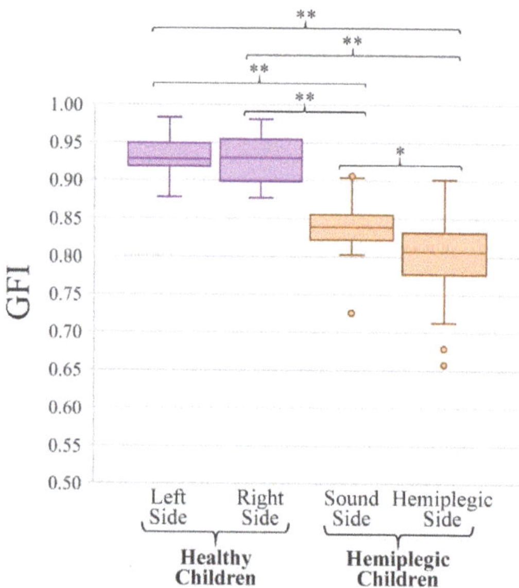

Figure 8. Boxplots of the GFI values of healthy and hemiplegic children of the test set. Asterisks highlight statistically significant differences between groups or sides (*: $p < 0.05$ and **: $p < 0.001$). Circles represent outliers.

5. Discussion

Since the 1980s, numerous studies demonstrated the utility of sEMG for investigating the muscle function in basic research as well as in clinics. Although numerous research studies relative to muscle physiology investigated the muscle function in healthy and pathological subjects at the level of basic research, the number of papers reporting of sEMG applications in clinics, which really improved the quality of patient management, is definitely more limited. Then, the body of knowledge available at this time, relative to research and clinical experiences carried out in the last 40 years, fully demonstrates the capabilities and outlines the limitations of sEMG.

A recent work [7] investigated the usage and barriers of sEMG in neurorehabilitation, by sending a 30-question survey to 52 experts on sEMG from different standpoints, countries, and backgrounds. Among the 18 questions for which a consensus higher than 75% was reached, some of them well relate to this work. Specifically, a consensus higher than 90% was reached on the following nine points:

1. "sEMG provides information on the neuromuscular function that is not provided by other assessment techniques/tools in neurorehabilitation" (91%);
2. "In clinical rehabilitation sEMG enhances the assessment and characterization of neuromuscular impairment in patients" (94%);
3. "sEMG allows to evaluate the effects of non-invasive interventions designed to impact muscle activity" (91%);
4. "sEMG may be useful to evaluate the appropriateness of the activation among muscles participating to a specific movement" (97%);
5. "sEMG allows to outline the sequential timing of muscular actions during given movements" (100%);
6. "sEMG allows to evaluate the appropriateness of the activation among muscles participating to a specific movement" (97%);
7. "sEMG assessment can be performed as a stand-alone technique to complement/optimize gait/motion analysis" (100%);
8. "Timing of muscle activations and their variability must be considered of utmost importance for clinical applications in neurorehabilitation among the EMG-derived variables" (100%);
9. "The difficulty of performing sEMG data analysis and interpretation without specific education/training is a potential barrier to the employment of sEMG in clinical neurorehabilitation" (97%).

From the cited work, which is very recent and, at this time, unique in the field of sEMG applied in clinics, we can summarize three statements on which the consensus is total:

- sEMG is a necessary tool to obtain a deep insight into the role of different muscles during any kind of movement;
- sEMG can be used as a stand-alone technique or it should be used as a complementary tool in gait/motion analysis, principally considering the timing of muscle activation;
- Performing sEMG data analysis and interpretation, with the tools currently available, is a complex task that requires specific training.

Hence, we can infer that for encouraging the spread of the usage of sEMG in clinics, scientists working on basic sEMG research should develop tools as much as possible that are user-independent, widely tested, and useful in clinics for facilitating the interpretation of multiple sEMG recordings. This is the purpose of the methodology herein presented.

Since the 1990s, some of the authors of this paper devoted a large part of their research activities developing user-independent methods to facilitate the application of sEMG signal analysis in clinics. Briefly, we developed, among others, the following tools: (i) a double threshold statistical detector of muscle activation [14] (1998); (ii) a comprehensive methodology for user-independent gait analysis, in which sEMG plays a major role [41] (2012); (iii) an improved algorithm for the user-independent segmentation and classification of gait cycles from foot-switch signals [13] (2014); (iv) an algorithm for quantifying the gait impairment score based on fuzzy logic [11] (2017); (v) an algorithm for hierarchical clustering of muscle activations [9] (2017); (vi) a user-independent index for quantifying asymmetry in muscle activations [12] (2019). This paper is the natural extension of our previous work since it describes the development of another two indices aimed at facilitating the usage of sEMG in clinics without requiring any user-dependent decisions. We stress that user-independent analysis is essential for assuring high repeatability of the obtained results among different laboratories.

We introduced two indices based on the Principal Activations, which numerically describe the muscle activations of a subject with respect to a reference population. We showed that the muscle activation of a subject may be quantitatively evaluated for a single muscle (muscle-specific index, MFI) as well as for a specific muscle pool (global index GFI). Moreover, to easily identify those muscles that are not activated in a physiological way, we proposed the representation of the MFIs by means of radar diagrams. This kind of representation may be easily adapted to any number of observed muscles.

The proposed indices allow for quantifying the similarity of the muscle activation of a specific subject to normality, which is defined as the behavior of a reference population. The present work uses as the reference population a group of 40 typically developing children studied during gait by investigating a group of five lower limb muscles (TA, LGS, RF, VM, and LH). The choice of these muscles assures having at least a flexor and an extensor muscle for each of the three joints usually investigated in gait (ankle, knee, and hip), which we consider as a solution generally satisfactory from a clinical point of view [41].

Notice that, since each MFI value is computed considering the sEMG signal generated only by the muscle it refers to, it does not depend on the number of observed muscles. Once the MFI values are obtained for any specific number of muscles considered, the GFI value may be obtained as the average of all the computed MFIs.

MFI and GFI find their most important application in clinics, when used to compare the activation and coordination of the muscle pool of interest of a specific subject to that of a reference population, for identifying possible deviations from the "normality", as Figure 5 shows. As an example, the left radar plot of Figure 5b shows that, on the sound side, the MFI value of TA is slightly lower than the threshold value. More interestingly from a clinical point of view, it is evident that knee extensors (RF and VM) show MFI values close to those of the affected side, while the LH shows a timing pattern compatible with that of 95% of the control population. A plausible clinical interpretation of this observation is that the compensatory effect on the sound lower limb, that is necessary to obtain an acceptable locomotion, principally involves knee extensors. This observation may play an important role in designing a rehabilitation protocol suitable to the needs of this specific subject, thus implementing a personalized-medicine-approach in rehabilitation. This specific capability of the proposed indices would help clinicians to develop a personalized rehabilitation program for each specific patient, thus improving the likelihood of success.

In this work, however, we compared rather extensively the behavior of the two indices when applied to a control population of typically developing children (different from that used to obtain the threshold values of the reference population) and a population of hemiplegic children. This was undertaken using two groups that are known to clearly differ in muscle activation patterns and coordination, only to give proof of the proper behavior of the indices. In fact, Table 2, Figure 6, Figure 7, and Figure 8 clearly show that MFI and GFI values differentiate the two sub-populations, demonstrating that the control population of typically developing children has a behavior that is always compatible with that of the reference population of typically developing children, while the group of hemiplegic children shows a clearly different behavior. These statements have been supported by proper statistics.

Notice that, even if 31 subjects out of 105 were discarded from the sample population, this does not limit the validity of the presented results. To the best of our knowledge, the data set used in this study is still the larger available database of sEMG and other gait-related signals describing the walking modalities of school-age children. Furthermore, several acquisitions were discarded solely due to poor adherence to the experimental protocol of the child, which is more likely to happen in younger subjects compared to older ones. Possible applications of objective indices describing the similarity of the activity of a single muscle or of a muscle pool belonging to a specific subject to the behavior of a reference normal population may have different applications in clinics. Indeed, indices can be used to demonstrate the anomalous function of a muscle during a specific task thus allowing the understanding of the causes of a motion abnormality, a necessary step for developing an effective rehabilitation program or to plan surgery. Moreover, indices quantifying the similarity of the muscle activation of a subject with reference to a matched normal population also allow for evaluating the effectiveness of a rehabilitation program over time, to document objectively the recovery of normal function at a single muscle level as well as at the level of a muscle pool.

Actually, when we refer to "normal population" in terms of gait, we must be aware that the concept of "normality" is often associated also to subjects that are affected by gait abnormalities that do not compromise noticeably their daily activities. It has already been reported that a fraction of typically developing children shows slight gait abnormalities that are not evident to the visual observation [21]. These abnormalities do not cause any specific limitation and hence are not reported to physical therapists or physicians for early correction. Nonetheless, it may not be ruled out that the possibility of such apparently negligible abnormalities, that may be already evident in childhood and in teenagers, could cause more severe problems to affected people in adulthood or in the elderly. These problems could range from an abnormal fall propensity to low back pain, and several other conditions. In this perspective, the availability of indices as MFI and GFI, as well as other indices quantifying the "quality" of gait [11,29,30,34,35], could allow for a relatively inexpensive and operator-independent screening of school-age children and teenagers, to identify slight gait abnormalities caused by poor muscle function or coordination and thus allowing for the definition of specific correction protocols.

Another important class of possible applications is represented by the follow-up of patients following a rehabilitation program after orthopedic surgery or for compensating gait abnormalities following acute, degenerative, and congenital conditions of the nervous system (i.e., stroke, multiple sclerosis, Parkinson's disease, cerebral palsy, etc.). Even in longitudinal evaluations, the availability of user-independent and reliable indices is crucial to allow for an objective patient assessment [42].

Finally, another very interesting application of these indices relates to sports training. By considering a specific movement performed by top-level athletes, one could build a "top reference population" to be used for scoring the performance of less talented athletes and, possibly, suggesting training programs to improve their performance. At this time, we do not have experience in this application, but we do believe it is worthy of being investigated.

6. Limitations of the Study

Although we believe that the indices we propose may be beneficial in clinics, we are aware of some limitations.

First, to apply the proposed indices to patients, from childhood to the elderly, we need three reference populations: typically developing children, normal adults, and elderly people. This is a limitation, but overcoming it only requires collecting and processing data from subjects belonging to the populations of interest. We already started collecting data from normal adults and shortly we will extend the study to elderly people.

A second limitation of this study is the number of investigated muscles, which is in this paper restricted to only five lower limb muscles. For improving the possible impact of the indices in clinical gait studies, it could be desirable to consider more than five muscles for each subject side. As already stated, we are currently working towards obtaining a reference adult population and we are considering twelve lower limb and trunk muscles. We foresee considering a set of at least twelve muscles also in the elderly reference population. Recording twelve or more muscles from typically developing children was not considered in our previous study mainly for two reasons. (i) The time needed to prepare a subject would have been definitely longer, and it is challenging to keep children concentrated for long periods of time. (ii) Especially for younger subjects, it may be difficult to place a large number of EMG probes on the limbs, due to the limited size of limbs.

However, the limitations of this study we are aware of, which we can easily overcome by extending the number of reference populations and considering a larger muscle pool, are well counterbalanced by the principal strength of this approach. This is the possibility of easily developing reference data for every cyclic movement, such as biking, running, stairs climbing, upper limb reaching tasks, swimming, and many others. In fact, although we tested the two indices and their computation in gait, the algorithms may be easily adapted to every cyclic movement, thus considerably enlarging the range of possible applications. In fact, the CIMAP algorithm, that is crucial for the extraction of the PAs on which GFI and MFI are based, was optimized for cyclic movements in general [10].

7. Conclusions

This work describes two quantitative indices for evaluating muscle activation in gait studies or several other cyclic movements. The MFI is relative to the activation of a single muscle, part of an observed muscle pool, and GFI is relative to the entire muscle pool.

In this study, we described how to compute the two indices and we demonstrated their proper performance in gait studies, considering a reference population of 40 typically developing children in which we detected sEMG from five lower limb muscles. The extension of the application of these indices to subjects from childhood to old age only requires the definition of another two reference populations, namely, one of normal adults and a second one of physiological elderly subjects. We are currently working on obtaining these two reference populations. We increased the number of observed muscles in adults and in the elderly from five to twelve, to extend the applicability of the methodology to larger muscle pools.

In conclusion, MFI and GFI values can provide a quantitative and reliable evaluation of muscle activation for identifying the abnormal function of single muscles involved in different movements and in various populations.

Given the importance of the availability of data describing various reference populations, we believe that experienced researchers working in this field should share through public data repositories their data, to make it possible to other research groups working in rehabilitation and sports medicine to benefit from the open access to reliable data sets.

Author Contributions: Conceptualization, S.R., M.G., V.A., M.K. and G.B.; methodology, S.R., M.G., V.A., M.K. and G.B.; software, M.G. and G.D.; validation, S.R., M.G., V.A., M.K. and G.B.; data curation, S.R., M.G., G.D. and D.F.; writing—original draft preparation, S.R., M.G. and M.K.; writing—review and editing, S.R., M.G., G.D., D.F., V.A., M.K. and G.B.; supervision, G.B., V.A. and M.K. All authors have read and agreed to the published version of the manuscript.

Funding: This research received no external funding.

Institutional Review Board Statement: Not applicable (this is a retrospective study carried out on a data set acquired several years ago, conforming to the Helsinki declaration, and gait analysis sessions were carried out under medical supervision).

Informed Consent Statement: Subject assent and signed parental informed consent were obtained for each subject.

Data Availability Statement: Data presented in this study are available on request from the corresponding author.

Conflicts of Interest: The authors declare no conflict of interest.

References

1. Wren, T.A.L.; Gorton, G.E.; Ounpuu, S.; Tucker, C.A. Efficacy of clinical gait analysis: A systematic review. *Gait Posture* **2011**, *34*, 149–153. [CrossRef] [PubMed]
2. Tao, W.; Liu, T.; Zheng, R.; Feng, H.; Tao, W.; Liu, T.; Zheng, R.; Feng, H. Gait Analysis Using Wearable Sensors. *Sensors* **2012**, *12*, 2255–2283. [CrossRef]
3. Chang, F.M.; Seidl, A.J.; Muthusamy, K.; Meininger, A.K.; Carollo, J.J. Effectiveness of Instrumented Gait Analysis in Children with Cerebral Palsy—Comparison of Outcomes. *J. Pediatr. Orthop.* **2006**, *26*, 612–616. [CrossRef] [PubMed]
4. De Luca, C.J. The Use of Surface Electromyography in Biomechanics. *J. Appl. Biomech.* **1997**, *13*, 135–163. [CrossRef]
5. Roetenberg, D.; Buurke, J.; Veltink, P.; Forner Cordero, A.; Hermens, H. Surface electromyography analysis for variable gait. *Gait Posture* **2003**, *18*, 109–117. [CrossRef]
6. Frigo, C.; Crenna, P. *Multichannel SEMG in Clinical Gait Analysis: A Review and State-of-the-Art*; Elsevier: Amsterdam, The Netherlands, 2009; Volume 24, pp. 236–245.
7. Manca, A.; Cereatti, A.; Bar-On, L.; Botter, A.; Della Croce, U.; Knaflitz, M.; Maffiuletti, N.A.; Mazzoli, D.; Merlo, A.; Roatta, S.; et al. A Survey on the Use and Barriers of Surface Electromyography in Neurorehabilitation. *Front. Neurol.* **2020**, *11*, 1137. [CrossRef]
8. Agostini, V.; Rosati, S.; Castagneri, C.; Balestra, G.; Knaflitz, M. Clustering analysis of EMG cyclic patterns: A validation study across multiple locomotion pathologies. In Proceedings of the 2017 IEEE International Instrumentation and Measurement Technology Conference (I2MTC), Turin, Italy, 22–25 May 2017; pp. 1–5.
9. Rosati, S.; Agostini, V.; Knaflitz, M.; Balestra, G. Muscle activation patterns during gait: A hierarchical clustering analysis. *Biomed. Signal. Process. Control.* **2017**, *31*, 463–469. [CrossRef]
10. Rosati, S.; Castagneri, C.; Agostini, V.; Knaflitz, M.; Balestra, G. Muscle contractions in cyclic movements: Optimization of CIMAP algorithm. *Proc. Annu. Int. Conf. IEEE Eng. Med. Biol. Soc. EMBS* **2017**, 58–61. [CrossRef]
11. Rosati, S.; Agostini, V.; Knaflitz, M.; Balestra, G. Gait impairment score: A fuzzy logic-based index for gait assessment. *Int. J. Appl. Eng. Res.* **2017**, *12*, 3337–3345.
12. Castagneri, C.; Agostini, V.; Rosati, S.; Balestra, G.; Knaflitz, M. Asymmetry Index in Muscle Activations. *IEEE Trans. Neural Syst. Rehabil. Eng.* **2019**, *27*, 772–779. [CrossRef]
13. Agostini, V.; Balestra, G.; Knaflitz, M. Segmentation and classification of gait cycles. *IEEE Trans. Neural Syst. Rehabil. Eng.* **2014**, *22*, 946–952. [CrossRef] [PubMed]
14. Bonato, P.; D'Alessio, T.; Knaflitz, M. A statistical method for the measurement of muscle activation intervals from surface myoelectric signal during gait. *IEEE Trans. Biomed. Eng.* **1998**, *45*, 287–299. [CrossRef]
15. Agostini, V.; Knaflitz, M.; Nascimberi, A.; Gaffuri, A. Gait measurements in hemiplegic children: An automatic analysis of foot-floor contact sequences and electromyographic patterns. In Proceedings of the IEEE MeMeA 2014—IEEE International Symposium on Medical Measurements and Applications, Lisboa, Portugal, 11–12 June 2014; pp. 1–4.
16. Agostini, V.; Nascimbeni, A.; Gaffuri, A.; Knaflitz, M. Multiple gait patterns within the same Winters class in children with hemiplegic cerebral palsy. *Clin. Biomech.* **2015**, *30*, 908–914. [CrossRef]
17. Di Nardo, F.; Maranesi, E.; Mengarelli, A.; Ghetti, G.; Burattini, L.; Fioretti, S. Assessment of the variability of vastii myoelectric activity in young healthy females during walking: A statistical gait analysis. *J. Electromyogr. Kinesiol.* **2015**, *25*, 800–807. [CrossRef]
18. Di Nardo, F.; Ghetti, G.; Fioretti, S. Assessment of the activation modalities of gastrocnemius lateralis and tibialis anterior during gait: A statistical analysis. *J. Electromyogr. Kinesiol.* **2013**, *23*, 1428–1433. [CrossRef]
19. Di Nardo, F.; Fioretti, S. Statistical analysis of surface electromyographic signal for the assessment of rectus femoris modalities of activation during gait. *J. Electromyogr. Kinesiol.* **2013**, *23*, 56–61. [CrossRef] [PubMed]
20. Benedetti, M.; Agostini, V.; Knaflitz, M.; Gasparroni, V.; Boschi, M.; Piperno, R. Self-reported gait unsteadiness in mildly impaired neurological patients: An objective assessment through statistical gait analysis. *J. Neuroeng. Rehabil.* **2012**, *9*, 64. [CrossRef] [PubMed]
21. Agostini, V.; Nascimbeni, A.; Gaffuri, A.; Imazio, P.; Benedetti, M.G.; Knaflitz, M. Normative EMG activation patterns of school-age children during gait. *Gait Posture* **2010**, *32*, 285–289. [CrossRef]
22. Winter, D.A.; Yack, H.J. EMG profiles during normal human walking: Stride-to-stride and inter-subject variability. *Electroencephalogr. Clin. Neurophysiol.* **1987**, *67*, 402–411. [CrossRef]
23. Ghislieri, M.; Agostini, V.; Knaflitz, M. How to Improve Robustness in Muscle Synergy Extraction. *Annu. Int. Conf. IEEE Eng. Med. Biol. Soc.* **2019**, *2019*, 1525–1528. [CrossRef]
24. Ghislieri, M.; Agostini, V.; Knaflitz, M. Muscle Synergies Extracted Using Principal Activations: Improvement of Robustness and Interpretability. *IEEE Trans. Neural Syst. Rehabil. Eng.* **2020**, *28*, 453–460. [CrossRef] [PubMed]
25. Mazzetta, I.; Zampogna, A.; Suppa, A.; Gumiero, A.; Pessione, M.; Irrera, F. Wearable Sensors System for an Improved Analysis of Freezing of Gait in Parkinson's Disease Using Electromyography and Inertial Signals. *Sensors* **2019**, *19*, 948. [CrossRef] [PubMed]
26. Martínez, M.; Villagra, F.; Castellote, J.M.; Pastor, M.A. Kinematic and kinetic patterns related to free-walking in parkinson's disease. *Sensors* **2018**, *18*, 4224. [CrossRef]
27. Khoury, N.; Attal, F.; Amirat, Y.; Oukhellou, L.; Mohammed, S. Data-driven based approach to aid Parkinson's disease diagnosis. *Sensors* **2019**, *19*, 242. [CrossRef]

28. Lai, D.T.H.; Levinger, P.; Begg, R.K.; Gilleard, W.L.; Palaniswami, M. Automatic Recognition of Gait Patterns Exhibiting Patellofemoral Pain Syndrome Using a Support Vector Machine Approach. *IEEE Trans. Inf. Technol. Biomed.* **2009**, *13*, 810–817. [CrossRef]

29. Alaqtash, M.; Sarkodie-Gyan, T.; Yu, H.; Fuentes, O.; Brower, R.; Abdelgawad, A. Automatic classification of pathological gait patterns using ground reaction forces and machine learning algorithms. In Proceedings of the 2011 Annual International Conference of the IEEE Engineering in Medicine and Biology Society, Boston, MA, USA, 30 August–3 September 2011; pp. 453–457.

30. Kamruzzaman, J.; Begg, R.K. Support Vector Machines and Other Pattern Recognition Approaches to the Diagnosis of Cerebral Palsy Gait. *IEEE Trans. Biomed. Eng.* **2006**, *53*, 2479–2490. [CrossRef]

31. Zhang, B.; Zhang, Y.; Begg, R.K. Gait classification in children with cerebral palsy by Bayesian approach. *Pattern Recognit.* **2009**, *42*, 581–586. [CrossRef]

32. Cui, C.; Bian, G.-B.; Hou, Z.G.; Zhao, J.; Su, G.; Zhou, H.; Peng, L.; Wang, W. Simultaneous Recognition and Assessment of Post-Stroke Hemiparetic Gait by Fusing Kinematic, Kinetic, and Electrophysiological Data. *IEEE Trans. Neural Syst. Rehabil. Eng.* **2018**, *26*, 856–864. [CrossRef]

33. Chakraborty, S.; Nandy, A. Automatic diagnosis of cerebral palsy gait using computational intelligence techniques: A low-cost multi-sensor approach. *IEEE Trans. Neural Syst. Rehabil. Eng.* **2020**, *28*, 2488–2496. [CrossRef]

34. Schwartz, M.H.; Rozumalski, A. The gait deviation index: A new comprehensive index of gait pathology. *Gait Posture* **2008**, *28*, 351–357. [CrossRef] [PubMed]

35. Schutte, L.M.; Narayanan, U.; Stout, J.L.; Selber, P.; Gage, J.R.; Schwartz, M.H. An index for quantifying deviations from normal gait. *Gait Posture* **2000**, *11*, 25–31. [CrossRef]

36. Baker, R.; McGinley, J.L.; Schwartz, M.H.; Beynon, S.; Rozumalski, A.; Graham, H.K.; Tirosh, O. The Gait Profile Score and Movement Analysis Profile. *Gait Posture* **2009**, *30*, 265–269. [CrossRef]

37. Kugler, P.; Jaremenko, C.; Schlachetzki, J.; Winkler, J.; Klucken, J.; Eskofier, B. Automatic recognition of Parkinson's disease using surface electromyography during standardized gait tests. In Proceedings of the 2013 35th Annual International Conference of the IEEE Engineering in Medicine and Biology Society (EMBC), Osaka, Japan, 3–7 July 2013; pp. 5781–5784.

38. Bojanic, D.M.; Petrovacki-Balj, B.D.; Jorgovanovic, N.D.; Ilic, V.R. Quantification of dynamic EMG patterns during gait in children with cerebral palsy. *J. Neurosci. Methods* **2011**, *198*, 325–331. [CrossRef] [PubMed]

39. Infarinato, F.; Romano, P.; Goffredo, M.; Ottaviani, M.; Galafate, D.; Gison, A.; Petruccelli, S.; Pournajaf, S.; Franceschini, M. Functional Gait Recovery after a Combination of Conventional Therapy and Overground Robot-Assisted Gait Training Is Not Associated with Significant Changes in Muscle Activation Pattern: An EMG Preliminary Study on Subjects Subacute Post Stroke. *Brain Sci.* **2021**, *11*, 448. [CrossRef] [PubMed]

40. Bogey, R.A.; Barnes, L.A.; Perry, J. Computer algorithms to characterize individual subject EMG profiles during gait. *Arch. Phys. Med. Rehabil.* **1992**, *73*, 835–841. [PubMed]

41. Agostini, V.; Knaflitz, M. Statistical gait analysis. In *Distributed Diagnosis and Home Healthcare (D2H2)*; American Scientific Publishers: Stevendon Ranch, CA, USA, 2012; Volume II, pp. 99–121.

42. Agostini, V.; Ganio, D.; Facchin, K.; Cane, L.; Moreira Carneiro, S.; Knaflitz, M. Gait parameters and muscle activation patterns at 3, 6 and 12 months after total hip arthroplasty. *J. Arthroplast.* **2014**, *29*, 1265–1272. [CrossRef]

 *sensors*

Article

An Algorithm for Choosing the Optimal Number of Muscle Synergies during Walking

Riccardo Ballarini [1], **Marco Ghislieri** [1,2], **Marco Knaflitz** [1,2] and **Valentina Agostini** [1,2,*]

[1] Department of Electronics and Telecommunications, Politecnico di Torino, 10129 Turin, Italy; riccardo.ballarini@studenti.polito.it (R.B.); marco.ghislieri@polito.it (M.G.); marco.knaflitz@polito.it (M.K.)
[2] PoliTo^{BIO}Med Lab, Politecnico di Torino, 10129 Turin, Italy
* Correspondence: valentina.agostini@polito.it; Tel.: +39-011-0904136

Abstract: In motor control studies, the 90% thresholding of variance accounted for (VAF) is the classical way of selecting the number of muscle synergies expressed during a motor task. However, the adoption of an arbitrary cut-off has evident drawbacks. The aim of this work is to describe and validate an algorithm for choosing the optimal number of muscle synergies (ChoOSyn), which can overcome the limitations of VAF-based methods. The proposed algorithm is built considering the following principles: (1) muscle synergies should be highly consistent during the various motor task epochs (i.e., remaining stable in time), (2) muscle synergies should constitute a base with low intra-level similarity (i.e., to obtain information-rich synergies, avoiding redundancy). The algorithm performances were evaluated against traditional approaches (threshold-VAF at 90% and 95%, elbow-VAF and plateau-VAF), using both a simulated dataset and a real dataset of 20 subjects. The performance evaluation was carried out by analyzing muscle synergies extracted from surface electromyographic (sEMG) signals collected during walking tasks lasting 5 min. On the simulated dataset, ChoOSyn showed comparable performances compared to VAF-based methods, while, in the real dataset, it clearly outperformed the other methods, in terms of the fraction of correct classifications, mean error (ME), and root mean square error (RMSE). The proposed approach may be beneficial to standardize the selection of the number of muscle synergies between different research laboratories, independent of arbitrary thresholds.

Keywords: gait; locomotion; motor module; number of synergies; VAF

Citation: Ballarini, R.; Ghislieri, M.; Knaflitz, M.; Agostini, V. An Algorithm for Choosing the Optimal Number of Muscle Synergies during Walking. *Sensors* **2021**, *21*, 3311. https://doi.org/10.3390/s21103311

Academic Editor: Marco Iosa

Received: 31 March 2021
Accepted: 7 May 2021
Published: 11 May 2021

Publisher's Note: MDPI stays neutral with regard to jurisdictional claims in published maps and institutional affiliations.

1. Introduction

Muscle synergies are a valuable tool to understand the mechanisms behind motor control in a quantitative and non-invasive way. Applications range from the medical field (e.g., monitoring of patients suffering from neurological/neurodegenerative diseases [1–3] or joint disorders [4]), to the rehabilitation field (e.g., pre/post-treatment comparisons [5,6]), to the robotic field (e.g., control of robotic devices or exoskeletons [7,8]), to the sport field [9].

The hypothesis of muscle synergies provides an insight into how the central nervous system (CNS) is able to manage a highly complex system with many muscles and joints. Indeed, the basis of this hypothesis is the ability of the CNS to reduce a large number of degrees of freedom of the movement thanks to the combination of a few discrete elements [10]. In other words, to generate movements, the CNS would not control the different muscles individually, but through functional groups, called muscle synergies.

Muscle synergies are usually extracted from surface electromyography (sEMG) signals, properly pre-processed, using the non-negative matrix factorization (NMF) algorithm [11,12]. This factorization algorithm requires the number of muscle synergies (n) as an input, which is not known a priori. Therefore, the factorization is typically run several times, considering different numbers of synergies ($n_i = [n_{min}, n_{max}]$). The only constraint is that the number of synergies must not exceed the number of muscles (m) considered in the sEMG acquisition ($n_{max} \leq m$); otherwise, the meaning of "synergy" itself would be lost. Afterward, in post-processing, one has to choose the "correct" number of synergies n_c (with $n_{min} \leq n_c \leq n_{max}$), i.e., the input that feeds the factorization algorithm providing "good" results, with a "small-enough" reconstruction error. In recent years, the correct number of muscle synergies (n_c) has been proposed as a meaningful feature for the analysis of motor control strategies in pathological populations [13–17]. A decreased neuromuscular complexity during gait has been assessed in post-stroke patients with respect to a healthy population [13]. Similar results were also found in another work [14], in which a reduced number of muscle synergies (two to four muscle synergies) were observed in the affected side of post-acute stroke patients with respect to a healthy population (four muscle synergies) while executing cycling training. These studies suggest that the number of muscle synergies and their composition could be correlated with motor control capacity and its reduction in pathological conditions [13–17].

Here lies one of the main issues of the muscle synergy extraction process: currently there is a lack of reliable methodologies for choosing the optimal number of muscle synergies. Most of the published studies choose n_c based on the reconstruction accuracy of the factorization, through the variance accounted for (VAF) [1,2,18–25]. To a lesser extent, the coefficient of determination R^2 [16,26,27] is also used, which is not conceptually different from VAF. However, this approach requires the selection of an arbitrary threshold for the VAF. The number n_c is defined as the smallest number of synergies that ensures a VAF value above the threshold. In literature, the VAF threshold is commonly set at 90% [1,2,4,7,18,20–24] and less frequently at 95% [19,25]. This method is very simple to implement, but it has several drawbacks: the threshold is arbitrary, it is set without an objective motivation, and there is not a single threshold value shared by all researchers. A few works have explored alternatives to VAF-based criteria. In particular, a statistical approach uses unstructured sEMG signals generated by randomly shuffling the original data across time and muscle [16], while other works consider the variability of muscle synergies between task cycles [28], or a task decoding-based metric [29,30].

The aim of this work is to overcome VAF-based methods using a data-driven approach. We designed and validated an algorithm for choosing the optimal number of muscle synergies (ChoOSyn), based on two parameters directly extracted from muscle synergies during locomotion: (1) the consistency within the motor task epochs (to identify synergies that are stable over the duration of the walking task), (2) the intra-level dissimilarity between synergies (to identify a base of information-rich synergies, avoiding unnecessary redundancy). Both a simulated and a real dataset were used to compare the performance of ChoOSyn against VAF-based methods.

2. Materials and Methods

2.1. Real Dataset

The real dataset originates from the retrospective analysis of sEMG signals previously recorded at Biolab (Politecnico di Torino, Italy) during gait analysis sessions [22–24]. The dataset contains gait signals from 20 healthy adults: 9 males (age: 56.9 ± 9.8 years, height: 1.71 ± 0.10 m, weight: 79.1 ± 22.0 kg) and 11 females (age: 51.5 ± 10.1 years, height: 1.66 ± 0.09 m, weight: 74.5 ± 24.0 kg).

Subjects walked at a self-selected speed for approximately 5 min. SEMG signals were acquired using the multi-channel recording system STEP32 for Statistical Gait Analysis (Medical Technology, Turin, Italy) [31]. The electrodes were positioned over the following 12 muscles of the dominant lower limb and over the trunk (bilaterally): Gluteus Medius (GMD), Tensor Fasciae Latae (TFL), Rectus Femoris (RF), Vastus Medialis (VM), Lateral Hamstring (LH), Medial Hamstring (MH), Lateral Gastrocnemius (LGS), Peroneus Longus (PL), Soleus (SOL), Tibialis Anterior (TA), and both right and left Longissimus Dorsii (LD_R and LD_L) muscles.

The volunteers enrolled in this work signed a written informed consent to participate in a study concerning muscle synergies adopted by healthy subjects during locomotion. The experimental protocol conformed to the principles of the Helsinki declaration.

2.2. Simulated Dataset

Similar to previous studies [21,32,33], pseudo-real sEMG data have been generated from the real dataset to simulate the muscle activity during gait (simulated dataset). The following steps were used to generate simulated data:

- From the dataset of 20 subjects, 15 subjects were extracted, showing $n = 4$ (5 subjects), $n = 5$ (5 subjects), and $n = 6$ (5 subjects) clearly recognizable muscle synergies, as assessed by expert operators (V.A. and M.G.). Hence, for each subject, activation coefficients (C) and weight vectors (W) were obtained. Figure 1A shows an example of muscle synergies ($n = 5$) representative of a specific subject.
- For each group of 5 subjects, data augmentation was performed to obtain 25 "simulated subjects", considering all the possible combinations of W and C. In other words, the matrix of weight vectors of the first subject (W_{subj1}) was combined with the coefficient matrix of every subject in the group ($W_{subj1}\,C_{subj1}$, $W_{subj1}\,C_{subj2}$, $\cdots$ $W_{subj1}\,C_{subj5}$), and the same was performed for the other weight matrixes (W_{subj2}, $\cdots$ W_{subj5}), obtaining 25 sets of muscle synergies. Overall, 25 sets were obtained with $n = 4$, 25 sets with $n = 5$, and 25 sets with $n = 6$, for a total of 75 sets.
- For each set of W and C, each muscle's envelope was reconstructed as the product $W_{muscle} * C$, where W_{muscle} is the weight vector of a specific muscle. Figure 1B provides an example for the LGS muscle.
- For each muscle's envelope, a simulated sEMG signal (S) was generated by multiplying the envelope by a zero-mean Gaussian process (G_S) with standard deviation $\sigma = 1$ a.u. (Figure 1C). At this step, no additive noise was superimposed on the signals. This does not mean that there was "no noise", but rather that additional noise to the noise originally present in the envelope was not introduced.
- Then, different levels of background noise were added to obtain different SNR values (15 dB, 20 dB, 25 dB, and 30 dB), through a zero-mean Gaussian process (G_N) with a standard deviation $\sigma = 1/10^{SNR/20}$ a.u. [21,34]. Figure 1D shows an example in which SNR was equal to 20 dB. The formula below (1) summarizes how each simulated sEMG signal was generated:

$$S = W_{muscle} \times C \times G_S + G_N \tag{1}$$

Therefore, a total of 375 simulated sets were obtained, since we introduced both signals with no additive noise (75 sets) and signals with 4 different SNR values (75 $\times$ 4 sets).

2.3. Muscle Synergy Extraction and Sorting

After sEMG pre-processing [21–24], muscle synergies were extracted and properly ordered as outlined in Figure 2.

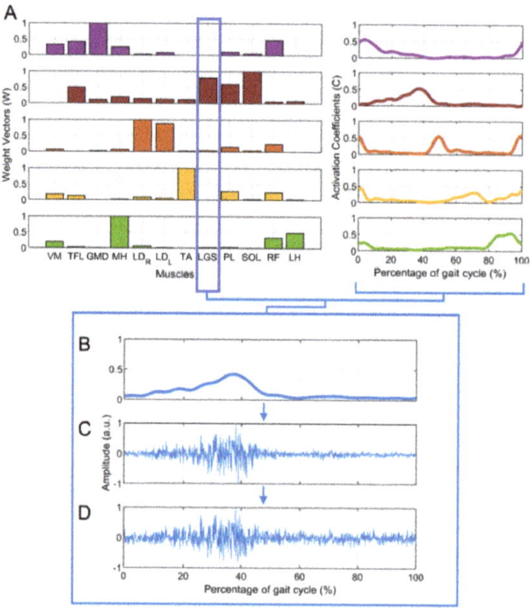

Figure 1. Example of the generation of a simulated sEMG signal for the lateral gastrocnemius (LGS) muscle: the first step is (**A**) the extraction of muscle synergies (W and C) from the real data of a representative subject with 5 muscle synergies, the second is (**B**) the reconstruction of the LGS envelope (obtained as $W_{LGS} * C$). Then, (**C**) a simulated sEMG signal without additive noise is generated. Finally, noise is added to the previous signals. An example of a simulated sEMG signal with SNR = 20 dB is shown in (**D**).

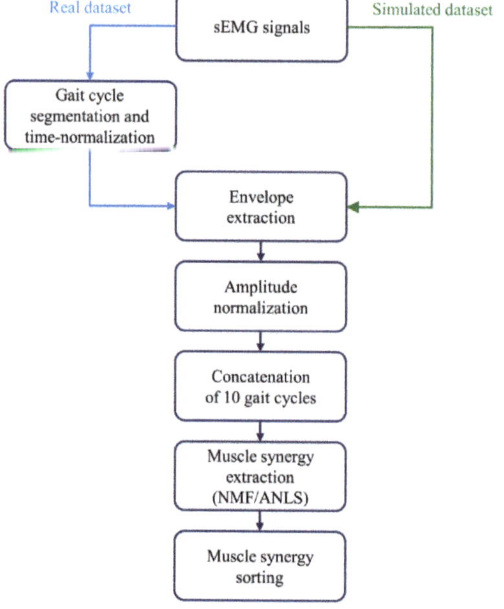

Figure 2. Data processing: steps to extract muscle synergies from the sEMG signals.

First, gait cycles were segmented and time-normalized to 1000 samples. Second, signals were high-pass filtered at 35 Hz through an 8th order Butterworth digital filter to attenuate slow movement artifacts and baseline wandering [22,35,36]. Third, signals were demeaned and rectified. Fourth, rectified signals were low-pass filtered at 12 Hz through a 5th order Butterworth digital filter to obtain the sEMG envelopes [22,35,36]. Fifth, each envelope was normalized in amplitude with respect to its global maximum. Then, we concatenated 10 adjacent gait cycles [37]. If the walk contained N gait cycles, the total number of subgroups was calculated rounding down $N/10$ to the smallest integer. As an example, if the walk contained 152 gait cycles, we considered 15 subgroups. Afterward, muscle synergies were extracted for each 10-cycle subgroup [21–24] through non-negative matrix factorization (NMF) [11,12]. The NMF algorithm models the original sEMG data as a linear combination of weight vectors (W) and activation coefficients (C), whose dimensions depend on the selected number of muscle synergies. In particular, the former models the time-independent contribution of each muscle to a specific muscle synergy, while the latter describes the time-dependent modulation of each muscle synergy. Instead of using the multiplicative update rule of the standard NMF approach [12], we chose to use another version of the algorithm, NMF with alternating non-negative least-squares (NMF/ANLS) [38], due to its advantages in terms of reduced computational time [38]. For the NMF/ANLS we set the following parameters: maximum iterations = 1000 [22], reruns = 5, residual error $<10^{-6}$ [22], and output variation $<10^{-6}$ [22]. To explore different solutions, the NMF algorithm was run several times on the same original sEMG data by changing the number of muscle synergies n in the range [1,8].

Finally, the 10-gait-cycle activation coefficient (10,000 samples) was time-averaged across windows of 1000 samples. Thus, we obtained an average activation coefficient for each subgroup.

The factorization returns W (and C) in a different order for each subgroup, and, therefore, proper sorting was required to average the correspondent W (and C) between the subgroups. For each number of synergies (n), we applied a k-means algorithm to reorder the weight vectors across the different subgroups (number of clusters: n, distance metric: cosine similarity, max iterations: 10^5, replicas: 15) [22]. Activation coefficients were then sorted accordingly.

2.4. Choosing the Optimal Number of Synergies (ChoOSyn)

The algorithm for choosing the optimal number of muscle synergies (ChoOSyn algorithm) is based on the two following muscle synergy features:

- High consistency across time [22,23], that supports the possibility of finding a solution that is as stable as possible among the various 10-cycle subgroups of the motor task
- Low similarity across synergies, to avoid selecting muscle synergies containing redundant information.

We chose these criteria after considering the characteristics of muscle synergies extracted from the real dataset.

In the following sections, we introduce the mathematical description of the parameters used to quantify the features described above. These parameters are a function of the number of synergies, so they assume a specific value for each number of synergies. They are also applicable for $n \geq 2$, because the similarity parameter cannot be extracted at $n = 1$ (since there is only one synergy).

2.4.1. Intra-Cluster Variability

The intra-cluster variability (ICV) quantifies the possible inconsistency of weight vectors (ICV_W) and activation coefficients (ICV_C) across time, i.e., between the different subgroups of 10 gait cycles. Its purpose is to quantify the level of variability of a given synergy during the considered task.

More specifically, for each number of synergies n ($2 \leq n \leq 8$), for each synergy i (with $i = 1, \ldots, n$), and for each subgroup j, we calculate the distance between each "cluster element" (W^{ij} and C^{ij}) and the "cluster centroid" ($\overline{W}^i$ and $\overline{C}^i$), through cosine similarity [22,23,39]. Then, ICV is defined as:

$$ICV_W = max\left(1 - \frac{W^{ij} \cdot \overline{W}^i}{\|W^{ij}\| \, \|\overline{W}^i\|}\right) \tag{2}$$

$$ICV_C = max\left(1 - \frac{C^{ij} \cdot \overline{C}^i}{\|C^{ij}\| \, \|\overline{C}^i\|}\right) \tag{3}$$

for the weights and the coefficients, respectively. Notice that the average operator is always applied across subgroups [22,23]. The "max" function is used to select the most variable muscle synergy ("worst" condition), obtaining a single ICV value for each n value. The ICV value ranges from 0 (i.e., perfectly repeatable muscle synergy between the different subgroups) to 1 (i.e., completely different muscle synergy across subgroups).

2.4.2. Weight Similarity

The parameter weight similarity (WS) is introduced to select the two most similar weight vectors ("worst-case") belonging to different muscle synergies.

For each number of synergies n ($2 \leq n \leq 8$), and for each synergy i (with $i = 1, \ldots, n$), the average weight vector across subgroups is considered ($\overline{W}^i$), representing the weights of a specific synergy over the entire locomotion task. Then, "cosine similarity" is introduced to quantify the degree of correlation between each couple of weight vectors $\left(\overline{W}^i 1 \, \overline{W}^k\right)$, and the WS parameter is defined as in (4):

$$WS = max\left(\frac{\overline{W}^i \cdot \overline{W}^k}{\|\overline{W}^i\| \, \|\overline{W}^k\|}\right) \tag{4}$$

where $\overline{W}^i$ and $\overline{W}^k$ represent the average weight vectors computed across subgroups for the i- and k-synergy, respectively. The WS value ranges from 0 (i.e., completely dissimilar muscle synergies) to 1 (i.e., completely similar muscle synergies).

2.4.3. Coefficient Similarity

The coefficient similarity (CS) parameter is introduced to select activation coefficients that limit, as much as possible, any redundant information between different muscle synergies. In this case, the correlation of muscle synergies is evaluated between levels n and $n - 1$, to check if the splitting of a specific synergy (at level $n - 1$) into two synergies (at level n) really provides new information.

For each number n of synergies ($2 \leq n \leq 8$), and for each synergy i (with $i = 1, \ldots, n$), the average activation coefficient across subgroups is considered ($\overline{C}^i$), representing the coefficients of a specific synergy over the entire locomotor task. Then, we identify the two (out of n) synergies that originated from a specific synergy belonging to the $n-1$ level. These synergies are obtained (except for $n = 2$) by clustering the weights at level n into $n-1$ clusters (with the weights of level $n-1$ as centroids), through k-means [28]. In this way, the coefficients of the two synergies of interest ($\overline{C}^i$ and $\overline{C}^k$) will belong to the same cluster (having "forced" n elements to cluster into $n-1$ clusters). Finally, "cosine similarity" is introduced to quantify the degree of correlation between the two activation coefficients just identified, and the CS parameter is defined as in (5):

$$CS = \frac{\overline{C}^i \cdot \overline{C}^k}{\|\overline{C}^i\| \, \|\overline{C}^k\|} \tag{5}$$

The *CS* value ranges from 0 (i.e., high information content provided by the new muscle synergy introduced in the level n) to 1 (i.e., low information content provided by the new muscle synergy introduced in the level n).

2.4.4. ChoOSyn

The ChoOSyn algorithm combines the parameters described above to determine the optimal number of muscle synergies. In particular, for each number of synergies n $(2 \leq n \leq 8)$, we define:

$$ChoOSyn_W(n) = WS(n) + ICV_W(n) \tag{6}$$

$$ChoOSyn_C(n) = CS(n) + ICV_C(n), \tag{7}$$

for the weights and the coefficients, respectively. The above formulas include the quantification of the synergy similarity through WS and CS to avoid redundant information and the quantification of the synergy consistency across time through ICV to discourage the choice of unstable muscle synergies (see Section 2.4.5—ChoOSyn rules).

Figure 3 shows, as bar diagrams, the values of the $ChoOSyn_W$ and $ChoOSyn_C$ parameters obtained from the data of two representative (real) subjects. Average bar diagrams of these two parameters were also obtained for the whole simulated and real datasets (and reported in the Results section).

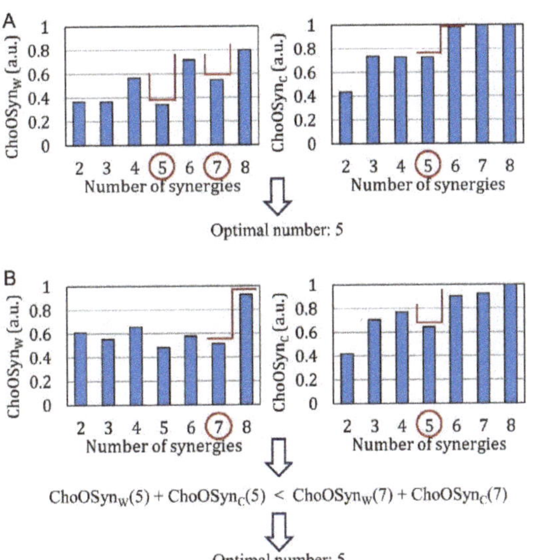

Figure 3. Examples of $ChoOSyn_W$ and $ChoOSyn_C$ values calculated on muscle synergies extracted from the data of two representative real subjects. "Steps" and local minima are highlighted by red segments. These examples show how the optimal number of synergies is chosen when the outputs of the two parameters are (**A**) the same or (**B**) different.

While for the real dataset, we do not know, a priori, the correct number of synergies, this information is known for the simulated dataset. Therefore, through analyzing the bar diagrams of $ChoOSyn_W$ and $ChoOSyn_C$ extracted from the simulated dataset, it can be seen that, in correspondence with the correct number of synergies ($n = n_c$), there is always a "step" and/or a local minimum. In the following section, this observation will be used to empirically introduce selection rules for obtaining the correct number of synergies. The term "step" refers to a "sharp" increase in the value of the parameter, preceded and followed by "stable" values. The term "local minimum" refers to a situation in which there is an abrupt decrease followed by an abrupt increase in the parameter values [28]. Figure 3 shows examples where both steps and local minima are highlighted (red lines).

To choose the correct number of synergies using both $ChoOSyn_W$ and $ChoOSyn_C$ parameters we used specific rules detailed below.

2.4.5. ChoOSyn Rules

The presence of a step reveals that muscle synergies maintain an almost steady consistency and similarity as n increases ($n \leq n_c$), but after exceeding n_c ($n > n_c$) they become highly variable and with redundant information. Instead, the local minimum represents a condition in which there are low values of the ChoOSyn parameters at the level n, but if n increases or decreases by 1 ($n-1$ and $n+1$ levels), the muscle synergies "get worse".

To identify steps and local minima, the ChoOSyn algorithm must be able to recognize cases where there is an increase in the value of the parameter from cases where the parameter is almost stable. We introduce the change in the ChoOSyn parameters as n increases:

$$\Delta ChoOSyn(n) = |ChoOSyn(n+1) - ChoOSyn(n)| \tag{8}$$

$$with\ 2 \leq n \leq 7 \tag{9}$$

Every variation of the parameter value greater than the average of $\Delta ChoOSyn$ is defined as an increase, and every variation smaller than $\overline{\Delta ChoOSyn}$ is defined as "stability". In this way, the algorithm can identify the previously introduced steps and local minima.

If there are multiple steps and local minima in the same bar plot, the algorithm selects only the two highest values of n (Figure 3A, left panel).

On this basis, the two parameters $ChoOSyn_W$ and $ChoOSyn_C$ make two separate selections (Figure 3). Finally, a single optimal value of n is chosen as follows:

- There is at least a common choice in the selection(s) provided by the two parameters (Figure 3A). In this case, the common number of synergies is selected.
- The two parameters provide a different selection for the number of synergies (Figure 3B). The number is chosen as the one providing the lowest sum of $ChoOSyn_W(n)$ and $ChoOSyn_C(n)$ (i.e., with the lowest similarity and highest consistency).

2.5. VAF-Based Methods

As already mentioned in the Introduction, the variance accounted for (VAF) is widely used in the literature to quantify the reconstruction accuracy after the factorization, and it is defined as the uncentered Pearson's correlation (in percentage) [1,2,18–25]:

$$VAF = \left(1 - \frac{\sum_{i=1}^{m}(M_i - R_i)^2}{\sum_{i=1}^{m} M_i^2}\right) \times 100, \tag{10}$$

where M is the matrix before the factorization, R is the reconstructed matrix obtained as the product between W and C, and m is the number of muscles (12 in this work).

We compare the performance of ChoOSyn with the three main VAF-based methods (Figure 4):

- T-VAF (Threshold VAF) (Figure 4A): this method is the most widely used in the literature [1,2,18–25]. It involves the setting of an arbitrary threshold and the subsequent choice of the first number of synergies with VAF above the threshold. The threshold is commonly set at 90% and less frequently at 95%: therefore, we chose to test both 90% and 95% thresholds.
- E-VAF (Elbow VAF) [11] (Figure 4B): this method requires finding the "elbow" of the VAF curve, i.e., the highest curvature point. It is the only VAF-based method that does not use arbitrary thresholds.
- P-VAF (Plateau VAF) [40] (Figure 4C): this method requires finding the point beyond which the VAF curve reaches a plateau. It uses an arbitrary threshold: the mean-square error obtained by fitting the VAF-curve through a straight line must be smaller than

10^{-2}. Cheung et al. [40] used a threshold equal to 10^{-5}, but in our simulated dataset 10^{-2} provided the best performance. The first point satisfying this condition is chosen.

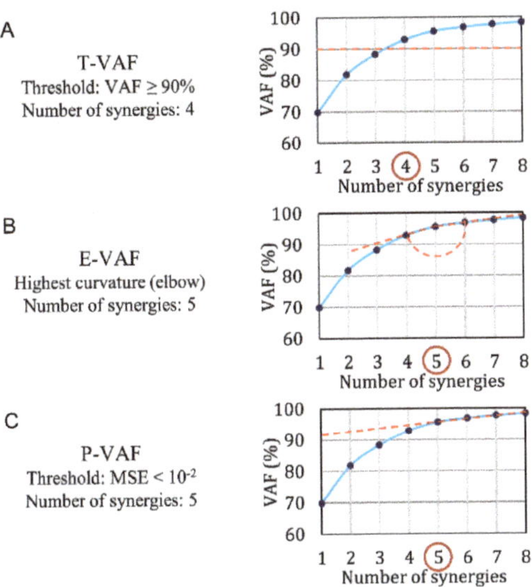

Figure 4. VAF-based methods used in the literature: (**A**) T-VAF, the most used method, (**B**) E-VAF, based on the curvature, and (**C**) P-VAF, based on the plateau. The reported VAF curve is calculated from the data of a real representative subject (the same for Figures 3 and 4).

2.6. Performance Evaluation

The performances of the ChoOSyn algorithm were compared to those of T-VAF, E-VAF, and P-VAF methods by means of the fraction of correct classifications, mean error (ME), and root mean square error (RMSE), both for the simulated and real datasets. RMSE is used to quantify how far a given method deviates from the correct number of synergies, while ME provides information about the sign, to know whether the method goes wrong by defect or excess. ME and RMSE are defined as follows:

$$ME = \frac{\sum_{i=1}^{NS}(n_i - n_{c,i})}{NS} \tag{11}$$

$$RMSE = \sqrt{\frac{\sum_{i=1}^{NS}(n_i - n_{c,i})^2}{NS}} \tag{12}$$

where n is the number of synergies identified by a method, n_c is the number of correct synergies, and NS is the total number of subjects in the dataset.

To know which number of synergies should be considered as correct (n_c) in the real dataset, we developed a "ground truth" using the judgment of two expert operators. Their judgment was performed blind to the details of the ChoOSyn algorithm as well as to the results of the various methods tested. For each real subject, they analyzed the muscle synergy plots considering different numbers of muscle synergies n and they chose—separately—the number they considered as correct, based on their knowledge of motor control strategies, muscle synergy analysis, and gait biomechanics. It should be noted that expert judgment is subjective, at least to some extent. Cohen's kappa statistic [41] was used to compute the degree of agreement between the raters. In case of disagreement, the two expert operators discussed the discordant cases to achieve a common ground truth. For the simulated dataset, its own nature guarantees its objectivity, knowing a priori the correct number of muscle synergies.

3. Results

3.1. Simulated Data

Figure 5A–C show the two ChoOSyn parameters extracted from the simulated dataset. More specifically, data obtained simulating $n = 4$, $n = 5$, and $n = 6$ muscle synergies are displayed. The bar plots show a marked "step", in correspondence to the correct number of synergies. This is the main feature that allows the algorithm to identify the optimal number of synergies without thresholds. In addition, in some cases, the plots present a local minimum just before the step.

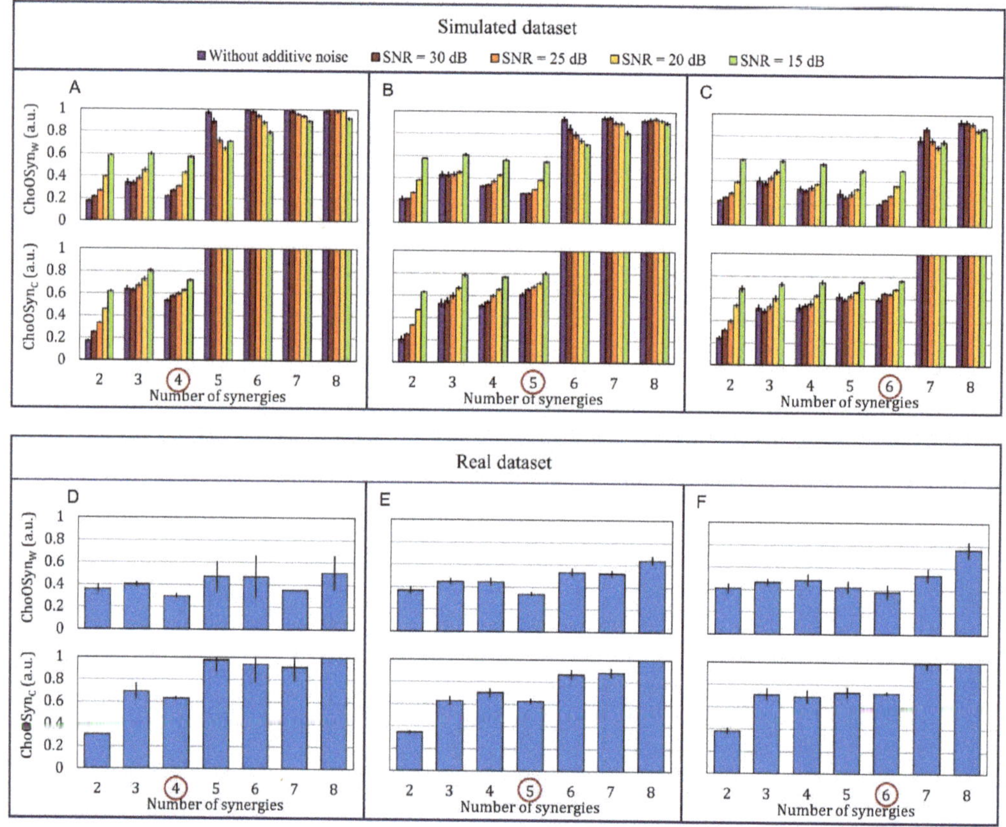

Figure 5. Bar plots representing the mean ± standard error of the two parameters $ChoOSyn_W$ and $ChoOSyn_C$. (**A–C**) Upper plots: simulated dataset. Each bar represents a different noise condition. The dataset is divided into three subsets with (**A**) 4, (**B**) 5, and (**C**) 6 muscle synergies, respectively. (**D–F**) Bottom plots: real dataset. It is divided into subjects that express (**D**) 4, (**E**) 5, and (**F**) 6 muscle synergies, respectively. We used the ground truth to divide the real dataset into three subsets.

The final results obtained applying the ChoOSyn rules are reported in Table 1. The row "no noise" shows the performance of the different methods tested without additive noise. The results obtained considering increasing levels of additive noise are also reported to evaluate the robustness of the methods at different SNR values.

Table 1. Simulated dataset—performance of the different methods in terms of the fraction of correctly classified, mean error (ME), and root-mean-squared error (RMSE).

Fraction of Correctly Classified	T-VAF (90%)	T-VAF (95%)	E-VAF	P-VAF	ChoOSyn
No noise	2/75	36/75	75/75	75/75	73/75
SNR = 30 dB	0/75	24/75	75/75	75/75	73/75
SNR = 25 dB	0/75	18/75	75/75	75/75	74/75
SNR = 20 dB	0/75	9/75	74/75	72/75	72/75
SNR = 15 dB	0/75	0/75	65/75	73/75	63/75
ME [1]	T-VAF (90%)	T-VAF (95%)	E-VAF	P-VAF	ChoOSyn
No noise	−1.29	−0.52	0.00	0.00	−0.03
SNR = 30 dB	−1.48	−0.68	0.00	0.00	−0.03
SNR = 25 dB	−1.57	−0.79	0.00	0.00	−0.01
SNR = 20 dB	−2.11	−1.04	0.01	0.04	−0.05
SNR = 15 dB	−3.07	−1.93	−0.15	0.03	0.04
RMSE [1]	T-VAF (90%)	T-VAF (95%)	E-VAF	P-VAF	ChoOSyn
No noise	1.39	0.72	0.00	0.00	0.16
SNR = 30 dB	1.56	0.82	0.00	0.00	0.16
SNR = 25 dB	1.65	0.92	0.00	0.00	0.12
SNR = 20 dB	2.19	1.17	0.12	0.20	0.28
SNR = 15 dB	3.15	2.00	0.53	0.16	0.53

[1] Unit of measure of ME and RMSE: number of synergies.

Overall, T-VAF methods fail to identify the correct numbers of synergies, while E-VAF, P-VAF, and ChoOSyn show comparable performances (except for SNR = 15 dB).

The synthetic signals are less complex to factorize, and, hence, the reconstruction accuracy (VAF) shows higher values already at lower numbers of synergies. Indeed, T-VAF identifies as the optimal number of synergies 1 to 4 units lower than the correct one. T-VAF also shows the highest ME and RMSE values.

The performance of T-VAF (95%) decreases with increasing additive noise, while, considering E-VAF, P-VAF, and ChoOSyn, the performance degradation is notable only in the worst condition (at 15 dB). For the ChoOSyn algorithm, this result was also predictable from the bar plots of Figure 5A–C. There is a marked step in the case without additive noise (purple bar) and, to a lesser extent, with high SNR values (red, orange, and yellow color bars), while the step becomes markedly shorter for the green bars representing SNR = 15 dB.

3.2. Real Data

Figure 5D–F show the two ChoOSyn parameters extracted from the real dataset. Overall, trends observed in the simulated dataset (Figure 5A–C) are also present in the real dataset (Figure 5D–F). In the latter case, we used the expert ground-truth to divide the population into subjects that express 4, 5, and 6 muscle synergies. The inter-rater agreement, computed by means of Cohen's kappa, was equal to 0.5, suggesting a moderate agreement between the two expert operators.

Considering the real dataset (Table 2), ChoOSyn achieved the best performance, with 17 out of 20 correct classifications and the lowest ME and RMSE. A slightly worse performance was observed for E-VAF, which obtained 12 out of 20 correct classifications. T-VAF and P-VAF achieved the worst performances.

Table 2. Real dataset—Performance of the different methods as the fraction of correctly classified mean error (ME), and root-mean-squared error (RMSE).

	T-VAF (90%)	T-VAF (95%)	E-VAF	P-VAF	ChoOSyn
Fraction of correctly classified	8/20	7/20	12/20	6/20	17/20
ME [1]	−0.90	0.55	0.70	0.90	0.20
RMSE [1]	1.30	0.98	1.18	1.18	0.55

[1] Unit of measure of ME and RMSE: number of synergies.

4. Discussion

Choosing the correct number of muscle synergies that control a motor task is fundamental to understand how the CNS drives the muscles. However, the method employed for selecting the correct number of synergies plays a critical role. The number of synergies characterizing a given activity, e.g., locomotion, varies within and across studies even for unimpaired individuals [28]. There is a lack of standardized methods for the precise identification of the number of synergies, making comparisons across studies and cohorts difficult.

The method currently accepted and used by the vast majority of researchers is based on VAF (variance accounted for) [1,2,18–25], which quantifies the reconstruction accuracy, i.e., how faithfully the muscle synergies represent the signals collected from the muscles. Among the various VAF criteria applied to select the optimal number of synergies, the threshold-VAF (T-VAF) is the most widely adopted, although, it relies on the definition of a fixed threshold T (i.e., T = 90%, Figure 4A). The first number of synergies that produces a VAF value equal to or greater than the threshold is selected as optimal. In the literature, the threshold T is commonly set at 90% [1,2,4,7,18,20–24], and less frequently at 95% [19,25].

The T-VAF method is very simple to implement, but, on the other hand, the presence of a fixed threshold is a well-known issue. First, the threshold is set arbitrarily and different research groups may use different T-values: globally, there is a lack of clear criteria to choose a specific value with respect to another. Second, a small variation in the T-value could also significantly change the results. Therefore, it would be necessary to test the robustness of the threshold itself.

In this work, we proposed and validated a method to choose the optimal number of muscle synergies (ChoOSyn), which is independent of the definition of arbitrary thresholds. ChoOSyn is an alternative to VAF and relies on two parameters directly estimated from data: consistency and similarity of muscle synergies.

Other research groups have introduced alternatives to VAF cutoff criteria. More specifically, Cheung et al. [16] proposed a statistical approach based on real and unstructured sEMG signals, generated by randomly shuffling the original sEMG signals across time and muscles, to select the correct number of muscle synergies in a more interpretable way with respect to standard threshold-based approaches. Ref. [28], instead, introduced intraclass and between-level correlation coefficients to discriminate "reliable" from "unreliable" synergies. Their approach was based on *k*-means clustering and was tested on 9 healthy subjects, considering eight leg muscles during treadmill walking. Delis et al. [29,30] developed a more "physiological" approach, introducing a task decoding-based metric during an arm pointing task.

The approach proposed in this work was validated both on a simulated and on a real dataset, considering an overground walking task. The performance of ChoOSyn was directly compared against VAF-based methods, in terms of the fraction of correct classification, mean error (ME), and root mean square error (RMSE).

Analyzing the simulated dataset, we found that ChoOSyn correctly identified the number of synergies in almost all cases with very small errors. E-VAF and P-VAF methods showed overall performances similar to ChoOSyn. On the contrary, the T-VAF method fails to identify the correct number of muscle synergies. This is probably due to the nature of the dataset: the simulated signals are less complex to factorize, and the VAF assumes higher values already at small numbers of synergies. Indeed, the ME values show that the T-VAF always goes wrong by defect.

When tested on the real dataset, ChoOSyn achieved the best performance, with a marked difference compared to the other methods. The ME and RMSE values of ChoOSyn are also much lower than those of the other VAF-based methods. The worst performing methods were T-VAF and P-VAF, which are based on arbitrary thresholds, further highlighting the problem mentioned above. E-VAF, which does not require thresholds, has the best performance of the VAF-based methods.

Therefore, we proved that ChoOSyn shows equal (simulated dataset) or even higher (real dataset) performance in the correct identification of the number of synergies with respect to the methods currently available in the literature. Indeed, the misclassifications are limited and the number of synergies obtained is close to the correct number, with ME and RMSE values comparable (simulated dataset) or smaller (real dataset) than those obtained with VAF-based methods. Moreover, ChoOSyn operates without thresholds.

The number of muscle synergies can also be strongly influenced by other steps of the muscle synergy extraction process, such as the sEMG pre-processing (e.g., low-pass filtering techniques) [42] and the number and choice of muscles acquired [18]. However, the focus of this contribution is on developing an approach that can be applied after a factorization algorithm, to select the correct number of muscle synergies, and not on evaluating the effect of different pre-processing techniques on the identification of the synergy number. We demonstrated that the ChoOSyn algorithm is more reliable than VAF-based methods. This suggests that the two newly introduced ChoOSyn parameters and the concepts behind them are relevant. It is desirable to obtain a high consistency of muscle synergies over the motor task duration, and low intra-level similarity between synergies (avoiding redundant information). Following these guidelines facilitates the proper selection of the correct number of synergies. We found that a method based on these concepts is more discriminative than the reconstruction accuracy (at the base of VAF-methods) in the search for the correct number of muscle synergies.

The proposed method was tested on 20 healthy subjects. It would be interesting to test. ChoOSyn both on a larger population and on different cohorts (for age or pathological condition). The need for long-lasting sEMG acquisitions to properly select the number of muscle synergies does not limit the feasibility and applicability of the proposed approach to pathological populations. Indeed, gait analysis is commonly used only in those patients that can independently walk, for at least some minutes, without external supports or walking aids. In the past, several studies demonstrated the feasibility of long-lasting gait data acquisition in patients suffering from different neurological conditions, such as normal pressure hydrocephalus [43], mild ataxia [44], and cerebral palsy [45]. Future work should focus on providing algorithm validation for patients affected by neurological disorders, such as patients affected by Parkinson's disease or stroke survivors, by also increasing the number of expert operators for the "ground truth" definition. Moreover, the dataset used includes signals acquired from the lower limb and the trunk while walking. However, the ChoOSyn method is not necessarily associated with the specific motor task considered and can be generalized to signals acquired during a different motor task or from different muscles. Indeed, since the proposed approach does not rely on arbitrary thresholds or task-dependent rules, it can be potentially extended to other cyclic motor tasks, such as running or cycling.

5. Conclusions

We described and validated an algorithm (ChoOSyn) to select the optimal number of synergies expressed during gait, which overcomes the limitations of VAF thresholding methods. The proposed approach may support the standardization of reports, in motor control studies, among different research laboratories. Moreover, ChoOSyn may be applied to different repetitive motor tasks (reaching movements of the upper limbs, etc.) without any specific need for adaptation to the motor task considered.

Author Contributions: Conceptualization, V.A. and M.K.; methodology, R.B.; formal analysis, R.B.; data curation, R.B. and M.G.; writing—original draft preparation, R.B. and V.A.; writing—review and editing, R.B., M.G., M.K., and V.A.; supervision, V.A. and M.K. All authors have read and agreed to the published version of the manuscript.

Funding: This research received no external funding.

Institutional Review Board Statement: Not applicable.

Informed Consent Statement: Not applicable.

Data Availability Statement: Data presented in this study are available on request from the corresponding author.

Conflicts of Interest: The authors declare no conflict of interest.

References

1. Barroso, F.O.; Torricelli, D.; Molina-Rueda, F.; Alguacil-Diego, I.M.; Cano-De-La-Cuerda, R.; Santos, C.; Moreno, J.C.; Miangolarra-Page, J.C.; Pons, J.L. Combining muscle synergies and biomechanical analysis to assess gait in stroke patients. *J. Biomech.* **2017**, *63*, 98–103. [CrossRef] [PubMed]
2. Banks, C.L.; Pai, M.M.; McGuirk, T.E.; Fregly, B.J.; Patten, C. Methodological Choices in Muscle Synergy Analysis Impact Differentiation of Physiological Characteristics Following Stroke. *Front. Comput. Neurosci.* **2017**, *11*, 1–12. [CrossRef]
3. Mileti, I.; Zampogna, A.; Santuz, A.; Asci, F.; Del Prete, Z.; Arampatzis, A.; Palermo, E.; Suppa, A. Muscle Synergies in Parkinson's Disease. *Sensors* **2020**, *20*, 3209. [CrossRef] [PubMed]
4. Feeney, D.F.; Capobianco, R.A.; Montgomery, J.R.; Morreale, J.; Grabowski, A.M.; Enoka, R.M. Individuals with sacroiliac joint dysfunction display asymmetrical gait and a depressed synergy between muscles providing sacroiliac joint force closure when walking. *J. Electromyogr. Kinesiol.* **2018**, *43*, 95–103. [CrossRef]
5. Mileti, I.; Zampogna, A.; Taborri, J.; Martelli, F.; Rossi, S.; Del Prete, Z.; Paoloni, M.; Suppa, A.; Palermo, E. Parkinson's disease and Levodopa effects on muscle synergies in postural perturbation. In Proceedings of the 2019 IEEE International Symposium on Medical Measurements and Applications (MeMeA), Istanbul, Turkey, 26–28 June 2019. [CrossRef]
6. Falaki, A.; Huang, X.; Lewis, M.M.; Latash, M.L. Dopaminergic modulation of multi-muscle synergies in postural tasks performed by patients with Parkinson's disease. *J. Electromyogr. Kinesiol.* **2017**, *33*, 20–26. [CrossRef] [PubMed]
7. Grazioso, S.; Caporaso, T.; Palomba, A.; Nardella, S.; Ostuni, B.; Panariello, D.; Di Gironimo, G.; Lanzotti, A. Assessment of upper limb muscle synergies for industrial overhead tasks: A preliminary study. In Proceedings of the 2019 IEEE International Workshop on Metrology for Industry 4.0 and IoT, Naples, Italy, 4–6 June 2019; pp. 89–92. [CrossRef]
8. Taborri, J.; Agostini, V.; Artemiadis, P.K.; Ghislieri, M.; Jacobs, D.A.; Roh, J.; Rossi, S. Feasibility of Muscle Synergy Outcomes in Clinics, Robotics, and Sports: A Systematic Review. *Appl. Bionics Biomech.* **2018**, *2018*, 1–19. [CrossRef]
9. Santos, P.; Vaz, J.; Correia, P.; Valamatos, M.; Veloso, A.; Pezarat-Correia, P. Intermuscular Coordination in the Power Clean Exercise: Comparison between Olympic Weightlifters and Untrained Individuals—A Preliminary Study. *Sensors* **2021**, *21*, 1904. [CrossRef]
10. D'Avella, A.; Saltiel, P.; Bizzi, E. Combinations of muscle synergies in the construction of a natural motor behavior. *Nat. Neurosci.* **2003**, *6*, 300–308. [CrossRef]
11. Tresch, M.C.; Cheung, V.C.K.; D'Avella, A. Matrix Factorization Algorithms for the Identification of Muscle Synergies: Evaluation on Simulated and Experimental Data Sets. *J. Neurophysiol.* **2006**, *95*, 2199–2212. [CrossRef] [PubMed]
12. Lee, D.D.; Seung, H.S. Learning the parts of objects by non-negative matrix factorization. *Nat. Cell Biol.* **1999**, *401*, 788–791. [CrossRef]
13. Clark, D.J.; Ting, L.H.; Zajac, F.E.; Neptune, R.R.; Kautz, S.A. Merging of Healthy Motor Modules Predicts Reduced Locomotor Performance and Muscle Coordination Complexity Post-Stroke. *J. Neurophysiol.* **2010**, *103*, 844–857. [CrossRef] [PubMed]
14. Ambrosini, E.; De Marchis, C.; Pedrocchi, A.; Ferrigno, G.; Monticone, M.; Schmid, M.; D'Alessio, T.; Conforto, S.; Ferrante, S. Neuro-Mechanics of Recumbent Leg Cycling in Post-Acute Stroke Patients. *Ann. Biomed. Eng.* **2016**, *44*, 3238–3251. [CrossRef]
15. Routson, R.L.; Clark, D.J.; Bowden, M.G.; Kautz, S.A.; Neptune, R.R. The influence of locomotor rehabilitation on module quality and post-stroke hemiparetic walking performance. *Gait Posture* **2013**, *38*, 511–517. [CrossRef] [PubMed]
16. Cheung, V.C.-K.; Piron, L.; Agostini, M.; Silvoni, S.; Turolla, A.; Bizzi, E. Stability of muscle synergies for voluntary actions after cortical stroke in humans. *Proc. Natl. Acad. Sci. USA* **2009**, *106*, 19563–19568. [CrossRef]
17. Sawers, A.; Allen, J.L.; Ting, L.H. Long-term training modifies the modular structure and organization of walking balance control. *J. Neurophysiol.* **2015**, *114*, 3359–3373. [CrossRef]
18. Steele, K.M.; Tresch, M.C.; Perreault, E.J. The number and choice of muscles impact the results of muscle synergy analyses. *Front. Comput. Neurosci.* **2013**, *7*, 105. [CrossRef]
19. Rodriguez, K.L.; Roemmich, R.T.; Cam, B.; Fregly, B.J.; Hass, C.J. Persons with Parkinson's disease exhibit decreased neuromuscular complexity during gait. *Clin. Neurophysiol.* **2013**, *124*, 1390–1397. [CrossRef]

20. Pale, U.; Atzori, M.; Müller, H.; Scano, A. Variability of Muscle Synergies in Hand Grasps: Analysis of Intra- and Inter-Session Data. *Sensors* **2020**, *20*, 4297. [CrossRef]
21. Ghislieri, M.; Agostini, V.; Knaflitz, M. The effect of signal-to-noise ratio on muscle synergy extraction. In Proceedings of the 2018 IEEE Life Sciences Conference, LSC 2018, Montreal, QC, Canada, 28–30 October 2018; pp. 227–230. [CrossRef]
22. Ghislieri, M.; Agostini, V.; Knaflitz, M. Muscle Synergies Extracted Using Principal Activations: Improvement of Robustness and Interpretability. *IEEE Trans. Neural Syst. Rehabil. Eng.* **2020**, *28*, 453–460. [CrossRef]
23. Rimini, D.; Agostini, V.; Knaflitz, M. Intra-Subject Consistency during Locomotion: Similarity in Shared and Subject-Specific Muscle Synergies. *Front. Hum. Neurosci.* **2017**, *11*, 586. [CrossRef] [PubMed]
24. Rimini, D.; Agostini, V.; Knaflitz, M. Evaluation of muscle synergies stability in human locomotion: A comparison between normal and fast walking speed. In Proceedings of the 2017 IEEE International Instrumentation and Measurement Technology Conference (I2MTC), Turin, Italy, 22–25 May 2017. [CrossRef]
25. Sy, H.V.N.; Nambu, I.; Wada, Y. The adjustment of muscle synergy recruitment by controlling muscle contraction during the reaching movement. In Proceedings of the 2016 IEEE International Conference on Systems, Man, and Cybernetics, SMC 2016, Budapest, Hungary, 9–12 October 2016; pp. 756–761. [CrossRef]
26. Hirashima, M.; Oya, T. How does the brain solve muscle redundancy? Filling the gap between optimization and muscle synergy hypotheses. *Neurosci. Res.* **2016**, *104*, 80–87. [CrossRef]
27. Yokoyama, H.; Kaneko, N.; Ogawa, T.; Kawashima, N.; Watanabe, K.; Nakazawa, K. Cortical Correlates of Locomotor Muscle Synergy Activation in Humans: An Electroencephalographic Decoding Study. *iScience* **2019**, *15*, 623–639. [CrossRef] [PubMed]
28. Kim, Y.; Bulea, T.C.; Damiano, D.L. Novel Methods to Enhance Precision and Reliability in Muscle Synergy Identification during Walking. *Front. Hum. Neurosci.* **2016**, *10*, 455. [CrossRef] [PubMed]
29. Delis, I.; Berret, B.; Pozzo, T.; Panzeri, S. Quantitative evaluation of muscle synergy models: A single-trial task decoding approach. *Front. Comput. Neurosci.* **2013**, *7*, 1–21. [CrossRef]
30. Delis, I.; Hilt, P.M.; Pozzo, T.; Panzeri, S.; Berret, B. Deciphering the functional role of spatial and temporal muscle synergies in whole-body movements. *Sci. Rep.* **2018**, *8*, 8391. [CrossRef]
31. Agostini, V.; Ghislieri, M.; Rosati, S.; Balestra, G.; Knaflitz, M. Surface Electromyography Applied to Gait Analysis: How to Improve Its Impact in Clinics? *Front. Neurol.* **2020**, *11*, 1–13. [CrossRef]
32. Soomro, M.H.; Conforto, S.; Giunta, G.; Ranaldi, S.; De Marchis, C. Comparison of Initialization Techniques for the Accurate Extraction of Muscle Synergies from Myoelectric Signals via Nonnegative Matrix Factorization. *Appl. Bionics Biomech.* **2018**, *2018*, 1–10. [CrossRef]
33. Ranaldi, S.; De Marchis, C.; Rinaldi, M.; Conforto, S. The effect of Non-Negative Matrix Factorization initialization on the accurate identification of muscle synergies with correlated activation signals. In Proceedings of the 2018 IEEE International Symposium on Medical Measurements and Applications (MeMeA), Rome, Italy, 11–13 June 2018; pp. 1–5. [CrossRef]
34. Agostini, V.; Knaflitz, M. An Algorithm for the Estimation of the Signal-To-Noise Ratio in Surface Myoelectric Signals Generated During Cyclic Movements. *IEEE Trans. Biomed. Eng.* **2011**, *59*, 219–225. [CrossRef]
35. Ghislieri, M.; Knaflitz, M.; Labanca, L.; Barone, G.; Bragonzoni, L.; Benedetti, M.G.; Agostini, V. Muscle Synergy Assessment During Single-Leg Stance. *IEEE Trans. Neural Syst. Rehabil. Eng.* **2020**, *28*, 2914–2922. [CrossRef]
36. Torricelli, D.; Barroso, F.; Coscia, M.; Alessandro, C.; Lunardini, F.; Esteban, E.B.; d'Avella, A. Muscle Synergies in Clinical Practice: Theoretical and Practical Implications. In *Emerging Therapies in Neurorehabilitation II*; Springer: Cham, Switzerland, 2016; Volume 10, pp. 251–272. [CrossRef]
37. Oliveira, A.S.; Egizzi, L.; Efarina, D.; Kersting, U.G. Motor modules of human locomotion: Influence of EMG averaging, concatenation, and number of step cycles. *Front. Hum. Neurosci.* **2014**, *8*, 335. [CrossRef] [PubMed]
38. Kim, H.; Park, H. Nonnegative Matrix Factorization Based on Alternating Nonnegativity Constrained Least Squares and Active Set Method. *SIAM J. Matrix Anal. Appl.* **2008**, *30*, 713–730. [CrossRef]
39. D'Avella, A.; Bizzi, E. Shared and specific muscle synergies in natural motor behaviors. *Proc. Natl. Acad. Sci. USA* **2005**, *102*, 3076–3081. [CrossRef]
40. Cheung, V.C.K.; D'Avella, A.; Tresch, M.C.; Bizzi, E. Central and Sensory Contributions to the Activation and Organization of Muscle Synergies during Natural Motor Behaviors. *J. Neurosci.* **2005**, *25*, 6419–6434. [CrossRef]
41. Cohen, J. A Coefficient of Agreement for Nominal Scales. *Educ. Psychol. Meas.* **1960**, *20*, 37–46. [CrossRef]
42. Scalona, E.; Taborri, J.; Del Prete, Z.; Palermo, E.; Rossi, S. EMG factorization during walking: Does digital filtering influence the accuracy in the evaluation of the muscle synergy number? In Proceedings of the 2018 IEEE International Symposium on Medical Measurements and Applications (MeMeA), Rome, Italy, 11–13 June 2018. [CrossRef]
43. Agostini, V.; Lanotte, M.; Carlone, M.; Campagnoli, M.; Azzolin, I.; Scarafia, R.; Massazza, G.; Knaflitz, M. Instrumented Gait Analysis for an Objective Pre-/Postassessment of Tap Test in Normal Pressure Hydrocephalus. *Arch. Phys. Med. Rehabil.* **2015**, *96*, 1235–1241. [CrossRef]
44. Benedetti, M.G.; Agostini, V.; Knaflitz, M.; Gasparroni, V.; Boschi, M.; Piperno, R. Self-reported gait unsteadiness in mildly impaired neurological patients: An objective assessment through statistical gait analysis. *J. Neuroeng. Rehabil.* **2012**, *9*, 64. [CrossRef]
45. Agostini, V.; Nascimbeni, A.; Gaffuri, A.; Knaflitz, M. Multiple gait patterns within the same Winters class in children with hemiplegic cerebral palsy. *Clin. Biomech.* **2015**, *30*, 908–914. [CrossRef]

MDPI
St. Alban-Anlage 66
4052 Basel
Switzerland
Tel. +41 61 683 77 34
Fax +41 61 302 89 18
www.mdpi.com

Sensors Editorial Office
E-mail: sensors@mdpi.com
www.mdpi.com/journal/sensors